D1226421

BIOPHYSICAL CHEMISTRY

BIOPHYSICAL CHEMISTRY

PART

I

THE CONFORMATION OF BIOLOGICAL MACROMOLECULES

Charles R. Cantor
COLUMBIA UNIVERSITY

Paul R. Schimmel
MASSACHUSETTS INSTITUTE OF TECHNOLOGY

W. H. FREEMAN AND COMPANY
San Francisco

Cover drawing created by Steve Levine of the Lawrence Livermore Laboratories, using the Atoms program written by Ken Knowlton and Lorinda Cherry of Bell Laboratories and coordinates provided by R. G. Shulman.

Sponsoring Editor: Arthur C. Bartlett
Project Editor: Pearl C. Vapnek
Manuscript Editor: Lawrence W. McCombs
Designer: Robert Ishi
Production Coordinator: Linda Jupiter
Illustration Coordinator: Cheryl Nufer
Artists: Irving Geis and Eric Hieber
Compositor: Syntax International
Printer and Binder: R. R. Donnelley & Sons Company

Library of Congress Cataloging in Publication Data

Cantor, Charles R 1942–
The conformation of biological macromolecules.

(Their Biophysical chemistry; pt. 1)
Includes bibliographies and indexes.
1. Macromolecules. 2. Chemistry, Physical organic—Technique. I. Schimmel, Paul Reinhard, 1940– joint author. II. Title.
QH345.C36 pt. 1 574.1'9283s [574.1'924] 79-22043
ISBN 0-7167-1042-0
ISBN 0-7167-1188-5 pbk.

Printed in the United States of America

9 8 7 6 5 4 3 2 1

To Louis and Ida Dianne Cantor
and Alfred and Doris Schimmel

Contents in detail of Part I

Contents in brief of Parts II and III

Preface

Biophysical Chemistry is concerned with biological macromolecules and complexes or arrays of macromolecules. The work deals with the conformation, shape, structure, conformational changes, dynamics, and interactions of such systems. Our goal is to convey the major principles and concepts that are at the heart of the field. These principles and concepts are drawn from physics, chemistry, and biology.

We have aimed at creating a multilevel textbook in three separately bound parts. The material covers a broad range of sophistication so that the text can be used in both undergraduate and graduate courses. It also should be of value to general scientific readers who simply wish to become familiar with the field, as well as to experienced research scientists in the biophysical area. For example, perhaps half of the material requires only the background provided by a one-semester undergraduate course in physical chemistry. A somewhat smaller fraction necessitates the use of concepts and mathematical techniques generally associated with a more sophisticated background, such as elementary statistical thermodynamics and quantum mechanics.

Biophysical Chemistry is organized into three parts. The first part deals with the structure of biological macromolecules and the forces that determine this structure. Chapter 1 introduces the fundamental questions of interest to biophysical chemists, Chapters 2–4 summarize the known structures of proteins, nucleic acids, and other biopolymers, and Chapters 5–6 treat noncovalent forces and conformational analysis.

Part II summarizes some of the techniques used in studying biological structure and function. The emphasis is on a detailed discussion of a few techniques rather than an attempt to describe every known technique. Chapters 7–9 cover spectroscopic methods, Chapters 10–12 deal with hydrodynamic methods, and Chapters 13–14 discuss x-ray and other scattering and diffraction techniques.

Part III demonstrates how techniques and principles are used in concert to gain an understanding of the behavior and properties of biological macromolecules. The focus is on the thermodynamics and kinetics of conformational changes and ligand interactions. New techniques are introduced as needed, and a few selected case histories or systems are discussed in considerable detail. The topics include ligand interactions (Chapters 15–17), the special theories and techniques used to study molecules that are statistical chains rather than definite folded conformations (Chapters 18–19), protein conformational changes (Chapters 20–21), nucleic acid conformational changes (Chapters 22–24), and membranes (Chapter 25).

We have made every effort to keep the chapters as independent as possible, so that the reader has a wide choice of both the material to be covered and the order in which it is to be treated. Extensive cross-references to various chapters are included to help the reader find necessary background material if the parts are not read in sequence. Where possible, examples are taken repeatedly from a small number of systems, so that the reader can have the experience of contrasting information gained about the same protein or nucleic acid from a variety of different approaches.

Within each chapter, we have attempted to maintain a uniform level of rigor or sophistication. Short digressions from this level are segregated into boxes; longer digressions are indicated by a bullet (•) preceding the section or subsection heading. Readers with a less rigorous background in physics, mathematics, and physical chemistry should find helpful the many boxes that review elementary material and make the text fairly self-contained; Appendix A provides a basic review of principles of matrix algebra. Other boxes and special subsections are aimed at advanced readers; in many cases, these discussions attempt to illuminate points that we ourselves found confusing.

In different sections, the level of mathematical sophistication varies quite significantly. We have tried to use the simplest mathematical formulation that permits a clear presentation of each subject. For example, hydrodynamic properties are treated in one dimension only. The form of a number of the fundamental equations is extracted by dimensional analysis rather than through lengthy (and not particularly instructive) solutions of hydrodynamic boundary-value problems. On the other hand, x-ray and other scattering phenomena are treated by Fourier transforms, and many problems in statistical mechanics are treated with matrix methods. These advanced mathematical techniques are used in only a few chapters, and numerous boxes are provided to assist the reader with no previous exposure to such methods. The remaining sections and chapters are self-contained and can be understood completely without this advanced mathematical formalism.

Some techniques and systems are not covered in any fair detail. This represents a biased choice by the authors, not necessarily of which techniques we feel are important, but simply of which are instructive for the beginning student in this field.

Each chapter concludes with a summary of the major ideas covered. In addition, each chapter is heavily illustrated, including some special drawings by Irving Geis. Certainly, much can be learned simply by reading the chapter summaries and by studying the illustrations. Also, we believe the illustrations convey some of the excitement of the field.

Problems are provided at the end of each chapter. These vary in difficulty from relatively simple to a few where the full answer is not known, at least to the authors. Answers to problems are provided in Appendix B.

Detailed literature citations are not included, except to acknowledge the source of published material reproduced or adapted here. However, a list of critical references for each chapter is included. In virtually all cases, these articles will provide an immediate entrée to the original papers needed for more detailed study.

The problem of notation and abbreviations in this field is a difficult one. In drawing together material from so many different types of research, we have had to adapt the notation to achieve consistency and to avoid confusion among similar symbols. Wherever possible, we have followed the recommendations of the American Chemical Society, but inevitably we have had to develop some conventions of our own. A glossary of some of the more frequently used symbols is provided.

At MIT some of this material has been used in an undergraduate course in biophysical chemistry. The course was designed to meet the needs of students wishing a second course in physical chemistry, but developed in a biochemical framework. The idea was to construct a course that covered much of the same material with the same rigor as a parallel, more traditional course. The only preparation required was a one-semester course in undergraduate physical chemistry, which at MIT is largely concerned with chemical thermodynamics.

Over the years graduate courses in biophysical chemistry at MIT and at Columbia have made use of much of the material presented here. In addition, a special-topics course in protein structure has used some of the material. Because a broad range of subjects is covered, its usefulness as a text will hopefully meet a variety of individual teaching tastes and preferences, as well as enable instructors to vary content as needs develop and change.

It is obvious that a work of this complexity cannot represent solely the efforts of its two authors. As we sought to master and explain the wide range of topics represented in biophysical chemistry, we learned why so few books have been written in this field in the past two decades. We owe a great debt to many who helped us in ways ranging from sharing their understanding to providing original research data.

We give special thanks to Irving Geis, for his effort on a number of complex illustrations and for his helpful advice on numerous other drawings; to Wilma Olson, for reading a major portion of the entire manuscript; to Robert Alberty and Gordon Hammes, for their influence, through teaching and discussions, on the material on biochemical equilibria and kinetics; to Richard Dickerson, for providing material and advice that were essential for the preparation of Chapter 13; to Paul Flory, for inspiring our treatment of conformational energies and configurational statistics of macromolecules; to Howard Schachman, whose course at Berkeley inspired parts of several chapters; to R. Wayne Oler, for bringing the authors together for this undertaking, and to Bruce Armbruster, for sealing the commitment; to the helpful people at W. H. Freeman and Company, including Ruth Allen, Arthur Bartlett, Robert Ishi, Larry McCombs, and Pearl Vapnek; to Kim Engel, Karen Haynes, Marie Ludwig, Joanne Meshna, Peggy Nelson, Cathy Putland, and Judy Schimmel, for typing and related work associated with the manuscript; and to Cassandra Smith and to Judy, Kathy, and Kirsten Schimmel, for their patience with the intrusion this work has made on the authors' lives.

Many people read and commented on specific chapters, provided figures, notes and materials, and spent much time with us in helpful discussions. We gratefully thank these people: Robert Alberty, Arthur Arnone, Struther Arnott, P. W. Atkins, Robert Baldwin, Larry Berliner, Bruce Berne, Richard Bersohn, Sherman Beychok, Victor Bloomfield, David Brandt, John Brandts, John Chambers, Sunney Chan, Patricia Cole, Robert Crichton, Francis Crick, Donald Crothers, Norman Davidson, Richard Dickerson, David Eisenberg, Robert Fairclough, Gerry Fasman, George Flynn, David Freifelder, Ronald Gamble, Robert Gennis, Murray Goodman, Jonathan Greer, O. Hayes Griffith, Gordon Hammes, John Hearst, Ellen Henderson, James Hildebrandt, Wray Huestis, Sung Hou Kim, Aaron Klug, Nelson Leonard, H. J. Li, Stephen Lippard, Richard Lord, Brian Matthews, Harden McConnell, Peter Moore, Garth Nicolson, Leonard Peller, Richard Perham, Michael Raftery, Alexander Rich, Frederick Richards, David Richardson, Wolfram Saenger, Howard Schachman, Harold Scheraga, Benno Schoenborn, Verne Schumaker, Nadrian Seeman, Robert Shulman, Mavis Shure, Louise Slade, Cassandra Smith, Hank Sobell, Thomas Steitz, Robert Stroud, Lubert Stryer, Serge Timasheff, Ignacio Tinoco, Jr., Richard Vandlen, Jerome Vinograd, Peter von Hippel, Christopher Walsh, James Wang, Gregorio Weber, Peter Wellauer, Barbara Wells, Robert Wells, William Winter, Harold Wyckoff, Jeffries Wyman, and Bruno Zimm.

November 1979

Charles R. Cantor
Paul R. Schimmel

Glossary of symbols

This glossary includes some of the symbols used extensively throughout the text. In many cases, the same or very similar symbols are used in certain contexts with other meanings; the meaning of a symbol is explained in the text where it is introduced.

Symbol	Meaning
A	Absorbance.
A_{ij}	Amplitude of kinetic decay.
Å	Angstroms.
a	Hyperfine splitting constant. Long semiaxis of ellipse. Persistence length.
$\mathbf{a}$	Unit cell basis vector.
$\mathbf{a}^*$	Reciprocal cell basis vector.
a_{ij}	Parameters composed of rate constants.
a_s	Exponent relating sedimentation to chain length.
a_η	Exponent relating viscosity to chain length.
b	Short semiaxis of ellipse.
$\mathbf{b}$	Unit cell basis vector.
$\mathbf{b}^*$	Reciprocal cell basis vector.
C	Molar concentration.
C_n	Rotational symmetry group element. Characteristic ratio.
C_∞	Limiting characteristic ratio.
ΔC_p^0	Standard constant pressure heat capacity change per mole.
c	Velocity of light in vacuum. Ratio of k_R/k_T. Weight concentration.
c_p	Plateau weight concentration.
$\hat{c}_i$	Weight concentration of ith species or component.
$\mathbf{c}$	Unit cell basis vector.
$\mathbf{c}^*$	Reciprocal cell basis vector.
D	Debye.
D	Translational diffusion constant.
D_n	Dihedral symmetry group element.
D_{rot}	Rotational diffusion constant.
$D_{20,\mathrm{w}}$	D extrapolated to 20° C, water.
E_a	Activation energy.
E_d	Interaction energy between two dipoles.
E_{kl}	Nonbonded pair interaction potential.
E_{tor}	Torsional potential energy.
$E(\Phi_i,\Psi_i)$, E_i	Total rotational potential for residue i.
$\mathbf{E}$	Electric field.
e	Exponential function. Unit of charge on electron.
F	Frictional coefficient ratio.
$F(\mathbf{S})$	Structure factor.
$F_\mathrm{H}(\mathbf{S})$	Structure factor, heavy-atom contribution.

Symbol	Meaning
$F_{\mathrm{Tot}}(\mathbf{S})$	Structure factor for an array.
$F_{\mathrm{m}}(\mathbf{S})$	Molecular structure factor.
$\mathbf{F}$	Force.
$\mathscr{F}$	The Faraday.
f	Translational frictional coefficient.
f_{app}	Apparent fractional denaturation.
f_{D}	Fraction in denatured state.
f_{N}	Fraction in native state.
f_{min}	Translational friction coefficient of anhydrous sphere.
f_{rot}	Rotational friction coefficient for sphere.
f_{sph}	Translational friction coefficient for sphere.
f_a, f_b	Rotational friction coefficient around a, b axis of ellipse.
G	Gibbs free energy.
ΔG^0	Standard Gibbs free energy change per mole.
$\Delta \bar{G}^0$	Intrinsic standard free energy change (with statistical component removed).
$\Delta G_{\mathrm{I},ij}$	Free energy of interaction between two ligands.
ΔG_{r}	ΔG per residue.
ΔG_{Tot}	Total free energy change per mole.
ΔG_{el}	Change in electrostatic free energy.
ΔG_{T}	Total free energy of formation of configuration.
$\Delta\Delta G_{\mathrm{T}}$	Difference in ΔG_{T} between two configurations.
$\Delta \bar{G}_{\mathrm{gr}}$	Average helix growth free energy change per residue pair.
g	g value for free electron, 2.00232.
g_{x}, etc.	Component of g-factor tensor.
H	Enthalpy.
H_{xy}	Magnetic field in xy plane.
ΔH	Enthalpy change per mole.
ΔH^0	Standard enthalpy change per mole.
ΔH_{r}	ΔH per residue.
ΔH_{D}	Enthalpy change for conversion from fully native to fully denatured state.
ΔH_{app}	Apparent enthalpy change for conversion from fully native to fully denatured state.
$\mathbf{H}$	Magnetic field.
$\mathbf{H}_{\mathrm{res}}$	Magnetic field at which resonance occurs.
$\underset{\sim}{\mathrm{H}}$	Hamiltonian operator.
$\Delta\mathbf{H}_{\mathrm{loc}}$	Magnetic field generated by local environment.
h	Planck's constant.
$\hbar$	$h/2\pi$
I	Intensity of radiation. Nuclear spin quantum number. Ionic strength.
$I(\mathbf{S})$	Scattering intensity relative to a single electron at the origin.
i	$\sqrt{-1}$
$\hat{\mathbf{i}}$	Cartesian unit vector.
J	NMR coupling constant.
$\mathbf{J}_2$	Solute flux.
$\hat{\mathbf{j}}$	Cartesian unit vector.
K_{D}	True equilibrium constant for conversion from fully native to fully denatured state.
K_{p}	Michaelis constant for product.
K_{S}	Michaelis constant for substrate.
K_{η}	Coefficient relating viscosity to chain length.
K_{s}	Coefficient relating sedimentation to chain length.
K_{app}	Apparent equilibrium constant for conversion from fully native to fully denatured state.
K_i	Macroscopic equilibrium constant. Equilibrium constant for forming ith configuration. Equilibrium constant for transition from native state to intermediate state i.
$\tilde{K}_i$	Apparent dissociation constant, one-ligand system.
$\tilde{K}_{ij}$	Apparent dissociation constant, two-ligand system.
k	Boltzmann's constant. Microscopic equilibrium dissociation constant.
k_R	Microscopic dissociation constant for R state.
k_T	Microscopic dissociation constant for T state.
k_i	Microscopic equilibrium constant.
$\hat{\mathbf{k}}$	Cartesian unit vector.

Symbol	Meaning
L_c	Contour length.
L, L'	Equilibrium constant for R_0 T_0.
$\mathbf{L}$	Angular momentum.
l	Length of one polymer bond.
l_e	Length of statistical segment.
M	Molecular weight.
$\bar{M}_n$	Number-average molecular weight.
$\bar{M}_w$	Weight-average molecular weight.
$\bar{M}_i$	Molecular weight of ith macromolecular species.
M_{ij}	Species with i bound L_1 and j bound L_2.
$M^{(j)}$	Set of all species with j bound L_2.
$\mathbf{M}$	Magnetization.
$\mathbf{M}_{xy}$	Magnetization in xy plane.
$\underset{\sim}{\mathbf{M}}$	Statistical weight matrix.
m	Colligative molality. Mass of object.
m_e	Mass of electron.
m_i	Molality of ith species.
m_s	Quantum number of electron spin along z axis.
m_I	Quantum number of nuclear spin along z axis.
m'	Total molality.
$\underset{\sim}{\mathbf{m}}$	Magnetic dipole operator.
N_0	Avogadro's number.
N_C	Number of carbons in amphiphile R chain.
N'_C	Number of carbons in amphiphile that are imbedded in hydrocarbon core of micelle.
N_e	Number of statistical segments.
N_{Ch}	Number of chains in micelle.
N_{hg}	Number of head groups in micelle.
n	Refractive index. Number of sites. Number of bonds in polymer.
n_i	Number of moles of component i. Number of sites of certain type.
n_w	Weight-average degree of polymerization.
P	Pitch of helix. Pressure. Patterson function.
P_0	Solvent vapor pressure.
P_v	Solvent vapor pressure in presence of solute.
P_r	Axial ratio.
pK_a	$-\log_{10} K_a$
pO_2	Partial pressure of oxygen.
$(pO_2)_{1/2}$	Partial pressure of oxygen at half saturation.
$\underset{\sim}{\mathbf{p}}$	Momentum operator.
Q	Configurational partition function.
R	Gas constant.
R_G	Radius of gyration.
$\langle R_G^2 \rangle_0$	Unperturbed mean square radius of gyration.
$\bar{R}$	Fraction of molecules in R state.
$\underset{\sim}{\mathbf{R}}$	Nuclear position operator.
$\underset{\sim}{\mathbf{R}}(\alpha,\beta)$	Coordinate transformation matrix.
r	Distance of separation.
r_D	Donnan ratio.
r_e	Radius of equivalent sphere.
$\langle r^2 \rangle_0$	Unperturbed mean square end-to-end distance.
$\mathbf{r}$	Polymer end-to-end vector.
$\underset{\sim}{\mathbf{r}}$	Electron position operator.
S	Svedberg (unit of sedimentation coefficient).
S_A	Partial molal entropy.
S'_A	Unitary part of S_A.
ΔS_r	ΔS per residue.
ΔS^0	Standard entropy change.
ΔS^0_u	Unitary standard entropy change.
$\mathbf{S}$	Scattering vector.
s	Sedimentation coefficient. Statistical weight. Equilibrium constant for helix growth. Equilibrium constant for base-pair formation.
$s_{20,w}$	Sedimentation coefficient corrected to 20° C, water.
$\hat{\mathbf{s}}$	Unit vector along scattered radiation.
$\hat{\mathbf{s}}_0$	Unit vector along incident radiation.
T	Temperature (in degrees Kelvin usually).
T_m	Melting temperature.
T_1	Longitudinal relaxation time.
T_2	Transverse relaxation time.
$\underset{\sim}{\mathbf{T}}_i$	Transformation matrix.
t	Time.

Symbol	Meaning
U^0_{mic}	Attractive part of μ^0_{mic}.
u	Component of M_{xy} in phase with H_{xy}. Electrophoretic mobility.
V	Volume.
V_h	Hydrated volume.
$\bar{V}_i$	Partial specific volume of component i.
V_p	Maximum reaction velocity in reverse direction.
V_s	Maximum reaction velocity in forward direction.
v	Speed (also called velocity). Component of M_{xy} out of phase with H_{xy}.
v_i	Initial reaction velocity.
$\langle v_2 \rangle$	Effective average solute velocity.
$\bar{v}$	Partial molar volume.
$\bar{v}_s$	Partial molar volume of pure solvent.
$\mathbf{v}$	Velocity.
$W(r)$	Radial distribution function of end-to-end distance.
$W(x,y,z)$	End-to-end distance distribution function.
W^0_{mic}	Repulsive part of μ^0_{mic}.
$(\bar{X}_i)$	Equilibrium concentration.
$\Delta(X_i)$	Difference between temporal and equilibrium concentration.
x_b	Bottom of cell.
x_m	Meniscus position.
y	General physical property.
y_D	Physical property of denatured state.
y_N	Physical property of native state.
$\bar{y}$	Fractional saturation of site.
$\bar{y}_F$	Fractional saturation with ligand F.
z	Charge on macromolecule or ion in units of e.
z_i	Ionic valence of ith ion.
α	Degree of association. Dimensionless binding parameter like $(F)/k_R$.
α_H	Hill constant.
β	Dimensionless binding parameter.
β_e	Bohr magneton.
β_n	Nuclear magneton.
β_s	Mandelkern–Flory–Scheraga parameter.
β'	Scheraga–Mandelkern parameter.
Γ	Parameter affecting relaxation amplitudes.
γ	Magnetogyric ratio. $(A)/K_{AR}$ binding parameter. Velocity gradient dv_x/dz.
δ	Chemical shift parameter. Phase shift.
$\delta(x)$	Dirac delta function of argument x.
δ_1	Hydration (in grams per gram).
δ_{ij}	Kronecker delta.
ε	Dielectric constant. Molar decadic or residue extinction coefficient.
$\Delta\varepsilon$	Circular dichroism $(\varepsilon_L - \varepsilon_R)$.
η	Solution viscosity.
η_0	Solvent viscosity.
η_{rel}	Relative viscosity.
η_{sp}	Specific viscosity.
$[\eta]$	Intrinsic viscosity.
Θ_i	Fractional saturation of ith site.
θ	Scattering angle. Fractional helicity.
$[\theta]$	Molar ellipticity.
$\underset{\sim}{\Lambda}$	Matrix of λ_i's.
λ	Eigenvalue. Wavelength. Kinetic decay time.
λ_j	jth kinetic decay time of jth eigenvalue.
λ_1, λ_2	Parameters composed of rate constants.
μ_i	Chemical potential per mole.
μ_i^0	Standard chemical potential per mole.
$\hat{\mu}_i$	Chemical potential per gram.
$\hat{\mu}_i^0$	Standard chemical potential per gram.
μ^0_{mic}	Standard chemical potential of amphiphile in micelle.
μ^0_w	Standard chemical potential of amphiphile in aqueous phase.
$\boldsymbol{\mu}_m$	Magnetic moment.
$\underset{\sim}{\boldsymbol{\mu}}$	Electric dipole moment operator.
ν	Frequency. Simha factor in viscosity. Moles of ligand bound per mole of macromolecule.
ν_N	Saturation density for lattice with N units.
π	Osmotic pressure.
ρ	Mass density (in grams per cm^3).
$\rho(\mathbf{r})$	Electron density.
σ	Nucleation constant.

Symbol	Meaning
σ_h	Superhelix density.
τ	Number of supercoils.
τ_F	Fluorescence decay time.
τ_a, τ_b	Rotational relaxation time for *a*-, *b*-axis orientation.
τ_c	Rotational correlation time.
τ_r	Rotational relaxation time of sphere.
τ, τ_j	Reaction relaxation times.
Φ	Electrical potential. Voltage difference.
Φ_c	Universal constant for random coils $2 \cdot 1 \times 10^{23}$.
ϕ	N–C′ torsional angle. Phase of complex number.
ϕ_{1a}, ϕ_{20}, etc.	Monomer wave functions.
ϕ_F	Fluorescence quantum yield.
ϕ_p	Practical osmotic coefficient.
ϕ', ϕ''	Nucleic acid backbone torsional angles.
$[\phi]$	Molar rotation per residue.
χ	Mole fraction of all solute species.
χ_i	Mole fraction of *i*th component.
χ_A	Mole fraction of *A*th component.
χ_{gc}	Mole fraction G + C.
χ	Glycosidic bond torsional angle.
ψ	C′–C torsional angle.
ψ', ψ''	Nucleic acid backbone torsional angles.
Ω_{jk}	Number of ways of putting *k* helical units into *j* separated sequences.
Ω_k	$(n - k + 1)$ number of ways of placing *k* helical units in one sequence within chain of *n* residues.
$\Omega_{n,i}$	Number of ways of assorting *i* items (ligands) in *n* boxes (sites).
ω	Circular frequency or angular velocity.
ω_0	Larmor frequency.
ω', ω''	Nucleic acid backbone torsional angles.
$\Delta\omega_{1/2}$	Line width.
$\boldsymbol{\omega}$	Angular velocity.
imag	Imaginary part of.
$\langle\rangle$	Average.
$\langle\|\rangle$	Overlap integral.
$\langle\|\|\rangle$	Expectation value integral.
*	Superscript, complex conjugate, as in F^*.
$\|\|$	Amplitude of complex number or length of vector, as in $\|F\|$.
∇	Vector differential.
( )	Molar concentration, as in (A).
†	Superscript, transpose of matrix, as in $A^\dagger$.
$\frown$	Superscript, convolution product, as in $\overset{\frown}{AB}$.

General Rules

Symbol	Meaning
K	Macroscopic equilibrium constant.
k	Microscopic equilibrium constant or rate constant.
C	Molar concentration.
c	Weight concentration.
$\underset{\sim}{\mathbf{M}}$	All matrices and operators.
$\hat{\imath}$	All unit vectors.
R_G	Radius of gyration.
χ	Mole fraction.
Φ	Voltage or electrical potential.

PART I

THE CONFORMATION OF BIOLOGICAL MACROMOLECULES

The subject of this book is the use of physical chemistry to measure and understand the properties of large molecules of biological interest. More attention is given to proteins and nucleic acids than to other biopolymers such as polysaccharides and lipid-containing systems. This is because, apart from their obvious central role in all known life forms, the function and structure of proteins and nucleic acids are currently much better understood than are the corresponding characteristics of other biopolymers.

This first part meets several ends. It introduces the reader to the levels of structure in biopolymers, and it defines the kinds of fundamental questions these molecules challenge physical chemists to answer. Some mention also is made of basic strategy that often can assist in the resolution of these questions. Then we review the current state of knowledge about the detailed structure of proteins and nucleic acids. The principal aim is to see what generalities are possible from a large amount of available information.

Next we summarize what is known about other types of biopolymers such as polysaccharides and membranes. Here we highlight the ways in which these systems differ from proteins and nucleic acids. Finally we describe the forces and interactions that are responsible for creating the characteristics of organized biological structures. Knowledge about forces is essential if one is to understand or predict the three-dimensional structures assumed by proteins and nucleic acids.

1
An introduction to the strategy and tactics of biophysical chemistry

1-1 LEVELS OF STRUCTURES IN BIOLOGICAL MACROMOLECULES

The molecular weights of the systems discussed in this book range from 10^3 to 10^{12} daltons. In many cases, such a system is composed of many covalent species rather than only one. The whole system is called a single molecule when the parts are present in a well-defined stoichiometry and when there is little tendency for them to dissociate spontaneously under physiological conditions.

The molecules studied here can be divided into seven classes of progressively increasing size and complexity (Table 1-1). As the size of the system increases, it is convenient to divide it conceptually into increasingly larger units. As shown in Table 1-1, there is a tendency to choose these units so that each represents a tenth to a hundredth of the total system.

The first three classes contain molecules for which structural information is potentially available at atomic resolution. It is meaningful and useful to frame questions about the structure and function of such molecules at the level of the location or behavior of individual atoms or individual residues (recognizable groups of atoms).

The last four classes represent systems of such complexity that, even if atomic resolution structures were available, such information would be extremely cumbersome to use. For example, a 10^6-dalton structure would consist of about 10^5 atoms. Describing such a structure by naming each atomic type and its spatial position would be analogous to describing a city by reciting the names and addresses of each of its inhabitants. A more manageable level of description involves naming individual subunits or regions, and then concentrating on individual residues or atoms only when their function is particularly interesting. This is akin to breaking the city down

Table 1-1
Size and complexity of biopolymers

Class	Specific example*	Typical size	Typical mol wt	Subunits used for detailed description	Number of subunits
Oligomers[†]	Actinomycin D	20 Å sphere	10^3–10^4	Atoms (or residues)	10^2 (or 10)
Small proteins	Chymotrypsin	40 Å sphere	10^4–10^5	Amino acid residues (or atoms)	10^2–10^3 (or 10^3–10^4)
Nucleic acids	tRNA	100 Å rod	10^4–10^5	Nucleotide residues (or atoms)	10^2–10^3 (or 10^3–10^4)
Large proteins	Aspartate transcarbamoylase	70 Å sphere	10^5–10^7	Subunits or covalent chains	10–10^2
Small assemblies	Ribosome	200 Å sphere	10^5–10^7	Subunits or covalent chains	10–10^2
Large assemblies	Membranes, viruses	1,000 Å sphere	10^7–10^{12}	Regions, fragments, components	10–10^2
Intact DNA	*E. coli* DNA	0.1 cm rod	10^7–10^{12}	Regions, fragments, components	10–10^2

* All of these examples are discussed in detail in various parts of this book.
[†] The prefix *oligo-*, used frequently in this book, generally refers to a structure composed of 2 to 25 units.

into neighborhoods or zones and keeping track of only a few significant individuals, such as the mayor.

Ambiguities can easily arise when terms like subunit, residue, molecule, region, and moiety are used to describe biopolymers with any kind of generality. There is a hierarchy of structural units and structural features. Because the patterns of organization of proteins, nucleic acids, and complex noncovalent aggregates such as cell surfaces are somewhat different, no single nomenclature is optimal for describing all three types of structure. Furthermore, there is no universal agreement even on an exact terminology for proteins alone. Here, we shall introduce what we consider to be the most convenient way of classifying levels of macromolecular structure. Keep in mind that the following definitions cannot be viewed inflexibly because many biological structures have features that do not fall cleanly into one category or another. The levels of structure that must be considered are illustrated schematically in Figure 1-1.

Figure 1-1

Levels of structure in biological macromolecules. A bacteriophage head, or other discrete assembly of proteins or nucleoproteins is a quaternary (4°) structure. The three-dimensional structure of a single component of one of these assemblies is a tertiary (3°) structure. Local helical regions that may be contained in such a structure are called a secondary (2°) structure. The actual chemical structural formula, without regard to the particular arrangement of atoms except as required by chemical bonding, is the primary (1°) structure. [Drawing by Irving Geis.]

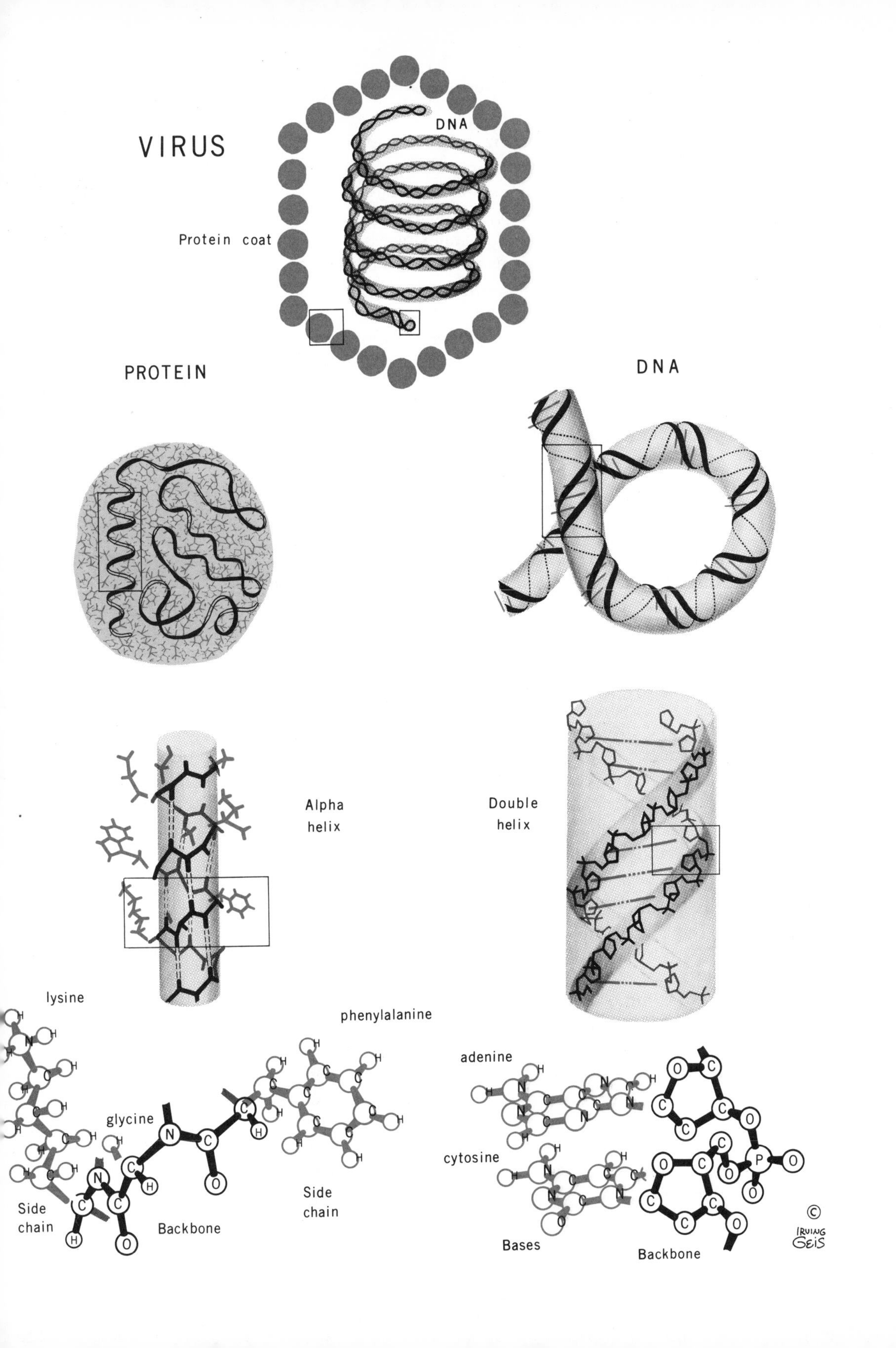
VIRUS
DNA
Protein coat
PROTEIN
DNA
Alpha
helix
Double
helix
lysine
phenylalanine
glycine
Side
chain
Backbone
Side
chain
adenine
cytosine
Bases
Backbone
©
IRVING
GEIS

The individual monomeric residues that form proteins and nucleic acids are amino acids and nucleotides, respectively. These generally have known *atomic structure*. That is, one can define the identity of every atom, the average location of its nucleus in space, and where, on the average, the electron density is concentrated. Such structures are not rigid or inflexible, but molecular vibrations and internal rotations about single bonds usually do not distort the picture too badly, especially in ordered crystals. The *configuration* of a monomer residue (as used in the language of organic chemistry) is its atomic structure, including known stereochemistry of asymmetric centers. The *conformation* is a more complete description and includes what is known about preferred orientations of groups that are, in principle, capable of movement by internal rotation. The conformation is generally an average over energetically accessible atomic structures and is the most complete description possible of a molecule in solution.

Primary structure is the sequential order of the residues

Proteins and nucleic acids are unbranched polymers. The covalent chain structure of either of these can be written $R_1R_2 \cdots R_i \cdots R_n$, where R_i gives the identity of the ith residue in the chain. A feature of proteins, nucleic acids, and most polysaccharides that immediately distinguishes them from many synthetic polymers is that the units are arranged in a head-to-tail fashion. Thus, the abbreviations R_aR_b and R_bR_a refer to molecules that have different covalent chemical structures. Also, the two ends of the polymer $R_1 \cdots R_n$ will have different chemical properties. To describe the covalent structure, the absolute *sequence* of residues must be written. There may be more than one polymer chain in the complete covalent structure and these, in turn, may contain intrachain or interchain covalent cross-links. The complete covalent structure is called the *primary structure* (1° structure). It must include a specification of the configuration of all asymmetric centers, both in the polymer chain backbone and on the side chains of each of the monomer residues.

The primary structure of hundreds of proteins and nucleic acids has been determined by elaborate chemical analysis. An up-to-date compilation of all known primary structures is published regularly (Dayhoff, 1975). It would be extremely time-consuming to write a chemical formula for the entire primary structure. Therefore, it is customary to abbreviate primary structures by using an alphabetic code. The most common version of this, (shown in Fig. 1-2) uses a one-letter code for each nucleotide and a three-letter code for each amino acid.[§] The terms *sequence* and *primary structure* are often used interchangeably and, in fact, they are identical for all single-chain non-cross-linked structures. Note that neither term implies knowledge of the conformation of any individual residues or the polymer backbone.

[§] The choice of this code, by a happy coincidence, anticipated the discovery that nature uses the same type of coding.

Type	*Example*	*Abbreviation*
Polystyrene	$\rightarrow CH_2—CH\phi—CH_2—CH\phi—CH_2—CH\phi \rightarrow$	
Polyamide	$\rightarrow NH—(CH_2)_6—NH—CO—(CH_2)_4—CO—NH—(CH_2)_6—NH—CO \rightarrow$	
Polypeptide	$\rightarrow NH—CH(CH_3)—CO—NH—CH(CH_2OH)—CO—NH—CH(H)—CO \rightarrow$	–Ala–Ser–Gly–
Polysaccharide	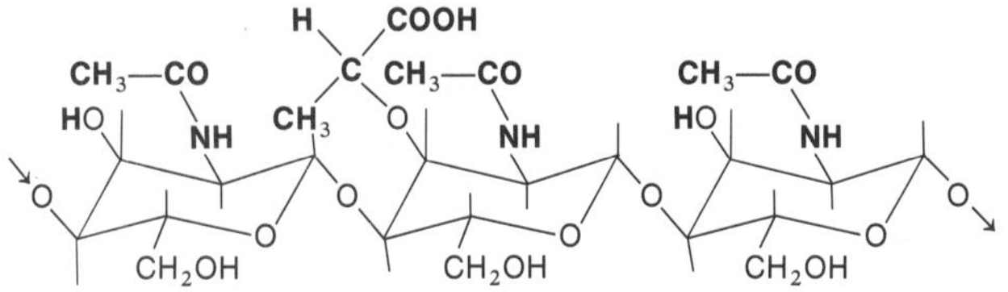	NAG–NAM–NAG
Polynucleotide	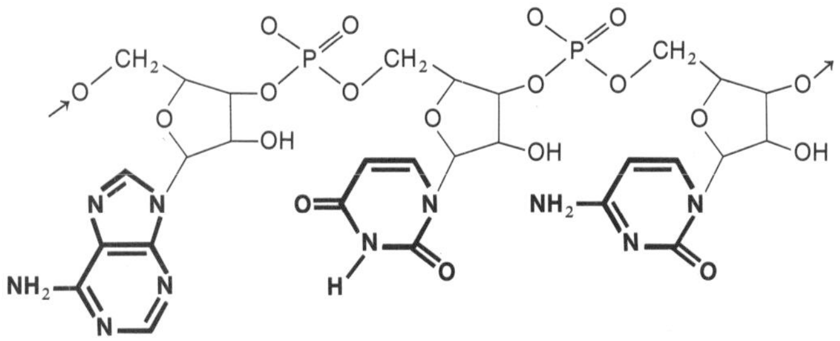	ApUpC

Figure 1-2

Typical linear polymers. Some synthetic materials (e.g., polyamides) have no preferred chain direction, whereas other synthetics and most biopolymers are head-to-tail arrangements of monomers. Backbone is shown in boldface; side chains are in lighter type.

Secondary structure describes helices of residues

Many biopolymer chains can form locally ordered, three-dimensional structures. For individual, linear, head-to-tail polymers, with asymmetric monomer units, the only kind of symmetric three-dimensional ordered structure possible is a helix. This contains a screw axis of symmetry. In cylindrical coordinates with the z axis equal to the screw axis, the position of the j_{th} backbone residue in a simple helical polymer of $n + 1$ residues is

$$z_j = j\, z_0 + \delta_z$$

$$x_j = r \cos(2\pi j z_0/P + \delta) \tag{1-1}$$

$$y_j = r \sin(2\pi j z_0/P + \delta)$$

where $j = 0, 1, \ldots, n$. As shown in Figure 1-3, r is the distance in the x–y plane from the z axis to a particular residue; z_0 is the vertical distance between adjacent residues; and P, the pitch of the helix, is the vertical distance that must be traveled for the helix to make one turn. The constants δ_z and δ define the position of the zeroth residue. In a more complex helix, each residue consists of several atoms. P and z_0 will be constant for all atoms. However, r and the phase factors δ_z and δ, although constant for every residue, are different for each atom within a residue.

A few simple limits of Equation 1-1 are worth noting. If $r = 0$, a simple linear structure results, generated by the translational symmetry operation $z \rightarrow z + z_0$, $x \rightarrow x$, $y \rightarrow y$. An exactly equivalent description of a linear structure occurs where $r \neq 0$ but $P = z_0$. Where $P = 2z_0$, the y coordinate of each corresponding atom of each residue alternates between $r \sin \delta$ and $-r \sin \delta$, whereas the x coordinate oscillates between $r \cos \delta$ and $-r \cos \delta$. Thus, the structure resembles an accordion or pleated sheet, as you can see easily by choosing $\delta = 0$.

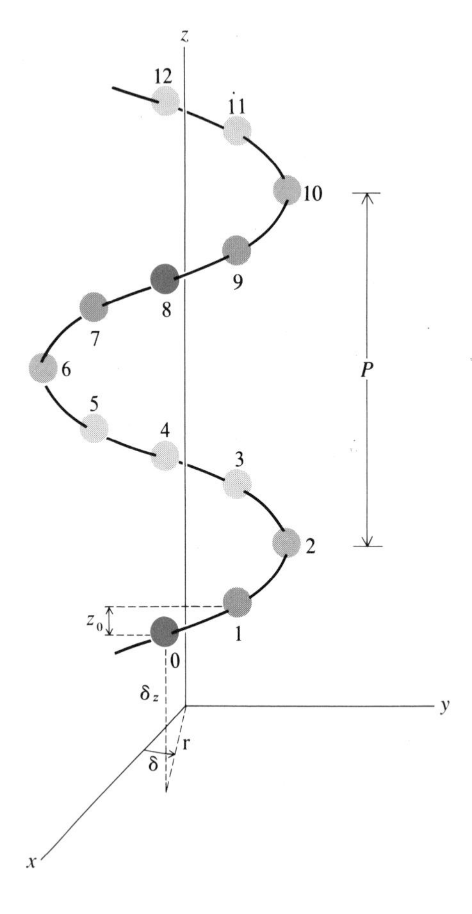

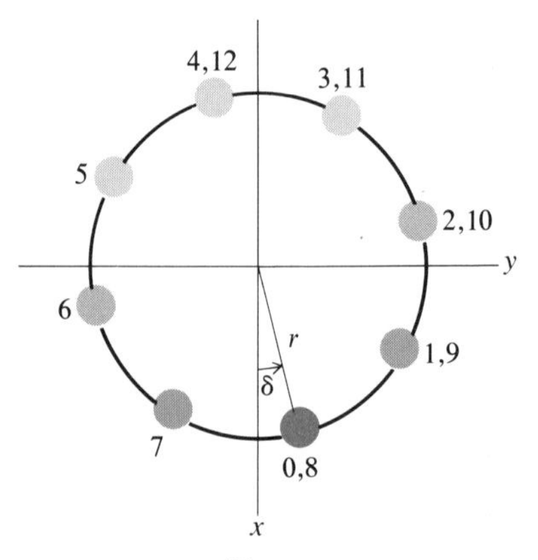

Figure 1-3

The geometry of a simple helix. The helix axis is coincident with the z-axis. P is the pitch, z_0 the vertical rise per residue. The location of the zeroth residue is further specified by δ (which defines its position in the x–y plane) and by δ_z (its distance above this plane). All identical repeating units are the same distance r from the helix axis; this distance is called the radius of the helix. The radius is shown more clearly in the top view. In the example illustrated, the helix is eightfold, so that the x and y coordinates of residues spaced eight residues ($8\ z_0$) apart are identical.

The quantity P/z_0 is the number of residues per turn of the helix. It need not be integral. A helix with n residues per turn is called n-fold, so the pleated sheet is a 2-fold helix; the polypeptide α helix is 3.6-fold.

Other helices can be formed by bringing together two or more individual helical strands. Double and triple helices are fairly common in nature. They can also be described by Equation 1-1, except that now one might have different fixed values of δ and δ_z for each atom type on each strand. In multiple-stranded helices, the individual covalent chains can run in parallel directions, but it is also possible to construct helices in which some strands run in opposite directions. The Watson–Crick DNA double-helical structure is a 10-fold double helix with antiparallel strands. Polypeptide β sheets are multistranded twofold helices in which some neighboring strands run parallel whereas others run antiparallel. In multiple-stranded helices, there will be symmetry operations that permit the coordinates of residues of one strand to be generated if the coordinates of another strand are known. Note that, because proteins and nucleic acids are copolymers that in general do not contain a periodic sequence of side chain types, all the symmetry operations we have been describing apply only to backbone atoms and not to the side chains.

The *secondary structure* (2° structure) of a biopolymer is defined as an enumeration of the particular regions of the primary structure involved in any kind of helix. More generally, secondary structure is a list of all three-dimensional regions that have ordered, locally symmetric backbone structures. For a protein, one would have to specify, for example, which particular residues are α helices and which are β sheets (and which polarity the strands have). The residues that are involved in neither of these are often called "random coil" for want of a better description, even though their structure may be far from random. Similarly, for a nucleic acid, one must identify the particular sequences involved in single, double, or triple helices, or no identifiable helical structure. About a dozen discrete secondary structure types are known for proteins and nucleic acids and a few more for polysaccharides. If a molecule is all one perfect secondary structure, such as an α helix or a DNA double helix, then a description of the secondary structure defines a precise conformation for the backbone and, if it is a periodic sequence, probably for the side chains also. For a molecule with a mixture of secondary structure types, the term conformation is also used to describe the totality of available knowledge about the overall structure.

Tertiary structure is the three-dimensional arrangement of residues

The *tertiary structure* (3° structure) of a protein or nucleic acid is the complete three-dimensional structure of one effectively indivisible unit. For a protein, this unit is usually one single covalent species, whether it contains a single polypeptide or more than one linked by covalent cross-links. For a nucleic acid, it is either a single covalent strand in the case of most RNAs, or two complementary double strands in the case of most DNAs. The tertiary structure includes a description not only of local symmetric structure (2° structure) but also of the spatial location of all residues insofar

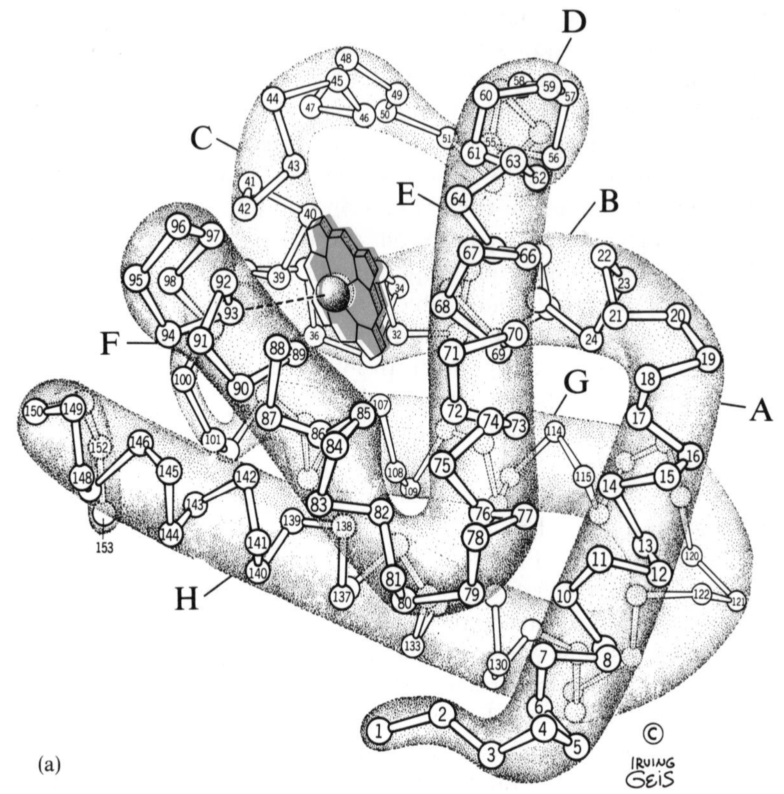

Figure 1-4

Three typical protein tertiary structures. (Only α-carbon positions are shown.) (**a**) Myoglobin consists mostly of α helices, labeled A through H, and contains a heme group (*colored*) that binds oxygen. [Drawing by Irving Geis.] (**b**) α-Chymotrypsin consists of three polypeptide chains and, at first glance, only a little α helix at the C-terminus of the C chain, but it also contains significant amounts of pleated sheet structure and four disulfides (*colored*). (**c**) Part of the collagen helix, which consists of three polypeptide chains, each of them in the helical form called poly-L-proline II. This is a tertiary structure because of the superhelix formed by interwinding these three separate helices.

as is possible. Three representative protein tertiary structures are shown in Figure 1-4. It is difficult for most observers to make much sense out of drawings of tertiary structures because too much detailed information is presented at once. Therefore, it is often desirable to present a schematic drawing of the tertiary structure by showing only the polymer backbone or sometimes only a representative location for each residue, such as the α-carbon for a protein. We have done this in Figure 1-4 to indicate three very different features that can contribute to tertiary structure. In myoglobin,

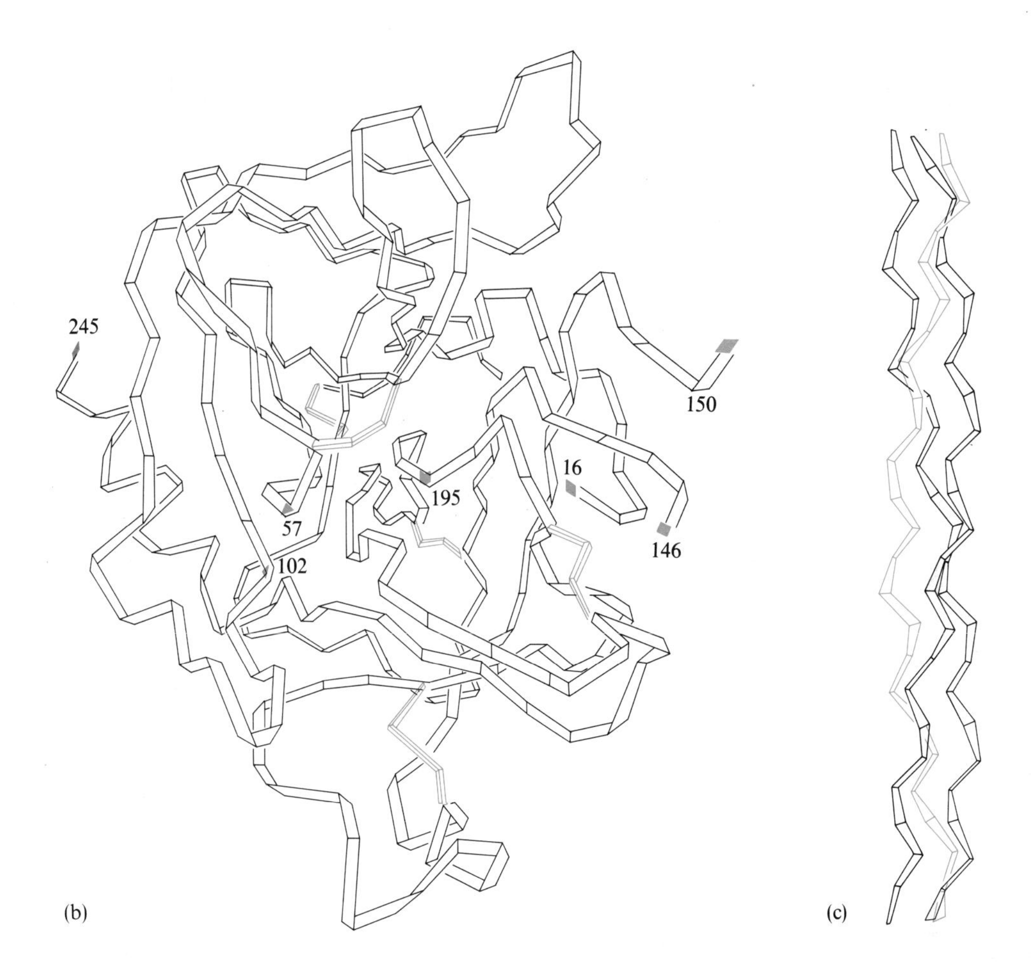

almost all residues exist in helical secondary structure, and so the tertiary structure is simply a description of the spatial orientation of these helical regions. Chymotrypsin has very little easily recognizable secondary structure, and here even a rough description of the tertiary structure must identify individually where each residue is located. A third tertiary structure feature is long-range symmetric organization. For example, superhelices can be formed by periodic perturbation of a simple helix. Superhelices can be single- or multiple-stranded. An example is the three-stranded collagen superhelix shown in Figure 1-4.

The term *conformation* is used synonymously with tertiary structure. This is because, if one knowns the conformation (i.e., the angle of rotation around all single bonds in a polymer), one indeed also knows the tertiary structure, and vice versa. Problems at the tertiary structure level do not necessarily imply that one knows or is interested in the *entire* three-dimensional structure. For example, the gross shape of the molecule is one aspect of its tertiary structure. So also is the question of whether a given residue is inside or outside of a compact globular protein or nucleic acid. At the present time, there are a few dozen proteins and one nucleic acid where the tertiary structure is known in sufficient detail to justify the construction of an approximate molecular model of the whole structure. New ones are appearing at the rate of about 6 to 10 per year.

Quaternary structure is the arrangement of subunits

The highest level of structure we shall consider is *quaternary structure* (4° structure). This is formed by the noncovalent association of independent tertiary structure units. The subunits of the quaternary structure may or may not be identical, and their arrangement in the quaternary structure may or may not be symmetric. An example of a simple quaternary structure is provided by vertebrate hemoglobins. These consist of four subunits—two each of two types (α and β), each of which is a single polypeptide chain folded into a compact globular tertiary structure. Each chain contains a bound heme group. The intact 66,000 dalton $\alpha_2\beta_2$ hemoglobin tetramer can be dissociated fairly easily into dimers.

$$\alpha_2\beta_2 \rightleftarrows 2\alpha\beta$$

Further dissociation into monomers requires more rigorous disruptive conditions. The actual geometry of the hemoglobin quaternary structure depends on whether the hemes in each individual subunit contain bound oxygen. However, as shown in Figure 1-5, each $\alpha\beta$ dimer is always related to the other by a C_2 symmetry axis (180° rotation). In addition, because α and β subunits themselves have quite similar tertiary structure, pseudosymmetry axes also exist relating their positions.

A typical more complex quaternary structure is the *E. coli* ribosome. This contains three RNA molecules and approximately 55 different protein chains. All but one of the latter are present in single copies; one protein occurs in four copies. The intact (2.6×10^6)-dalton 70S ribosome is easily separated into two subunits of unequal size: 30S and 50S particles.[§] As shown in Figure 1-6, these can each be broken down further into a range of intermediate particles, and finally into individual components. For structures such as the ribosome that are really a hierarchy of quaternary structure levels, it is sometimes convenient to use terms such as "quinternary structure" to define increasing stages of association. However, we shall avoid these terms.

The quaternary structure of a few multisubunit proteins is known fairly accurately from x-ray diffraction studies. These structures range from proteins as simple as superoxide dismutase (with two identical subunits) to molecules as complicated

[§] It is quite customary to name large structures by their sedimentation velocity (in Svedberg units, S) where actual enumeration of their contents either is unknown or else is too lengthy to bother with.

as aspartate transcarbamoylase (which has the subunit stoichiometry r_6c_6). A number of much more complex quaternary structures are known approximately through the use of electron microscopy, sometimes assisted by low-resolution x-ray measurements. Figure 1-7 shows an example for a very large assembled protein, the *E. coli* pyruvate dehydrogenase complex. This complex has a molecular weight of about 5×10^6 daltons and consists of three types of chains: t, p, and f. The latter two are associated into the dimeric units p_2 and f_2. The overall subunit composition of pyruvate dehydrogenase probably is $t_{24}(p_2)_{12}(f_2)_{12}$. Another example is tobacco mosaic virus, which has 2,130 identical protein subunits arranged in a helix around a single coiled RNA strand 6,400 nucleotides long. There is some information available about quaternary structure of even more complex systems—e.g., protein assemblies in striated muscle fibers, and protein substructures within rather complex viruses such as T2 and T4 bacteriophages.

1-2 SOME CENTRAL QUESTIONS IN BIOPHYSICAL CHEMISTRY

Now that we have presented the terminology with which the structure of large molecules can be described, it is possible to pose a number of questions that biophysical chemists often ask about particular molecules or assemblies. Not all these questions are answerable, and not all of them are relevant to every system. It is a constant challenge to assess which are really the significant ones for a given system, and then to design effective strategies for finding the answers.

Questions about sample quality

Some basic questions must be asked about the molecule or system itself to see if it is suitable for any kind of detailed study.

1. *Is it pure? Is it homogeneous in composition?* For example, does the sample contain only a single high-molecular-weight species? This can often be confirmed by measuring the molecular size using techniques such as ultracentrifugation, electrophoresis, and exclusion chromatography (to be described in detail in Chapters 11 and 12). Such techniques can also be used to purify one component from a mixture of macromolecules. A more detailed analysis of homogeneity often can involve chemical analysis, frequently supplemented by spectroscopic measurements. For example, questions commonly asked are whether a protein is contaminated with nucleic acid, whether it contains covalently attached sugars (and, if so, how much), whether it consists of an unbroken polypeptide chain (and, if so, its length). Note that it does not always pay to ask whether a protein or nucleic acid is free from contaminating small molecules because these are inevitably present. Most biopolymers are polyelectrolytes and as such must be accompanied by a counterion atmosphere. It is sometimes

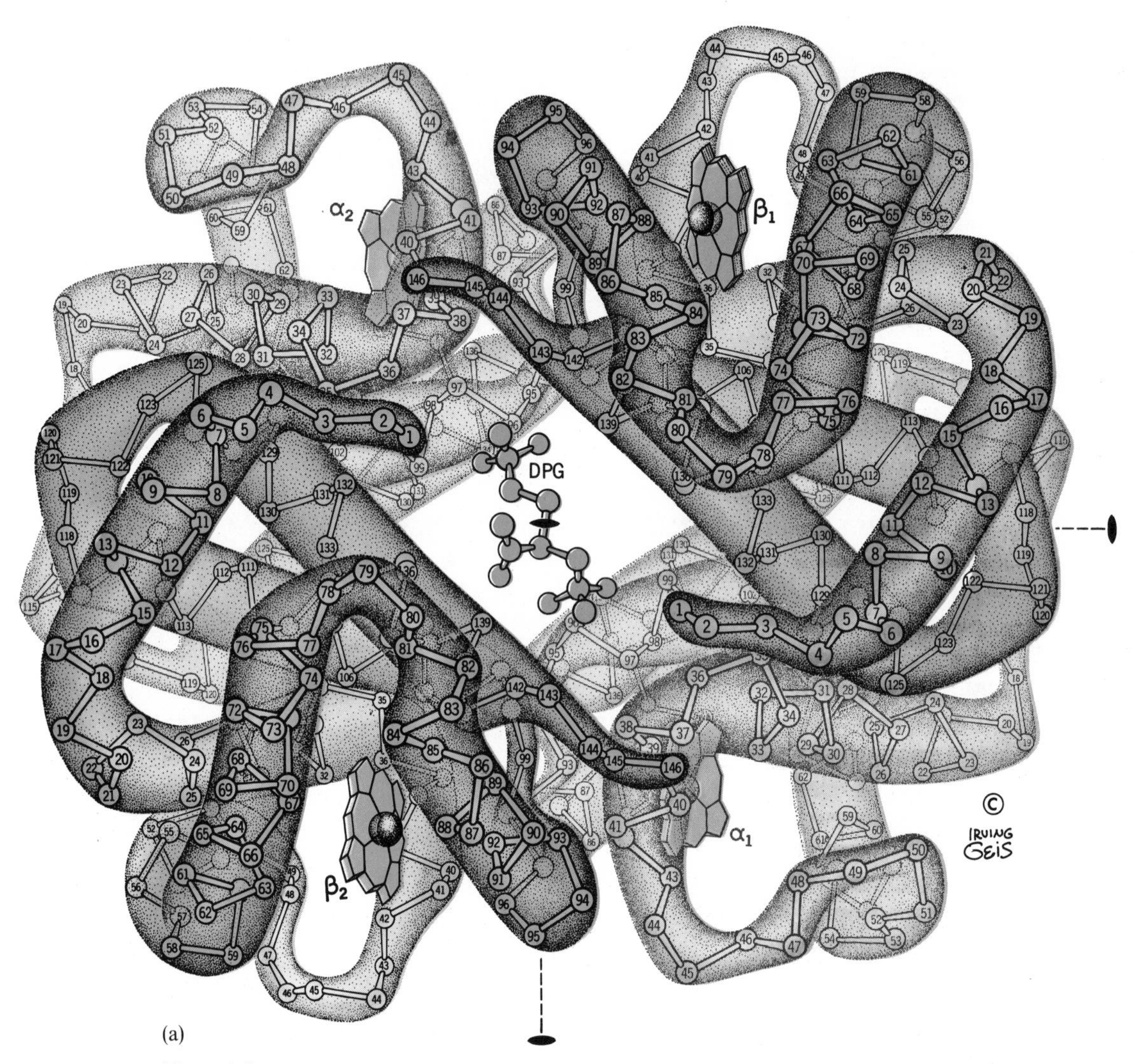

Figure 1-5

Quaternary structures of hemoglobin. There are four subunits, each with one bound heme. A twofold rotational symmetry axis exists in the center of the molecule perpendicular to the plane of the page. This 180° rotation interchanges α_1 with α_2 and β_1 with β_2. Because α and β chains are almost identical, two pseudotwofold axes exist as shown in the horizontal and vertical directions: one interchanges α_2

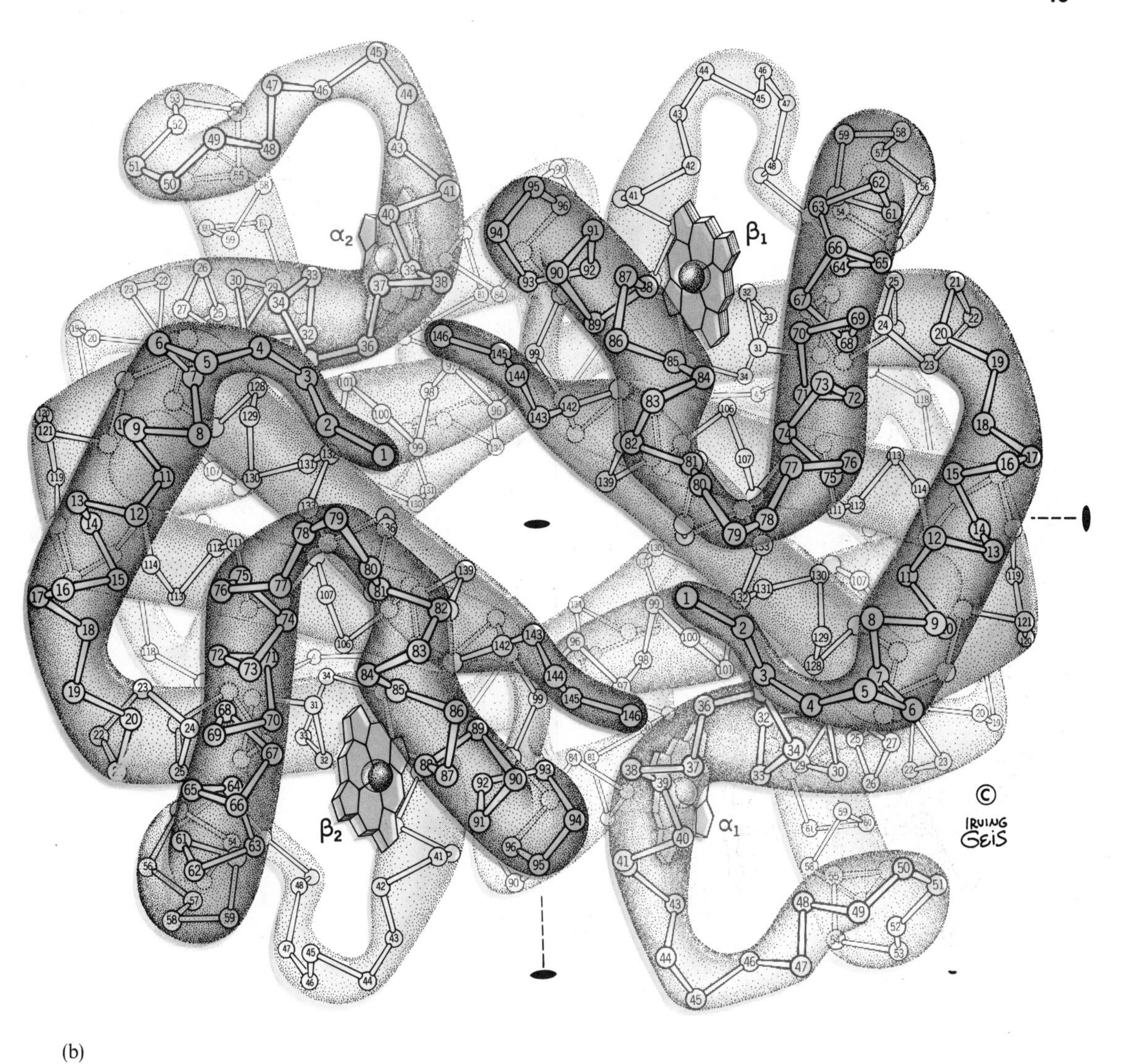

(b)

with β_1 and α_1 with β_2; the other interchanges α_1 with β_1 and α_2 with β_2. (**a**) Deoxyhemoglobin. Shown is a bound diphosphoglycerate molecule (DPG) that stabilizes this form. (**b**) Oxyhemoglobin. Note that the subunits have rotated to change the space between the β chains and destroy the tight, positively charged binding site of diphosphoglycerate. [Drawings by Irving Geis.]

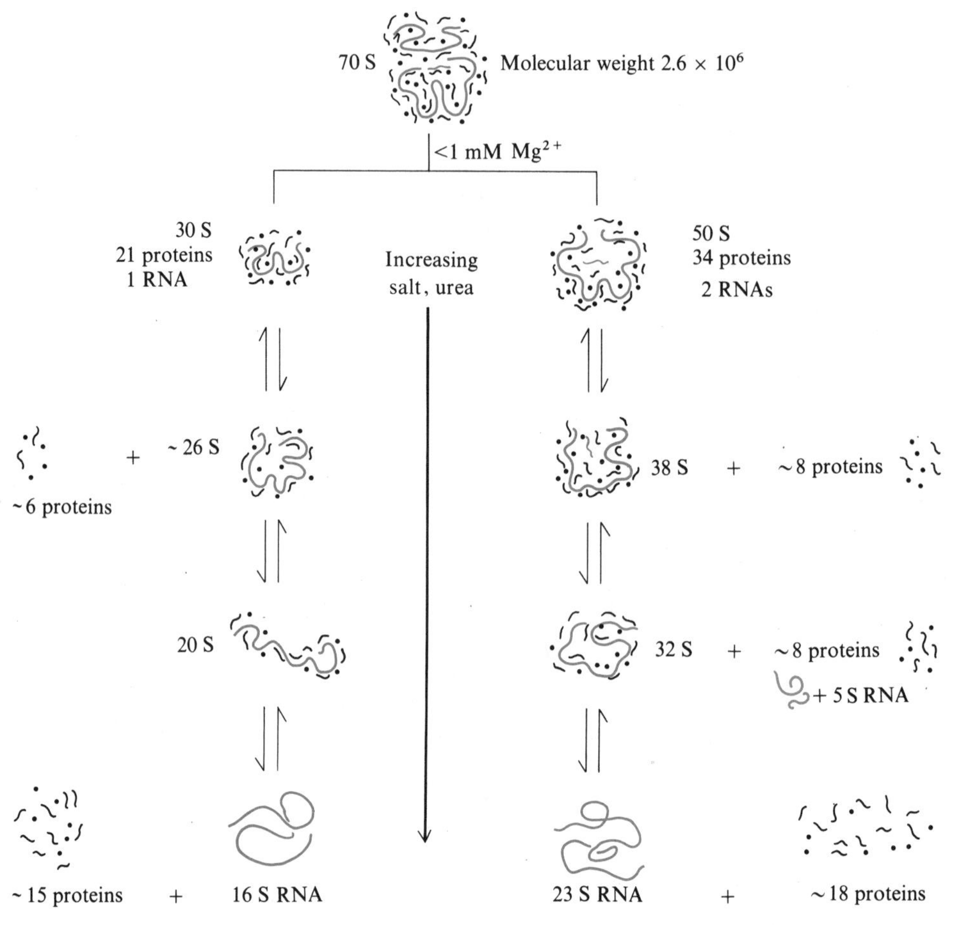

Figure 1-6

Some of the states known to occur when E. coli ribosomes are treated with progressively more severe disruptive conditions. The original 70S particle contains 55 different proteins (all but one in a single copy each) and three rRNAs.

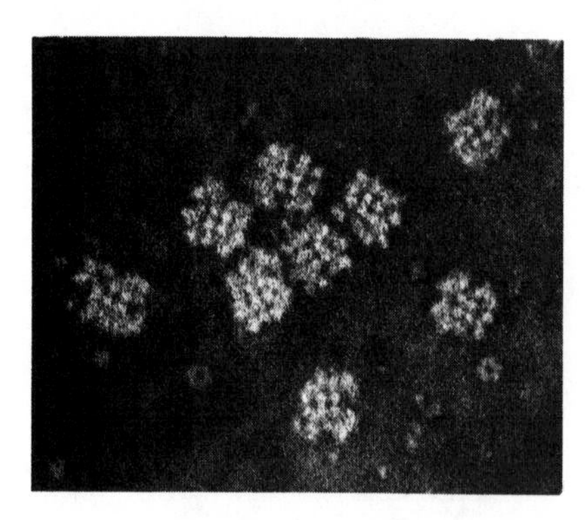

Figure 1-7

Electron micrograph of the E. coli pyruvate dehydrogenase complex, visualized through negative staining. [Courtesy of Dr. Lester Reed.]

important to gain control over which species of counterion are present. Very frequently, there are particular small molecules that either must be present or must be excluded for proper function of the macromolecule.

2. *Is it native? Is it complete?* It is important to know whether a molecule isolated in the test tube has the same structure and properties as it did inside the living cell where it was synthesized and where most biopolymers normally function. In many cases, we know that the answer emphatically is no. In vivo, many molecules are packaged or arranged into complex organizations. These are so complicated that studying them by physical means is virtually impossible. Furthermore, they are so fragile that most methods of opening the cell to get at the molecule disrupt the organized structure. For example, DNA is probably never free as an extended isolated molecule in vivo. It is always involved in interactions with proteins and seems to be packaged or organized into ordered forms, both in prokaryotes and in eukaryotes. Even the simplest viruses organize their DNA. The DNA inside a T2 bacteriophage is carefully packed and actually occupies most of the available free internal volume of the phage head. As shown in Figure 1-8, osmotic lysis to free the DNA causes a fantastic increase in the total volume in which the DNA finds itself, although the actual molecular volume of the DNA presumably remains essentially the same.

Because conditions in vivo cannot be precisely reproduced in vitro, one has several options. Sometimes, it is possible to study individual classes of molecules inside the cell. This requires either that the molecule of interest form a large fraction of the total cell weight (as, for example, hemoglobin inside erythrocytes) or that the molecule have some unique physical or chemical property that enables it to be seen above the background of the total cell constituents (as in the case of certain paramagnetic or intensely colored molecules). Alternatively, it is sometimes possible to increase the population of the desired molecule within the cell by chemical and biological manipulation or to incorporate some kind of probe or label, which then enables direct, physical observations to be made. These techniques are very important, but they are still in their infancy.

If the molecule cannot be studied in vivo, then one must be constantly aware that it could be different in vitro. The differences can be as dramatic as the phage DNA discussed above, or they can be very subtle (but no less important). For decades, hemoglobin was subjected to the most concentrated study any biopolymer has ever experienced. Much was learned, including the complete three-dimensional structure, but its functional behavior (oxygen binding) in vitro was perplexingly different from measurements in vivo. Furthermore, it varied from one laboratory to another. Finally, it was discovered that a small molecule, diphosphoglycerate, present inside the erythrocyte, dramatically alters the oxygen affinity of hemoglobin. In various laboratories, different amounts of diphosphoglycerate were lost during purification.

Damage to macromolecules during purification is also a considerable problem. Most cells contain enzymes capable of hydrolyzing or otherwise modifying proteins and nucleic acids. Normally, these are kept isolated or are otherwise restrained from attacking cell constituents, but they may be freed when the cell is disrupted.

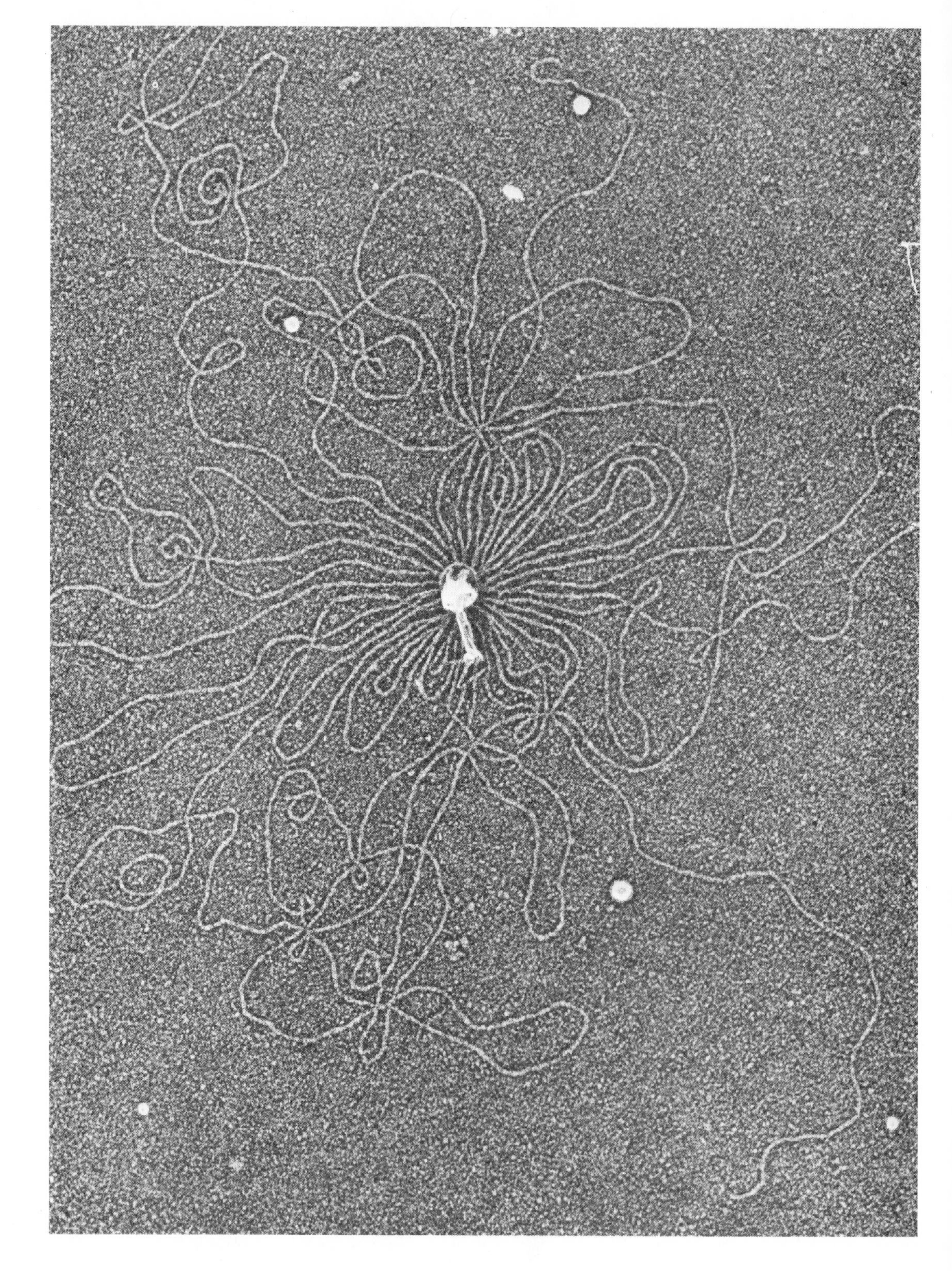

Figure 1-8

A bacteriophage T2 that has been osmotically shocked to free the DNA. It still is not known by what process the huge DNA was originally coiled up tight enough to be packed into the head of the phage in the first place. [From A. K. Kleinschmidt, D. Lang, D. Jacherts, and R. K. Zahn, *Biochim Biophys. Acta* 61:857 (1962).]

3. *Is it consistent?* When working with different preparations, do the molecules have identical properties? Variations in the state of the living organism from which the molecules were isolated, or unavoidable slight experimental variations in handling particularly delicate systems, can lead to altered structures. Functional assays provide one critical test of consistency. A number of physical measurements are sufficiently sensitive and nondestructive that they can often be used for quality control. When the source of variation can be identified, it is sometimes possible to circumvent it. For example, damage to RNAs by enzymatic cleavage can be minimized by isolating them from mutant bacterial strains containing defective or inactive degradative enzymes; the intrinsic variability of whole living organisms can be minimized by using animal cells grown in culture under standardized conditions. Genetic variations can be eliminated by using cloned cells that derive from a single genetically pure precursor.

Questions about structure

Once one is satisfied that a clean, reproducible system is at hand, a wide variety of potentially interesting questions can be asked about its molecular properties.

1. *What is the structure?* This is an obvious question and it can be answered on many levels, depending on the nature of the system and the amount of effort and material available. Determining the primary structure is primarily a biochemical rather than a physical problem. However, a knowledge of the primary structure is a prerequisite to many physical studies (see introduction to Part II). Occasionally, physical methods are helpful in sequence determination. Examples will be given later.

Information about the secondary structure is most easily obtained from various spectroscopic studies. Optical methods, discussed in Chapters 7 and 8, can tell reliably the kinds and amounts of secondary structure in a protein or nucleic acid. They generally cannot reveal which specific regions of the primary structure participate in various aspects of the secondary structure. NMR, when assisted by other techniques, is capable of yielding a fairly detailed assignment of individual residues to particular secondary structure regions of nucleic acids (Chapter 9).

The complete tertiary structure of a biological macromolecule is currently accessible only by x-ray crystallography and a few closely related diffraction techniques (Chapters 13 and 14). All of these techniques require that the molecule can be coaxed into forming a well-ordered crystalline array. This seems attainable for only a fraction of the biopolymers of interest at any time. Indeed, certain biological molecules or systems are disordered by their very nature, and these cannot be studied at high resolution by diffraction techniques.

In the absence of a crystalline structure, a considerable amount of useful information about the tertiary structure of a biopolymer can be obtained by combining many less informative techniques. Methods that can provide a picture of the size and shape of the molecule include electron microscopy and hydrodynamic techniques. These will be discussed in Chapters 10 through 12. A number of physical and chemical studies can tell us whether particular residues in the primary structure are relatively accessible to molecules in the solvent—and thus presumably on the surface of the structure—or whether they are buried. A few spectroscopic techniques exist that can provide higher-grade tertiary structure data. With much effort, distances between specific points within the structure can be measured by spectroscopic or scattering techniques. The difficulty is that all of these method provide only a single distance at a time.

If the macromolecule can be aligned, other spectroscopic techniques can reveal how certain secondary structure regions or individual residues are oriented within the tertiary structure. For quaternary structure, many different approaches allow the number and types of subunits to be measured. Usually these boil down to a careful quantitative analysis of the intact system and molecular weight measurements of all the subunits. If the system is large enough, the geometric arrangement of subunits can be seen by electron microscopy. Otherwise, inferences are possible from chemical techniques, such as cross-linking, or from some of the physical methods that provide distance information. Where only small numbers of subunits are involved, quaternary structure patterns sometimes are directly available by a careful analysis of hydrodynamic data.

2. *Is the structure predictable?* An affirmative answer to this question has been one of the major goals of biophysical chemists for years. First, consider secondary structures. There are two aspects of prediction. Can one predict or rationalize what the most stable geometry should be for any given polymer backbone by simply applying what one knows about covalent and noncovalent forces operating within it? As you will see in Chapters 5 and 6, the answer to this question is largely yes. A separate question is, given a particular primary structure and several choices of secondary structures, can one predict the overall pattern of secondary structure that will form under a given set of conditions? This means, for a protein, which residues will be found in helices or sheets and, for a nucleic acid, which residues will be in double helices? To make such predictions requires a knowledge of how all side chains affect the energy of each kind of secondary structure and also requires a way of searching among all possible structural arrangements.

The problem is easier for nucleic acids, and substantial progress has been made. The fundamental dominant interactions are Watson–Crick base pairs. Much of the potentiality for this is immediately visible by inspection of the primary structure.

Calculation of the energetics of particular base-pairing patterns requires a knowledge of less than 40 thermodynamic interactions, as will be demonstrated in great detail in Chapter 23. The problem becomes much harder with proteins. The sequence gives few immediate clues as simple as base-pairing complementarity. The number of thermodynamic interactions that must be considered is much larger because each residue can be one of twenty amino acids instead of only one of four nucleotides. In spite of these difficulties, moderately reliable qualitative schemes do exist for predicting the secondary structure of proteins, as we shall show.

A second stage for the prediction of a structure is to try to go from the primary or secondary structure to the tertiary structure. The mathematical complexity of this problem is fairly formidable, as will become apparent in later chapters. A rigorous solution seems beyond the range of any computers, existing or contemplated. However, by clever approximations and simplifications, substantial progress has been made in the past few years in developing methods of predicting how a protein or nucleic acid chain folds in three-dimensional space. However, a truly successful prediction of the tertiary structure of a polymer has not yet been demonstrated. A similar problem, but probably a much simpler one, is predicting the quaternary structure of an assembly, given the tertiary structure of the components. If these were rigid, the problem would be almost within the range of a brute force, trial-and-error energy calculation of all plausible arrangements. Such attempts have been made to explore possible structures of the fibers formed by aggregating sickle-cell hemoglobin. Here, a small amount of data on the structure of the fiber greatly facilitates calculations by restricting the number of geometric variables. In general, one can see an enormous role for structure-prediction methods in the future—perhaps not so much for *a priori* calculation, but rather as a means of incorporating all of the structural information available in an attempt to refine it to as accurate and detailed a structure as possible.

A question that is sometimes experimentally accessible is the extent of coupling between various parts of the structure. For example, can one helical region of a folded globular protein unfold without affecting any of the others? If you actually remove that region from the structure—say, by enzymatic cleavage—is it still helical as an isolated moiety? What effect does this removal have on the remaining structure? Suppose a protein or nucleic acid is subjected to environmental perturbation such as an increase in temperature? If a small perturbation can cause a large change in tertiary and secondary structure, it is called a cooperative structural transition. This is demonstrated in Chapters 20 and 21. Such a transition implies that different and (perhaps) distant regions of the molecule may have thermodynamically significant interactions. Hence, an *a priori* prediction of the structure will be difficult because it may be impossible to subdivide the structure into smaller pieces for the purpose of calculations.

3. *Is it rigid or flexible?* All molecules are at least slightly flexible. In a large molecule, one would like to know whether or not movements of whole side chains or larger units can take place. Some macromolecules, such as hemoglobin, have several fairly discrete, accessible conformations, as shown in Figure 1-5. This multi-

plicity is often exploited in the regulation of biological function. Other macromolecules exist in a whole ensemble of conformations, such as a typical denatured protein or a random polymer chain. Then one can only describe the average structure of the system. If one wants a molecule to recognize and bind only a very specific target, the simplest design would be to make it a rigid template that matches the target pattern very precisely. Sometimes, however, specific interactions are assisted by flexibility.

A few molecules, such as many immunoglobulins, exist as a set of relatively rigid units joined together by a few flexible links. It seems that inherent flexibility of this kind is associated with a specific, functional role of the molecule. Each immunoglobulin G (IgG) has two binding sites for the particular class of antigens against which it is specific (Fig. 1-9). If IgG were a rigid molecule, it would bind two antigens on a surface or in an array only if they were correctly spaced to match the geometry of the two sites. If the binding free energy per site were ΔG^0, the apparent macroscopic association constant between IgG and a single antigen would be $2\exp(-\Delta G^0/RT)$, where the factor of two enters because either arm of the IgG can match up with one of the antigens. However, because the IgG is flexible, it is much more likely that, once one site is bound, the second will also become attached. This is a manifestation of the "chelate effect." Although the enthalpy for filling the second site is the same as the first, the entropy loss is much less because the second site is already held near an antigen by the attachment of the first site. A rigid, bivalent immunoglobulin would be able to exploit the chelate effect for enhanced binding affinity only if by accident its structure permitted both sites to be occupied simultaneously.

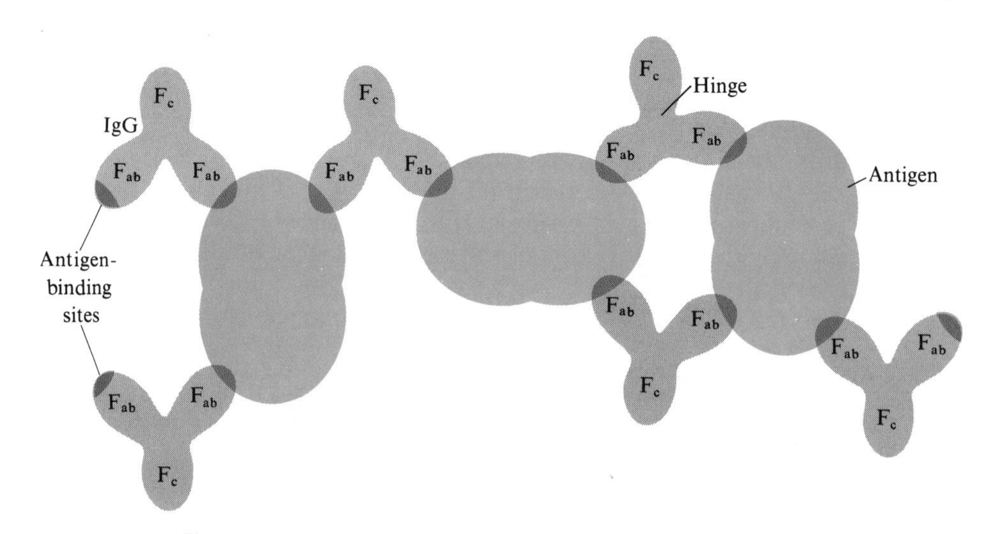

Figure 1-9

Schematic structures of antibodies. Several immunoglobulin G molecules in a typical antigen–antibody precipitate.

One way to explore the range of accessible conformations of a macromolecule is to vary the environment. The noncovalent forces that stabilize macromolecular structures are discussed in Chapter 5. From the properties of these forces, one can infer that alterations in temperature, pH, salt concentration, and solvent all have the potential to cause alterations in structure.[§] Studies as a function of these thermodynamic variables can often uncover a range of discrete structural forms and can also permit the enthalpy and entropy differences between these forms to be measured, as discussed in Chapters 20 through 23.

It is generally of interest to try to understand the effect of environmental perturbation on the secondary, tertiary, and quaternary structural levels. Preferential sensitivity to a particular variable often provides an important clue to the type of noncovalent interaction critically responsible for maintaining one structure in preference to others. A more specific kind of environmental perturbation is the coupling between conformational changes at different structural levels. For example, do alterations in the quaternary structure (such as removing one subunit) induce changes in the tertiary structure of any of the other subunits? Conversely, how does altering the tertiary structure of a subunit affect the nature or stability of the quaternary structure in which it participates?

4. *Is there anything familiar or unusual about the structure?* We have come to expect certain general features about each structural level of proteins or nucleic acids. Some of these are the consequences of the forces that determine structure. Even if we cannot predict the structure accurately, it is sometimes possible to recognize elements of it that appear to violate our current approximate level of understanding. Such structural features often are clues to functionally important questions. For example, with globular proteins, an inside and an outside can be roughly defined. Certain amino acid side chains are virtually never found on the inside. Charged groups, such as lysine or arginine, are normally solvated and have nearby counterions. Not only would burying an isolated side chain of this type cause a loss of the solvation energy, but there would also be an enormous electrostatic price to pay. For example, the attraction between two oppositely charged ions is roughly given by $e^2/\varepsilon r$, where e is the unit of charge, r is the distance between the ions considered as point charges, and ε is the dielectric constant of the medium between them. Outside the protein, ε is about 80, whereas it is probably a factor of 20 smaller on the inside. Thus, if one must bury a charge, there is an enormous energetic advantage in burying a suitable opposite charge as close to it as possible. Finding buried charges in a protein is like finding oases in a desert. There is probably some underlying reason why they are there.

Among the proteins or nucleic acids where considerable structural information is available, there is an impressive tendency for some different molecules to have surprisingly similar structures at various levels. Thus, it is always worthwhile to compare a new structure with the body of preexisting ones. If it falls into a particular

[§] Pressure and electric field are also important but are less frequently studied because experimental variation is more difficult.

family, such as the serine esterases, the information available from previous studies on members of the same family greatly expedites learning about the new structure. Observed structural differences are an immediate and crucial clue to the origin of any functional differences.

The origins of these structural families, and just how widespread they are, still are not clear. Certainly the major cause is divergent evolution. Primitive organisms must have contained far fewer proteins or nucleic acids than current ones do. An easy response to a new selection pressure would have been to duplicate a gene and then modify one of the daughters slightly to create a new function. In fact, primary or higher levels of structure can be used to trace the path of the evolution of organisms in much the same way that bone structures or other anatomical features are used. An alternative and more speculative explanation for some of the observed structural homologies among biopolymers is convergent evolution. For this, one argues that certain biological structures represent such an optimal solution to a functional problem that molecules originally distant from this optimum will drift toward it with time in response to selection pressure. It is difficult to prove that convergent evolution was a significant contribution to the pattern of structures visible today, but it is also difficult to rule it out completely.

Sometimes a structure is so unusual that it immediately suggests a function. For example, when the protein ferritin is viewed under the electron microscope, it appears as a protein shell surrounding an electron-dense core that chemical analysis indicates is predominantly Fe_2O_3. This makes it clear that ferritin is an iron-storage protein. Iron oxides are essentially insoluble in water and would be rather difficult for an organism to handle or transport. Ferritin solves this problem neatly by packaging the iron inside a protein.

5. *Do individual residues have altered properties?* In general, we understand fairly well the ionic equilibria, covalent chemistry, and physical properties of all the common side chains that make up proteins and nucleic acids. However, these properties frequently become changed when the residues are incorporated into ordered secondary or higher-level structures. This alteration can be a critical source of information about the local structure near a particular residue. The change in properties can be analyzed with some physical techniques at a much higher level of resolution than is possible when the structure is viewed as a whole with x-ray crystallography. As an example, certain spectroscopic techniques are highly sensitive to the exact bonding or electronic distribution around a given atom. If this atom has unique properties, it serves as a probe and can be viewed against the background of the rest of the macromolecule.

Anomalous chemical properties also are sometimes a clue to function. Hydroxyl groups such as the serine amino acid side chain are normally not very reactive toward electrophilic reagents, except at high pH when they are deprotonated. However, it has long been known that a particular serine residue in some proteins is highly reactive even at neutral pH. This is exploited by serine esterases that use the activated serine as a nucleophile in their catalytic site. We now know that the neighbors of this

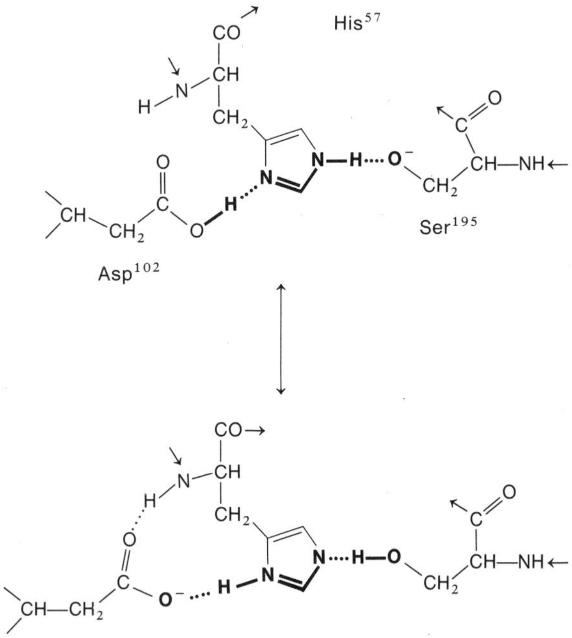

Figure 1-10

A model for hydrogen bonding between three residues at the active site of chymotrypsin and other serine esterases. Two tautomeric structures are shown. These constitute a charge-relay system that gives the OH of serine some partial negative character. However, recent evidence challenges the idea of a hydrogen bond between Ser^{195} and His^{57}; see D. A. Matthews et al., *J. Biol. Chem.* 252:8875 (1977).

serine in the tertiary structure interact with it in a way that is strongly activating. At first the activation was believed to proceed by a charge-relay system that stabilizes an ionized serine (Fig. 1-10). However, the best x-ray structural data now available suggest that much of the apparent reactivity of the serine arises from its favorable positioning next to the electrophilic carbonyl of a bound substrate, not from its increased nucleophilicity.

6. *How is the structure attained?* A fundamental problem in biology is morphology, the development and maintenance of form. For biopolymers, the mechanisms that are responsible for the formation of a structure may be different for the various levels. For primary structure, except for cross-links, this is essentially a biochemical problem.[§] The mechanism of the synthesis of specific protein and nucleic acid sequences is understood in considerable detail and will not be discussed here.

§ However, higher orders of structure may be involved in recently discovered RNA-splicing mechanisms.

Secondary structures, especially those involving the interaction of only nearby residues in the primary structure, form rapidly and spontaneously when enviromental conditions are favorable. Much is known about the detailed steps involved, and these are discussed for proteins and nucleic acids in Chapters 20 through 23.

The mechanisms by which tertiary and quaternary structures are formed are more complicated and more controversial. A few things are clear. Quite a number of protein or nucleic acid tertiary structures can form spontaneously in the test tube. These experiments are performed by starting with a native tertiary structure, denaturing as much of it as possible, removing the denaturing conditions, and monitoring the reappearance of the structure or its functional activity. As far as one can tell, the refolded molecules are identical in all respects to the native species with which one began. In some cases, it is possible to detect discrete steps along the refolding pathway, making details of the mechanism of refolding more accessible. However, it is not certain in most cases whether the attainment of tertiary structure is a thermodynamically or kinetically driven process. The former means that the final structure is thermodynamically the most stable of all possible structures and that the molecule has surveyed the ensemble of these structures in the process of selecting it. The kinetically driven process means that only a limited set of intermediate structures is attempted by the macromolecule as it sweeps through paths with low kinetic barriers to reach the final structure. As a result, the final structure could be a metastable one rather than the global thermodynamic minimum for the system.

Exactly the same kinds of considerations and uncertainties are applicable to the formation of quaternary structure. It is a monument to the experimental art of many talented investigators that quite a few of nature's most complicated quaternary structures—such as ribosomes, nucleosomes, and even whole viruses—can be reassembled from separated components in vitro. Hence, as in tertiary structures, the problem of greatest significance biologically is whether or not assembly in vivo proceeds in the same way as assembly in vitro. It is very difficult to answer this question rigorously in the affirmative. The most powerful approach is to find assembly-defective mutants and compare the intermediate states that pile up in vivo with stable intermediates that can be created in vitro. However, by their very nature, these mutants have altered assembly in vivo, and the intermediates seen may not correspond to states that are bona fide intermediates in normal organisms.

In some cases, it is possible to show that assembly in vivo must have proceeded differently from any process that simply involves a mixing of the components. There are two critical questions to ask. Are all of the molecules that participate in assembly present in the final structure? Do all of the molecules present in the final structure have the same primary structure that they had when they first joined the assembly in vivo? If the answer to either is no, then simple-minded direct assembly in vitro may not correspond to the precise process in vivo. For example, in the assembly in vivo of some viruses, scaffolding proteins appear to be necessary to help organize the viral shell or package the DNA within. However, these proteins leave the virion before assembly is complete.

Another very common occurrence is enzymatic modification of the tertiary or quaternary structure after folding. A striking example is insulin. As isolated, this

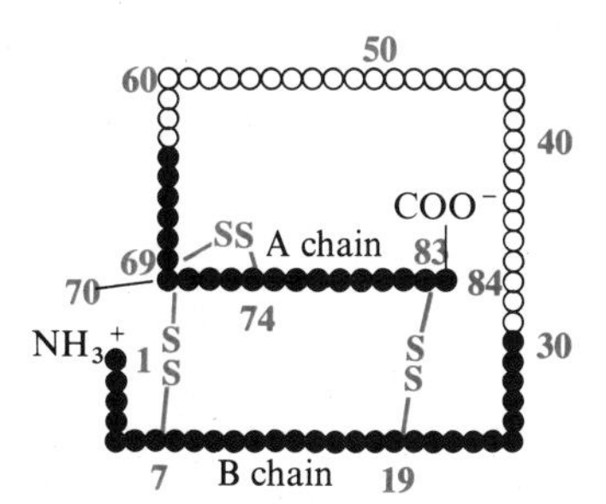

Figure 1-11

Primary structure of porcine proinsulin. Residues 31 to 63 are cleaved away to make the A and B chains of mature insulin.

small protein consists of two polypeptide chains joined by two disulfide cross-links (Fig. 1-11). There were numerous attempts to recombine the two separated chains in vitro, all of which were frustrated by extremely poor yields of regenerated insulin. Then, it was discovered that insulin is synthesized in vivo as a single polypeptide chain. After it folds, a large central piece is enzymatically cleaved away to produce the two-chain structure that accumulates in the cell. Where such covalent modification accompanies assembly, understanding the mechanism depends upon unraveling the temporal order of the various folding and modification steps. It is challenging to try to decide whether a particular modification is required as a thermodynamic prerequisite for a given folding step, or whether it occurs simply because of kinetic factors.

Questions about function

Knowing the structure of a macromolecule without any information about its function is like observing the structure of a building without knowing anything about its use or its inhabitants. It can be aesthetically pleasing but intellectually frustrating. Some questions about function are clearly in the realm of biochemistry, or biology, but here are a few that demand the attention of biophysical chemists.

1. *Can the function be predicted or rationalized?* Given only the structure of a molecule, it would be an enormous feat if all ramifications of its biological function could be predicted with no additional experimental information about that particular molecule, but with only our general knowledge about cellular biochemistry and the structure of other molecules. One can call this question molecular archeology because it is akin to the problem of trying to reconstruct the lives and social structure of a society from the buildings and other artifacts it left behind. For example, suppose one had deduced the active site structure of chymotrypsin, shown in Figure 1-10. Could one predict from this information that chymotrypsin is a hydrolytic enzyme? that it has very different rates of hydrolysis for esters and amides?

Perhaps a more realistic and approachable question is what additional information besides the structure is needed for a thorough understanding of the function. Generally, to construct a plausible mechanism for action of an enzyme, kinetic and thermodynamic data are needed in addition to structures. There is a big difference between knowing that a process is structurally plausible and claiming that it is energetically or kinetically feasible enough to be biologically significant.

2. *Is all of it necessary?* This question pertains to each level of structure in every particular function. For example, is it only the intact quaternary structure that can perform a certain function, or will isolated subunits do the job identically? If so, which subunits serve that function? One of the most interesting cases in which the properties of individual subunits can be compared with a native quaternary structure is the protein tryptophan synthetase from *E. coli.* This has a subunit composition $\alpha_2\beta_2$ and, upon dissociation, isolated α and β_2 units can be purified. These catalyze the following reactions:

α subunit: Indole-3-glycerol phosphate ⇄ Indole + Glyceraldehyde-3-phosphate

β_2 subunit: Indole + Serine ⇄ Tryptophan + H_2O

When the intact $\alpha_2\beta_2$ enzyme is used, the overall reaction observed is

Indole-3-glycerol phosphate + Serine ⇄ Tryptophan + Glyceraldehyde-3-phosphate + H_2O

With this last reaction, no evidence for the production or utilization of free indole can be found. This can be tested, for example, by adding radiolabeled indole to the reaction mixture and seeing whether it incorporates into tryptophan.

The rate of tryptophan synthesis by the intact enzyme is an order of magnitude faster than what would be expected from the properties of the isolated subunits. Is this due simply to the failure of indole to equilibriate with the solution? To examine this, the kinetic properties of mutant forms of the enzyme were studied. From the known genetics of the system, proteins could be isolated containing either inactive α chains, $\alpha_2^i\beta_2$, or β chains, $\alpha_2\beta_2^i$. These are capable of carrying out only the partial reaction catalyzed by the isolated active subunit. However, the rate of catalysis of $\alpha_2^i\beta_2$ is considerably faster than β_2. The rate of $\alpha_2\beta_2^i$ is faster than that of isolated α. Therefore, in the native quaternary structure, each subunit somehow alters the properties of the other to enhance its intrinsic catalytic rate.

What features of the tertiary structure are required for a particular function, and which if any are irrelevant? In general, a fairly intact tertiary structure seems necessary for the function of most proteins and nucleic acids. By comparing the structure of native and denatured forms, it is sometimes possible to discern the features of the tertiary structure that must be properly arranged for biological activity. For a polyfunctional molecule, dissection by progressive structural disruption often permits regions with different functional roles to be assigned. It is a fairly safe biological speculation that all of the primary structure serves some role biologically. If it didn't, it is hard to see why the selective processes of evolution would not have caused the unessential part to become deleted from the structure and no longer synthesized. The role may be fairly indirect. For example, part of the structure may be needed just to facilitate proper folding of other parts. In practice, it is sometimes found in vitro that part of a structure can be dispensed with for all known functions. This probably means that we simply do not yet understand the range or subtlety of the capabilities of that macromolecule or the factors in vitro that regulate its production, activity, and destruction.

3. *With what other molecules does it interact?* Usually the first hints of interaction between a macromolecule and any of the myriad of small or large biological molecules comes from functional studies. Then one often needs to know how many of a given type of molecule can interact simultaneously, how strongly they bind, and whether all appear to bind identically. These questions can be answered by direct thermodynamic measurements of binding (such as equilibrium dialysis) or by many less direct spectroscopic methods. If the molecules that bind have some convenient spectroscopic property (for example, if they are colored), they can be monitored directly. If not, then spectroscopic properties of the molecule that does the binding can sometimes be followed. More rigorous studies examine the thermodynamics and kinetics of the binding process.

For small-molecule interactions, systematic modification of the structure of the small molecule is often possible, and this can reveal exactly which features are recognized by a protein or nucleic acid and for what purposes. As a simple example, consider the protein transferrin. It binds Fe^{2+} very strongly to two equivalent sites. Its binding affinity for many other common biological divalent cations such as Ca^{2+} and Mg^{2+} is much lower. It turns out that transferrin is an iron-transport protein.

If a macromolecule interacts specifically with a variety of other molecules, then one wants to know how the presence of one affects the binding of others. Competitive binding suggests that either the same (or overlapping) sites are involved, or mutually exclusive separate sites are involved; the latter might arise, for example, when the binding of one molecule causes a structural change that eliminates the binding site of another molecule. Alternatively, binding may be cooperative. The presence of one small molecule could stimulate binding of others, either of the same type or of entirely different types. These effects can become very complicated if several different kinds of interactions occur simultaneously. We shall show later how a number of such cases can be analyzed.

4. *What is the structure of the functional site?* Most of the functions of large molecules can be localized at certain regions of the structure. Then all of the questions about structure that we raised before become of critical importance in these regions. In addition, one would like to know which residues of the macromolecule actually contact the molecule that is bound to it. Are any positioned in such a way as to suggest a catalytic role? A particularly important question is whether function or binding is accompanied by a structural change, either in the host macromolecule or in the bound molecule. For example, many proteins exist as a mixture of several discrete conformations. Only one of these may be found once the protein is saturated by binding some other molecule. Many small molecules that bind to double-stranded nucleic acid helices cause the strands to untwist somewhat, lengthening the whole

structure. All of this indicates that it is not necessarily safe to look at the structure of a passive macromolecule and assume that its structure will be identical when it is functioning.

If there are structural changes important for function, one of the first objectives is to identify the level of structure at which these occur and the residues that are involved. In the case of oxyhemoglobin and deoxyhemoglobin (Fig. 1-5), the major structural change that accompanies oxygenation appears to be a rearrangement of the quaternary structure. But this can be misleading. The oxygen binds to heme groups located nowhere near the contact points of the four subunits of the protein. Yet, for a quaternary structural change to occur, somehow the properties of residues at the interface must be altered. In fact, when oxygen is bound to a heme group in hemoglobin, the iron atom of the heme moves. This induces a series of subtle tertiary structure changes that produce an altered interface. These changes are no less important to the understanding of the mechanism by which hemoglobin binds oxygen cooperatively than is the more dramatic quaternary structural change.

Sometimes the macromolecule shows no evidence of a structural change when another molecule binds to it, but that molecule instead undergoes a structural change. A beautiful example of this occurs when the enzyme lysozyme binds a hexasaccharide. As shown in Figure 1-12, one of the six sugar-binding sites on lysozyme cannot accommodate a sugar in the normal chair conformation. The sugar ring must distort to a half-chair form in order to bind. This is energetically unfavorable. However, it occurs because the binding energy of the other sugars at their sites more than compensates. Looking at the whole process in a simplified way, lysozyme is capable of taking some binding energy and concentrating it at one point in the structure of the saccharide complex. The result of this is to assist the breakage of a carbon–oxygen bond in one sugar as part of the catalytic mechanism of the enzyme.

In lysozyme, structural changes at a functional site are large enough to be seen fairly clearly. However, much smaller changes can be important in facilitating chemical reactions at an active site. So are rather intimate details of the local environment. A major goal of biophysical chemistry is a quantitative explanation of the effectiveness of enzymes as catalysts. One of the obstacles in such explanations is the subtlety of many of the contributing factors. It can be difficult to obtain measurements that are sufficiently accurate to reveal them.

5. *How is function regulated?* A basic feature of most processes associated with living organisms is homeostasis, the tendency of the organism to resist or compensate for change. The ability of biological molecules to self-replicate and to catalyze vast networks of reactions carries with it an inherent risk of chemical instabilities. After all, an explosive detonation is simply an autocatalytic reaction. Clearly, biological organisms have learned to control their chemistry so that a correct balance of materials is synthesized, available energy is used efficiently, and fluctuations in a wide variety of environmental parameters are smoothed out. This demands that the function of many biological molecules must be amenable to control in response to the needs of the organism. This regulation can take place at many different levels.

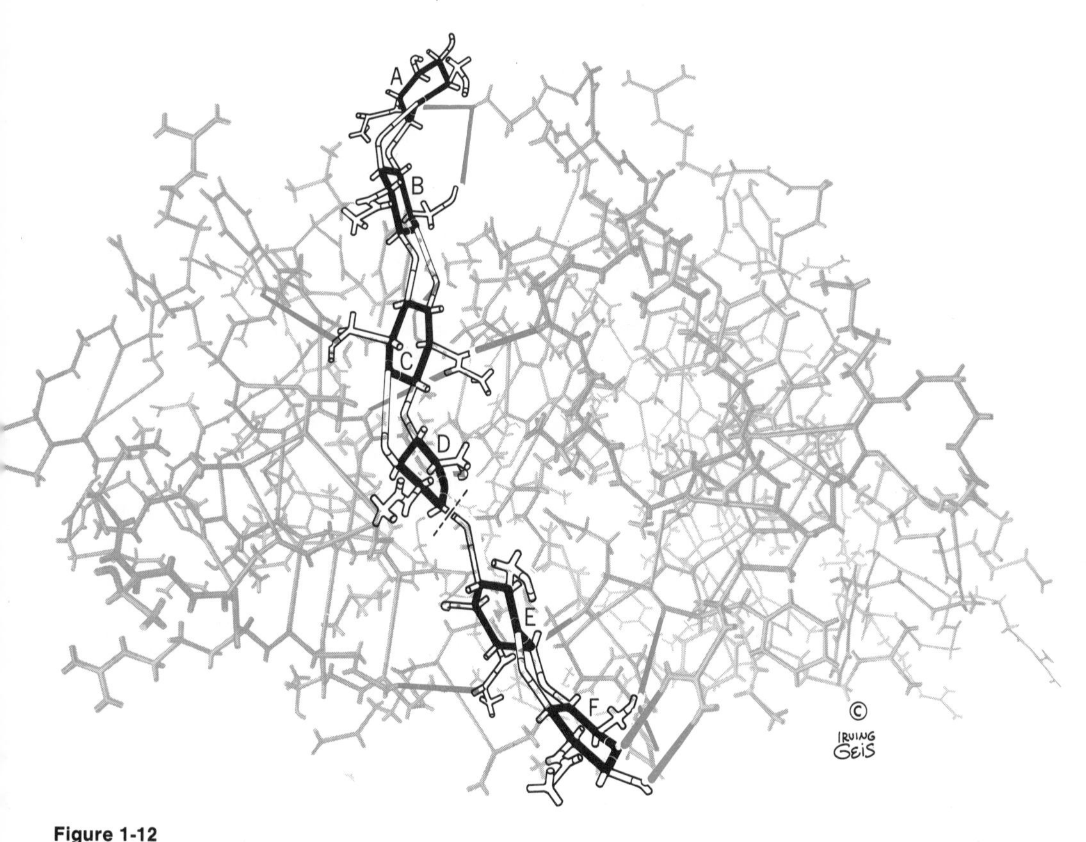

Figure 1-12

Tertiary structure of hen egg-white lysozyme containing the bound substrate (NAG–NAM)$_3$. See Figure 1-2 for the chemical structure of this substrate. Colored rods show hydrogen bonds between the enzyme and the substrate. Note that the sugar labeled D is distorted so that four atoms lie in a plane. The bond cleaved by lysozyme is shown by a dashed line. [Drawing by Irving Geis.]

Consider the activity of an enzyme in the cell. One level of control is the total amount of enzyme present. This is adjusted by altering the rates of its synthesis and degradation. Another control point is the fraction of the total enzyme that is active. This can be adjusted, for example, by quaternary equilibrium between active and inactive forms. These equilibria may be regulated by effectors, molecules that do not necessarily participate in the reaction catalyzed by the enzyme but simply shift it between active and inactive conformations.

The activity of an enzyme can also be controlled through altering the number of available catalytic sites by filling some with inhibitor molecules. Activity can be adjusted by changing the binding strength of the molecules that are destined for catalysis, by changing their rate of association or release, or by altering the intrinsic catalytic rate of the enzyme. The effective activity can also be adjusted at a more complicated level, by regulating the amount of substrate molecules on which the enzyme acts, or by controlling their location in the cell and therefore determining whether they can reach the enzyme.

Exactly the same kinds of considerations apply to proteins and nucleic acids that have no role in catalysis. Their functions are also regulated at a variety of levels. An excellent example is hemoglobin. We mentioned before that diphosphoglycerate alters the properties of hemoglobin. Specifically, diphosphoglycerate binds to the deoxy form of hemoglobin (Fig. 1-5). It thus stabilizes this form relative to the oxy form, and this makes it more difficult to bind oxygen at a fixed oxygen concentration. The biological effects of altering the diphosphoglycerate level are shown in Figure 1-13. Hemoglobin delivers oxygen from the lungs to the capillaries in all body tissues. The amount of oxygen that hemoglobin binds in the lungs is determined by the partial pressure of oxygen in air. This is about 100 torr at sea level. In the capillaries, hemoglobin releases oxygen to attain a new equilibrium that is governed by the level of oxygen dissolved in the tissues, about 30 torr. The difference in the two levels of binding is the amount of oxygen each hemoglobin molecule can deliver to the tissues.

Diphosphoglycerate shifts the oxygen binding curve of hemoglobin inside the erythrocyte. Because of the shape of this binding curve, an increase in diphosphoglycerate causes hemoglobin to unload oxygen more efficiently at lower atmospheric

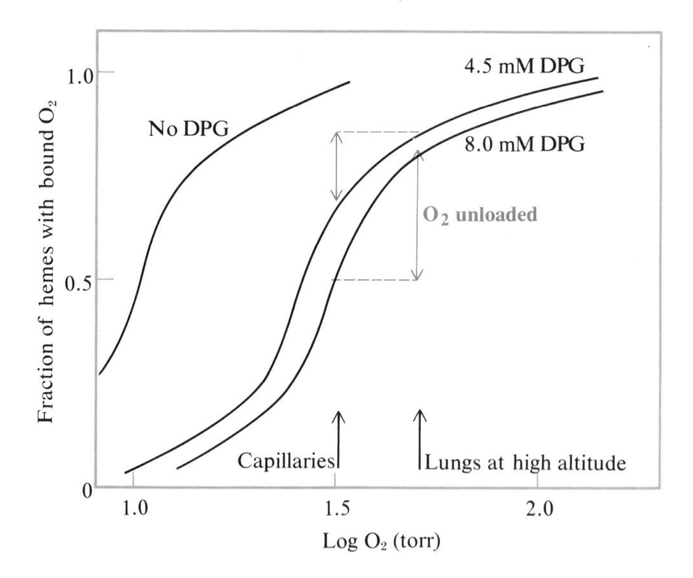

Figure 1-13

Oxygen binding of hemoglobin as a function of the partial pressure of oxygen. Shown are the approximate partial pressures in capillaries (30 torr) and in the lungs at high altitude (50 torr). Binding curves are shown for three different levels of diphosphoglycerate. The ability of hemoglobin to unload oxygen depends on the difference in amounts bound in the lungs and in the capillaries. Note how this difference is magnified at high diphosphoglycerate levels.

levels of oxygen. Therefore, one of the body's immediate responses to a decrease in oxygen pressure (such as an increase in altitude) is an increase in the level of diphosphoglycerate. This is a stopgap measure. A long-term solution is to synthesize more hemoglobin and increase the number of erythrocytes. This does occur, but it is a much slower process, taking several days whereas the diphosphoglycerate adaptation can occur in hours.

The process of regulation can clearly become quite complicated for molecules that have more than one biological function. Can the various functions proceed simultaneously? If so, can they be regulated independently? If not, what decides which function is dominant? In practice, biophysical chemists usually concentrate on the regulation of individual macromolecules. However, if a whole set of macromolecules must cooperate in some function, clearly there must be a way not only of monitoring the individual species but of regulating the whole ensemble as well. This task appears to be handled by more general regulatory processes—that is, large and small molecules designed and synthesized solely for the purpose of control. These include steroid and peptide hormones, cyclic nucleotides, and the proteins that synthesize or detect these. Such proteins are among the most interesting targets for biophysical studies because of their central role in biological processes.

The functions of other large molecules are the maintenance of cell structure and the control or transport of material inside or outside the cell. The study of regulation of these functions should prove fascinating because it may be necessary to coordinate on a molecular level events occurring at great distances from one another.

1-3 SOME BASIC STRATEGIES OF BIOPHYSICAL CHEMISTRY

Many of the questions we have posed are hard to answer. They would be difficult if only small molecules were involved, and the complexity of the large molecular structures imposes severe constraints on the types of possible experiments and magnifies the effort involved in any given experiment. Fundamentally, the most serious obstacle is the large number of variables present in the system. These include the enormous number of parameters needed to define the structure of the macromolecule and the large number of potential environmental variables relevant to its function. The large number of variables has two unpleasant consequences. The number of possible experiments increases geometrically with the number of variables. Therefore, the probability of choosing a significant experiment at random becomes hopelessly small. The second problem is that, even if a huge number of experiments have been done, it becomes quite difficult conceptually to visualize the results as a total unit.

Most people find it more comfortable to think about a response to a single variable that can be plotted as a graph in two dimensions. It is still feasible to handle two variables simultaneously because this can be handled as a surface in three dimensions. But beyond this, it is almost always necessary for people to group no more than two variables at a time. A protein structure may entail hundreds of significant variables, and few people can retain this much for immediate recall and correlation. Therefore, one must simplify the system for detailed biophysical studies. The essential strategy is a well-known one: divide and conquer. Isolate a portion of a structure, a variable, or a phenomenon. Control the rest or try to keep them constant, so that one aspect at a time can be analyzed and understood. There are a few basic ways to carry out this strategy.

Use a smaller system as a model

The available structural resolution of whatever techniques are used decreases with the size of a molecule. In principle, there is no reason why the structure of a large system could not be obtained at 1 Å resolution or better, but there are severe practical difficulties (see Chapter 13). It is far easier to study a small piece of the system at very high resolution and then to use these results to infer its structure in an intact system. For example, it is difficult to obtain accurate atomic coordinates for α-helical regions within entire globular proteins or for double helices within RNA tertiary structures. Therefore, it pays to study small model helical compounds as structural models. Once this is done, a way must be found to insure that the structures so obtained do not change significantly when they are part of a larger molecule. As another example, spectroscopic techniques are very powerful structural tools in molecules small enough that the contributions of every residue can be identified and monitored independently. Because spectra of larger molecules are an average over many residues, detailed interpretation is difficult. If, however, the properties of individual residues in specific environments or conformations are known from model studies, then it is sometimes possible to fit the spectrum of the polymeric system with specific hypotheses about the structure.

Observe only one part of the system

There is always a risk that working with a physical fragment of the system may be misleading because it need not behave the same way as the intact system. One alternative is to find a way to view only one part of the intact system at a time. This is the basic goal of probe or label studies. Sometimes the macromolecule contains a built-in probe such as a transition metal ion near or at a functional site or a colored prosthetic group like a heme. Alternatively, the investigator can introduce a probe at a fixed point in the structure by covalent attachment or by noncovalent binding. The type of probe used determines which features of the environment it reveals. These include polarity, flexibility, the presence of specific macromolecule residues or other bound molecules, the actual conformation of the probe, its electronic structure, and the nature of its binding to the macromolecule.

There is a risk inherent in all studies that rely on the insertion of a probe. The price of peeking is that the probe may perturb the structure. The region most likely to be perturbed is the area immediately adjacent to the probe. Unfortunately, this is the very region the probe is sensing.

Compare two systems that are almost identical

A straightforward approach to isolating individual structural variables is to make small perturbations in the structure and observe the results. How is this done? One approach is to compare the properties of closely related molecules. For example, some proteins (such as cytochrome *c* or hemoglobin) and some RNAs (such as transfer RNA) have been isolated from many different species. Because the resulting structures and functional properties are quite similar, one can learn what parts of the structure appear to be held constant. Presumably these regions are necessary for function. It can be more informative to compare structures that are similar but do not have the same function. Then, critically important structural features can sometimes be identified. For example, tRNA species involved in bacterial cell-wall peptidoglycan synthesis have certain distinct structural changes from the otherwise similar molecules that participate in protein synthesis on the ribosome (see Chapter 3). Sets of proteins exist that have homologous structures but different functions. One example is the serine esterases: chymotrypsin, trypsin, elastase, etc. There are even cases of molecules, such as superoxide dismutase and part of immunoglobulin G, that are completely unrelated in function (as far as we can tell) but that still have similar structures.

A particularly effective strategy for isolating structural and functional variables is the use of mutant proteins or nucleic acids. These have the intrinsic advantage that only a single or few amino acid or nucleotide changes have taken place. Yet this has produced the functional difference that allowed identification of the species as a mutant. Enormously clever tricks have been developed for bacteria; these techniques selectively generate a mutation in a specific protein or nucleic acid and persuade the organism to amplify the synthesis of the mutant species so that it is easy to purify it in quantity. For example, hundreds of mutants of the lactose repressor protein with individual amino acid substitutions have been accumulated. One is now in a position to change amino acids around almost at will by sophisticated genetic manipulation. It is not a simple task, but it is much easier than organic synthesis because, once one has constructed the necessary strain, it goes on making the mutant protein forever.

The state of the art of genetic manipulation is much less advanced with eukaryotes. Here one relies much more on naturally occurring mutations that result in a deleterious phenotype. This is spotted in the field or in a laboratory or hospital, sometimes as the result of systematic screening, but frequently as a by-product of other work or analysis. For example, many hemoglobin mutations lead to anemias, and these are readily recognized clinically. As a result, a vast number of mutants of human hemoglobin are known, the specific amino acid differences identified, and (in some cases) the functional effect understood.

For example, hemoglobin Sydney has a single valine → alanine change in the 67th position of the β chain. The resulting phenotype observed clinically is hemolytic anemia. This means the red blood cells are fragile and break easily. When molecules of hemoglobin are purified and studied, it is found that the heme groups from the β chains are very weakly held and tend to fall off. This will have two results. The heme-deficient protein is now much less effective as an oxygen carrier, both because it has two fewer hemes to bind oxygen and because the shape of its oxygen binding curve is altered, resulting in less efficient discharge of oxygen in the capillaries. Also, the heme that falls off is harmful. It probably interacts with the erythrocyte cell membrane and in some way results in the increased fragility of the erythrocytes. The molecular interpretation of the instability of heme binding to hemoglobin Sydney is straightforward. In normal hemoglobin, the valine helps form a pocket in which the heme binds. Its affinity for heme is due in part to the desire of heme to escape from water into the nonpolar environment of the pocket. With alanine present in the mutant, the pocket is too small for a good fit, and the empty space is presumably filled by water. This clearly weakens the nonpolar interactions.

Where genetic variation is not available or is insufficient, selective chemical modification is sometimes used. Certain amino acid residues can be selectively derivatized in the presence of others. Frequently, even if a protein has many copies of each type, only a limited number will be highly reactive. If the resulting modified protein has an altered function or structure, one can attempt to explain how the modification caused this. It is more difficult to obtain specific chemical modifications with nucleic acids because there are fewer types of residue and because the chemical properties of each type are distressingly similar. However, some tricks can simplify the problem somewhat. Suppose the result of chemical modification is a heterogeneous mixture of molecules derivatized at a few different sites. It is likely that some species have inactivated functions, whereas others are still normal. Then, physical separations based on the ability to function (such as affinity chromatography) can help resolve the mixture. In practice, although it is sometimes a laborious hit-and-miss procedure to obtain them, chemically modified macromolecules are a necessary prerequisite to many other techniques such as probe or label approaches—and to the heavy-atom isomorphous replacement required for almost all large molecule crystallography.

Isolate discrete states of the system

Many biological polymers can exist in a variety of different conformations that are stabilized differentially by some of the common environmental parameters such as temperature, pH, and salt concentration. If the goal is detailed structural studies, it is usually necessary to work with one pure form at a time. Therefore, it is helpful to survey a variety of conditions to determine the ranges where conformations are stable, and where conformational changes take place. Because many macromolecular structural changes occur in a very narrow range of conditions, each pure state can be likened to a phase of the system, and the conformational changes can be called phase transitions. All available knowledge about environmental effects can be assembled into a phase diagram, in much the same way as one would summarize states of water or alloys as a function of thermodynamic variables. We show several examples in Chapters 22 and 24 of how such diagrams are constructed. The important point is that, once such a phase diagram is available, it immediately lets one see what environmental ranges can be used to examine the structure of pure states, and which ranges should be used for convenient study of conformational changes. This can greatly simplify further study.

If the conformation changes gradually over a broad range of conditions, it is difficult to construct a phase diagram. Furthermore, it becomes unclear whether to describe the system as a few discrete states that interconvert gradually, or whether to postulate an ensemble of many states. Thermodynamic and kinetic tests that can help to resolve this issue are described in Chapters 20 and 21. It is a common tactic to attempt to treat the system as a mixture of the minimal number of discrete states needed to fit available experimental data. In simple cases, this number is often obvious by inspection. In complex cases, mathematical techniques exist that can reveal the number of variables needed to describe the population of discrete states of the system to within the accuracy of the experimental measurements (Box 1-1).

Box 1-1 A GENERAL SCHEME FOR MULTICOMPONENT ANALYSIS

Suppose that l different experimental parameters have been measured, each at r experimental conditions. Let P_{ji} be the value of the ith parameter under the jth condition. Assemble all the values of P_{ji} into an $l \times r$ matrix, $\underset{\sim}{\mathbf{P}}$. We can construct a symmetric square ($r \times r$) matrix $\underset{\sim}{\mathbf{Q}} = \underset{\sim}{\mathbf{P}}\underset{\sim}{\mathbf{P}}^{\dagger}$ by multiplying $\underset{\sim}{\mathbf{P}}$ by its transpose ($P^{\dagger}_{ij} = P_{ji}$). The matrix $\underset{\sim}{\mathbf{Q}}$ contains all of our knowledge of the system. The eigenvalues λ_{κ} and then the eigenvectors $\mathbf{\Psi}_{\kappa}$ of $\underset{\sim}{\mathbf{Q}}$ can be found in standard ways. If $\underset{\sim}{\mathbf{E}}$ is the unit matrix, then

$$|\underset{\sim}{\mathbf{Q}} - \lambda_{\kappa}\underset{\sim}{\mathbf{E}}| = 0 \quad \text{and} \quad \underset{\sim}{\mathbf{Q}}\mathbf{\Psi}_{\kappa} = \lambda_{\kappa}\mathbf{\Psi}_{\kappa}; \qquad \kappa = 1,r$$

It turns out that the $\mathbf{\Psi}_{\kappa}$ vectors form a basis set to describe the system in the same way that the wave functions of a molecular Hamiltonian can represent molecular states. The eigenvalues λ_{κ} turn out to be a measure of the importance of a state. What one does in practice is to look at the values of λ_{κ} and pick just the biggest ones. Statistical tests can be used to decide which and how many of the corresponding $\mathbf{\Psi}_{\kappa}$ must be included to account for the observed properties of the system. This is a terribly formal but quite general procedure. In practice, experimental data rarely are accurate enough to justify it, and more approximate treatments must be used. For an example of the use of this multicomponent analysis scheme, see R. H. Fairclough et al., *J. Mol. Biol.* 132:557 (1979).

Summary

There are four levels of structural organization of biological macromolecules with which we shall be concerned. Primary structure is the covalent chemical structure or sequence of residues in chains, plus a specification of any cross-links between or within chains. Secondary structure is the local ordered helical arrangement of residues. Tertiary structure is the complete three-dimensional structure of a single covalent or otherwise effectively indivisible unit. Quaternary structure is the arrangement of individual tertiary structure units within an assembly.

Among the most important basic questions asked in biophysical chemistry are the following. How can structures be determined? Can structure be predicted or rationalized? Are structures static or dynamic? How does an ordered structure alter chemical and physical properties? How is the structure utilized in biological function? How is biological function regulated?

The basic strategies used in studying biological polymers attempt to reduce the number of variables involved. It is usually necessary to simplify the system either by using a smaller system as a model, by using a technique that can observe only a part of the system, by comparing two systems that are almost identical so that effects due to the few differing moieties can be highlighted, or by controlling environmental variables to isolate discrete states of the system. Most of these approaches can be summarized in the notion "divide and conquer."

Problems

1-1. Consider a continuous helix with a radius r and a pitch P. Show that the length traveled along one turn of the helix (contour length) is $\sqrt{P^2 + 4\pi^2 r^2}$. HINT: Project the helix onto a cylinder enclosing it, and then unwrap the cylinder.

1-2. A particular protein can regulate its own biosynthesis. In solution, the protein exists as a monomer–octamer equilibrium. The monomer, which is unstable and tends to denature, stimulates the rate of synthesis; the octamer inhibits synthesis. Explain how this is advantageous to the organism.

References

GENERAL

Bloomfield, V. A., and R. F. Harrington, eds. 1975. *Biophysical Chemistry*. San Francisco: W. H. Freeman and Company, [Reprints of pertinent articles from *Scientific American*.]

Stryer, L. 1975. *Biochemistry*. San Francisco: W. H. Freeman and Company. [Includes a clear, up-to-date review of the biochemical roles of macromolecules discussed in this text.]

SPECIFIC

Dayhoff, M. A., ed. 1972–1976. *Atlas of Protein Sequence and Structure.* Silver Spring, Md.: National Biomedical Research Foundation.

Eisen, H. N. 1974. *Immunology.* New York: Harper & Row. [Chapters 15 and 16 describe antibody structure and antibody–antigen interactions.]

Kirschner, K., and H. Bisswanger. 1976. Multifunctional proteins. *Ann. Rev. Biochem.* 45:143.

Kraut, J. 1977. Serine proteases: Structure and mechanism of catalysis. *Ann. Rev. Biochem.* 46:331.

Magar, M. E. 1972. *Data Analysis in Molecular Biology.* New York: Academic Press. [A treatment of various kinds of multicomponent analysis.]

Nomura, M., A. Tissieres, and P. Lengyel. 1974. *The Ribosome.* Long Island, N.Y.: Cold Spring Harbor Press. [The most complete work on this subject.]

Perham, R. N. 1975. Self-assembly of biological molecules. *Phil. Trans. Roy. Soc. London Ser. B* 272:123. [An interesting and stimulating short article.]

Pogson, C. I. 1974. The protein chemistry of regulatory phenomena. In *Biochemistry Series One, 1: Chemistry of Macromolecules*, ed. H. Gutfreund (London: Butterworths), p. 287.

2
Structure of proteins

2-1 PROPERTIES OF AMINO ACIDS

The fundamental units of proteins are twenty α-amino acids. Their structures, names, and the common three-letter code used for abbreviations are given in Figure 2-1. In all twenty, except glycine, the α-carbon is an asymmetric center. For proteins or peptides synthesized on the ribosome, the configuration about this center must be the L-stereoisomer, shown in Figure 2-1. Because of its near universality, the prefix L will generally be omitted. Keep in mind that D-amino acids have occasionally been found in a few naturally occurring peptides. These, however, are not synthesized on the ribosome.

Two amino acids, isoleucine and threonine, have a second asymmetric center at the β-carbon. Its configuration is also constant, and the correct stereochemistry is shown in Figure 2-1. Amino acids with altered β-carbon stereochemistry are called alloisoleucine and allothreonine.

The basic covalent structure of a protein is a polypeptide chain:

$$NH_3^+—CHR_1—CO—NH—CHR_2—CO— \cdots —NH—CHR_{n-1}—CO—NH—CHR_n—COO^-$$

Unless otherwise indicated, peptide chains are always written so that the NH group in each peptide bond is on the right. Thus, Ile–Glu indicates a dipeptide in which Ile is the N-terminal residue and Glu is the C-terminal residue.

Ionization of amino and carboxyl groups in amino acids and peptides

The amino acids shown in Figure 2-1 are in the ionic form, which predominates at pH 7. The α-amino group is protonated at this pH; the carboxyl is deprotonated. Therefore, most amino acids at neutral pH have no net charge, but they exist as a zwitterion NH_3^+—CHR—COO^- and will clearly have a large dipole moment. Four amino acids have an additional charge on the side chain at pH 7: Asp and Glu are negative; Lys and Arg are positive.

Because pH is an important environmental variable in the study of proteins, one needs to know the pH at which various sites on amino acids titrate and how this pH is altered by incorporation of the amino acid into a protein. This information has been obtained for amino acids and small peptides by direct potentiometric titration. Protons are added or removed, and the pH of the solution is monitored. For intact proteins, a total potentiometric titration is a moderately simple experiment when sufficient quantities of protein are available. However, the interpretation

Figure 2-1

Structures of the twenty normal amino acids at pH 7. All are shown projected into a plane about the α-carbon. In actuality the amino group and hydrogen are above the plane of the page, whereas the carboxyl and the side chain are below it.

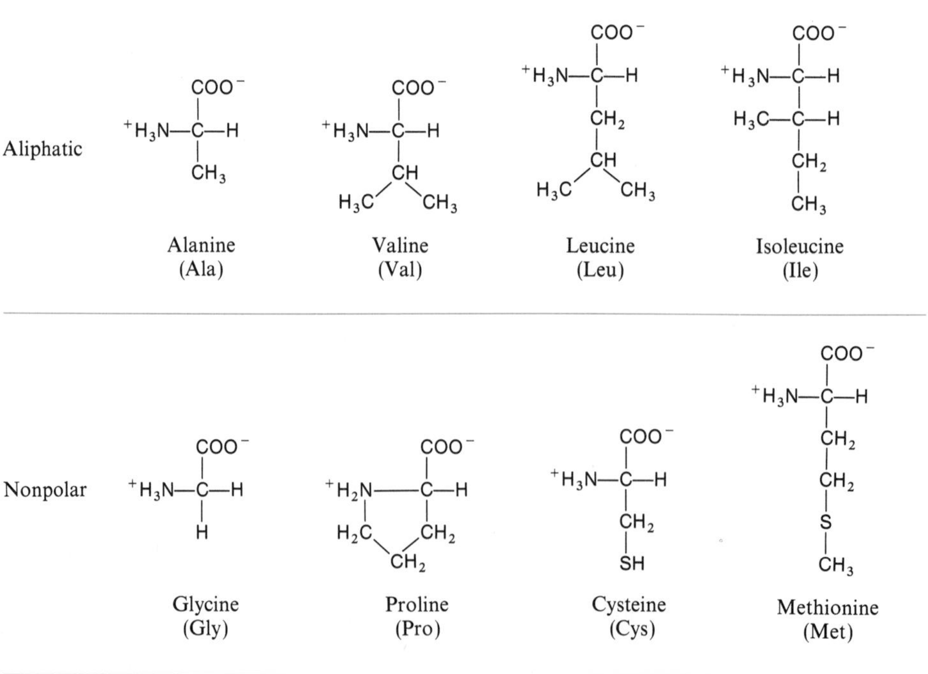

Aliphatic	COO^- ^+H_3N—C—H CH_3 Alanine (Ala)	COO^- ^+H_3N—C—H CH H_3C CH_3 Valine (Val)	COO^- ^+H_3N—C—H CH_2 CH H_3C CH_3 Leucine (Leu)	COO^- ^+H_3N—C—H H_3C—C—H CH_2 CH_3 Isoleucine (Ile)
Nonpolar	COO^- ^+H_3N—C—H H Glycine (Gly)	COO^- ^+H_2N——C—H H_2C CH_2 CH_2 Proline (Pro)	COO^- ^+H_3N—C—H CH_2 SH Cysteine (Cys)	COO^- ^+H_3N—C—H CH_2 CH_2 S CH_3 Methionine (Met)

of the results is difficult because it is often not clear to which residue (among many similar or identical residues) to assign a particular region of the titration curve. In these cases, other (more specific) ways of following the titration of individual residues must be selected. NMR is particularly convenient for some side chains, such as His; optical spectra are helpful for others, such as Tyr. (See Chapters 7 and 9.)

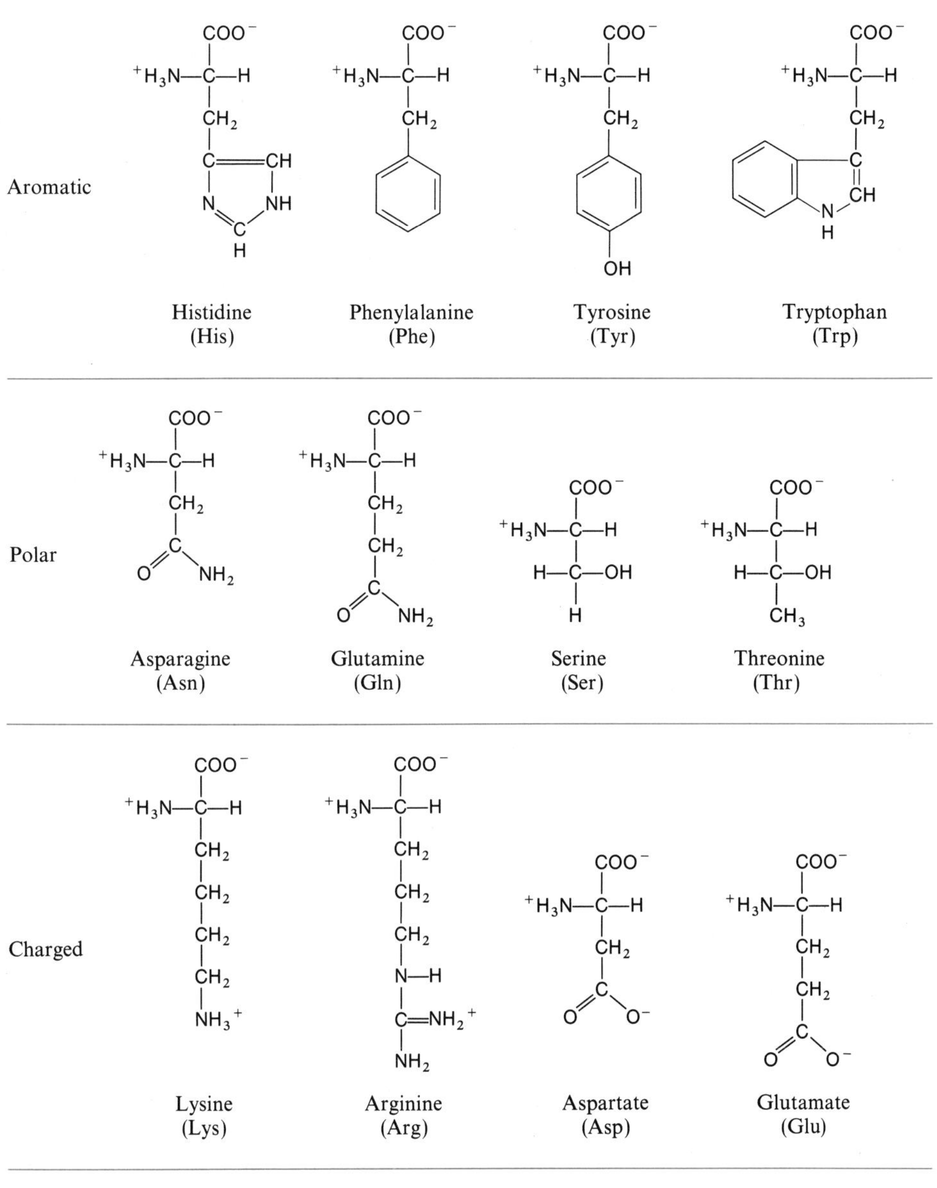

Aromatic

Histidine (His) — Phenylalanine (Phe) — Tyrosine (Tyr) — Tryptophan (Trp)

Polar

Asparagine (Asn) — Glutamine (Gln) — Serine (Ser) — Threonine (Thr)

Charged

Lysine (Lys) — Arginine (Arg) — Aspartate (Asp) — Glutamate (Glu)

In proteins or polypeptides, individual residues lose all their zwitterionic character in forming peptide bonds. One must be concerned only with the two chain termini and the amino acid side chains. Two effects will cause these to titrate differently than the corresponding sites in individual amino acids. These are (1) electrostatic interactions caused by the differences in the local chemical structure, and (2) thermodynamic interactions caused by coupling of ionization and conformational equilibria. Consider the difference between the N-terminus or C-terminus of a protein and the amino and carboxyl groups of free amino acids. A large electrostatic attraction between NH_3^+ and COO^- is present at pH 7 in the free amino acids. This makes it more difficult to remove a protein from NH_3^+ or to add one to COO^-, as shown by the ionic equilibria for oligoalanines in Table 2-1. Although charge effects are very important in Ala or $(Ala)_2$, by $(Ala)_4$ the interaction between terminal charges has ceased to be significant. Box 2-1 reviews the principles of ionization equilibria and illustrates the use of titration data to estimate electrostatic interaction energies.

Box 2-1 PRINCIPLES OF IONIZATION EQUILIBRIA

The ionization reaction $\text{H—A} \rightleftarrows \text{H}^+ + \text{A}^-$ will be governed by an acid dissociation constant:

$$K_a = (\text{H}^+)(\text{A}^-)/(\text{H—A})$$

It is convenient to express K_a in a form analogous to pH:

$$\text{pK}_a = -\log K_a$$

When conditions are adjusted such that the acid is half ionized, $(\text{H—A}) = (\text{A}^-)$, and $K_a = (\text{H}^+)$ or $\text{pK}_a = \text{pH}$. The dissociation constant is related to the standard free energy of dissociation by

$$\Delta G^0 = -RT \ln K_a$$

ΔG^0 is the free energy difference between products (H^+ and A^-) and reactant (H—A) when both are in their standard states (say, 1 M in an aqueous solution). The actual free energy change involved in carrying out the ionization reaction under some other conditions is

$$\Delta G_{\text{ioniz}} = \Delta G^0 + RT \ln [(\text{H}^+)(\text{A}^-)/(\text{H—A})]$$

Thus, whenever (H^+), (A^-), and (H—A) satisfy the conditions of the equilibrium constant K_a, then $\Delta G_{\text{ioniz}} = 0$.

Suppose that the ionization reaction is coupled to some other process or interaction—for example, binding of a proton to A^- changes the interaction of A^- with some other group in the molecule. Let the free energy change involved in this interaction be ΔG_c, where the subscript c refers to coupling. Then the total free energy change involved in the ionization reaction is

$$\Delta G_{\text{Tot}} = \Delta G_{\text{ioniz}} + \Delta G_c = \Delta G^0 + \Delta G_c + RT \ln [(\text{H}^+)(\text{A}^-)/(\text{H—A})]$$

Table 2-1
Ionic equilibria for some oligoalanines

Equilibria	pK_1	pK_2
$NH_3^+—CH(CH_3)—COOH \overset{1}{\rightleftarrows} NH_3^+—CH(CH_3)—COO^- \overset{2}{\rightleftarrows} NH_2—CH(CH_3)—COO^-$ ${}^+Ala \quad {}^+Ala^- \quad Ala^-$	2.34	9.69
$NH_3^+—CH(CH_3)—CO—NH—CH(CH_3)—COOH$ ${}^+Ala—Ala$ $\overset{1}{\rightleftarrows} NH_3^+—CH(CH_3)—CO—NH—CH(CH_3)—COO^-$ ${}^+Ala—Ala^-$ $\overset{2}{\rightleftarrows} NH_2—CH(CH_3)—CO—NH—CH(CH_3)—COO^-$ $Ala—Ala^-$	3.12	8.30
${}^+Ala—Ala—Ala \overset{1}{\rightleftarrows} {}^+Ala—Ala—Ala^- \overset{2}{\rightleftarrows} Ala—Ala—Ala^-$	3.39	8.03
${}^+Ala—Ala—Ala—Ala \overset{1}{\rightleftarrows} {}^+Ala—Ala—Ala—Ala^- \overset{2}{\rightleftarrows} Ala—Ala—Ala—Ala^-$	3.42	7.94

If the system is allowed to equilibrate, $\Delta G_{Tot} = 0$. The H^+ concentration at which the acid is half ionized is

$$(H^+)_{1/2} = e^{-(\Delta G^0 + \Delta G_c)/RT}$$

The apparent pK_a' is

$$pK_a' = -\log (H^+)_{1/2} = (\Delta G^0 + \Delta G_c)/2.303\ RT$$

If a model system is available in which there is no coupling, the pK_a determined in that system is

$$pK_a = \Delta G^0/2.303\ RT$$

Therefore, from the difference in the two pK_a values, the coupling energy can be evaluated as $\Delta G_c = 2.303\ RT\ (pK_a' - pK_a)$.

Let us use this result to analyze the electrostatic interaction between the COO^- and NH_3^+ groups in alanine, using data given in Table 2-1. The unperturbed pK_a for dissociation of an H^+ from COO^- can be estimated from the titration of $(Ala)_4$ as 3.42. The perturbed pK_a' for Ala is 2.34. Because RT is ~ 0.6 kcal mole^{-1} at room temperature,

$$\Delta G_c = (2.303)(0.6)(2.34 - 3.42) = -2.5 \text{ kcal mole}^{-1}$$

Thus, for the reaction

$$NH_3^+—CH(CH_3)—COOH \rightleftarrows NH_3^+—CH(CH_3)—COO^- + H^+$$

interactions between the NH_3^+ and COO^- lower the free energy of the molecule by 2.5 kcal mole^{-1}. This effect promotes easier dissociation, which is expressed as a lower pK_a.

A potentiometric titration does not directly reveal which position is titrating. This information usually can be found either through spectroscopic studies or through the use of chemical analogs. For example, the following analogs allow carboxyl and amino pK_a values of oligoalanines to be assigned unequivocally:

$$CH_3\text{—CO—NH—CH}(CH_3)\text{—COOH} \overset{1}{\rightleftarrows} CH_3\text{—CO—NH—CH}(CH_3)\text{—COO}^- \qquad pK_1 = 3.72$$

$$NH_3^+\text{—CH}(CH_3)\text{—CO—}NH_2 \overset{2}{\rightleftarrows} NH_2\text{—CH}(CH_3)\text{—CO—}NH_2 \qquad pK_2 = 8.02$$

Ionization of side chains

When ionizable side chains are present, the titration data can be quite complicated. At pH < 2, lysine exists predominantly in the form

$$NH_3^+\text{—CH(COOH)—}(CH_2)_4\text{—}NH_3^+$$

A titration of lysine shows three pK_a values. One value of 2.18, by analogy with alanine, clearly is the carboxyl ionization. It is shifted to even lower pH in Lys than in Ala, due to the extra electrostatic attraction between the COO^- and the side-chain NH_3^+. The other ionizations occur at pK_a values of 8.90 and 10.28. One of these values must represent the α-amino and the other the ε-amino, but it is not immediately obvious which is which. This can be determined by examining model compounds.

Three distinct ionization steps can be seen in Ala–Lys. These have pK_a values of 3.22, 7.62, and 10.70. The lowest clearly is the carboxyl, as you can see by comparing Ala–Ala and Ala–Lys. Two possibilities exist for the other ionizations: (1) the α-amino loses a proton before the side chain does, or (2) the side chain loses a proton first.

$$NH_3^+\text{—CH}(CH_3)\text{—CO—NH—CH}((CH_2)_4NH_3^+)\text{—COO}^-$$

7.62

(a) $NH_2\text{—CH}(CH_3)\text{—CO—NH—CH}((CH_2)_4NH_3^+)\text{—COO}^-$ ⇌ ⇌ $NH_3^+\text{—CH}(CH_3)\text{—CO—NH—CH}((CH_2)_4NH_2)\text{—COO}^-$ (b)

10.70

$$NH_2\text{—CH}(CH_3)\text{—CO—NH—CH}((CH_2)_4NH_2)\text{—COO}^-$$

Two arguments can be used to distinguish between schemes *a* and *b*. Primary alkyl amines have higher pK_a values than do α-carbonyl-substituted methyl amines. For example, the pK_a of *n*-butylamine is 10.59, whereas the amine of an α-amino acid unperturbed by electrostatic effects titrates near pH 8, as shown by the oligoalanine data in Table 2-1. This argues for scheme *a*. If this scheme is correct, the pK_a of the ε-amino group on lysine should be slightly higher than that of butylamine because of its electrostatic attraction to the COO^-. In contrast, the α-NH_3^+ would have to titrate in the presence of the charged ε-NH_3^+ and therefore its pK_a should be lower than the α-amino in Ala–Ala. These predictions are all consistent with the observed values. If scheme *b* were correct, one would have to explain why the α-amino in Ala–Ala titrates with a pK_a of 8.30, whereas in Ala–Lys, with no charge in the side chain, it titrates with a pK_a of 10.70. There is no way to rationalize this, and thus scheme *b* is not the preferred pathway.

To find out where the ε-amino of lysine should titrate in a polypeptide, it is necessary to examine compounds where the free amino and carboxyl termini are far enough away from the side chain to cause negligible electrostatic interactions. For example, in Ala–Lys–Ala–Ala, three pK_a values are observed at 3.58, 8.01, and 10.58. The first two are very close to the corresponding positions in $(Ala)_4$, and the ε-amino is not shifted significantly compared with *n*-butylamine. However, in compounds with a multiplicity of ionizable side chains, titration curves become nightmarishly complicated. For example, Lys–Lys–Lys shows pK_a values at 3.06, 7.34, 9.80, 10.54, and 11.32.

Ionization equilibria in proteins

The preceding examples show that the titration of individual residues in a protein depends on the sequence of the amino acids. However, we have been dealing with short peptides that have a relatively flexible conformation. Once ordered secondary and tertiary structures are present, one must consider electrostatic interactions with all nearby charges. Attempts to calculate these interactions quantitatively have been moderately successful in some cases, but it is a truly formidable problem. As shown later in the book, every polyelectrolyte attracts a counterion atmosphere. This tends to weaken all direct ionic interactions. The density of the ion atmosphere depends on the total concentration of electrolyte in the solution, but it also is a function of the geometry of nearby charged groups on the polymer. In principle, titration measurements at high ionic strength predominantly reflect only electrostatic interactions between nearby charged groups. Thus, some information is available about local secondary and tertiary structure from such titration measurements. In practice, however, if all one measures is the overall potentiometric titration, the data are not very informative.

The hydrogen ion titration of a typical protein is shown in Figure 2-2. Even a simplistic interpretation requires a knowledge of the ionization equilibria of all amino acid side chains present. Data for the side chains that are not perturbed by

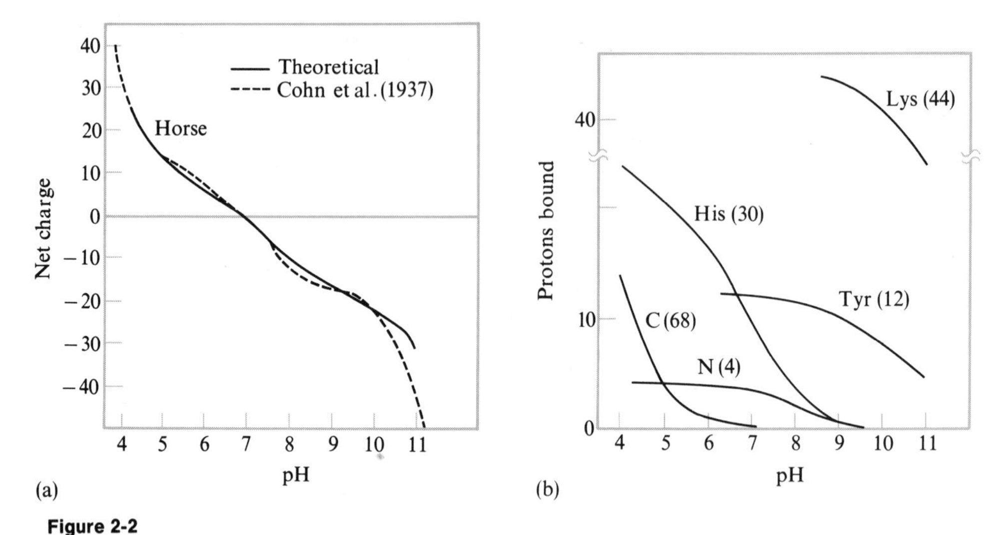

Figure 2-2

The pH titration of horse oxyhemoglobin at 0.02 ionic strength and 29°C. (**a**) The experimental data have been shifted up by 3.0 charge units to give the correct isoionic pH. The calculated curve attempts to account for the ionization properties of all side chains as perturbed by electrostatic interactions within the known quaternary structure of the protein. (**b**) Calculated number of protons bound by each ionizable moiety as a function of pH. Abbreviations used are c (Glu plus Asp) and n (N-terminus). Numbers in parentheses are the numbers of each type of group. [After W. R. Orttung, *Biochemistry* 9:2394 (1970).]

nearby charges can be obtained from measurements on peptides, as we have discussed earlier. A summary of the pH range in which each side chain can normally be expected to titrate is shown in Table 2-2. In practice, where the titration of individual side chains of a protein can be assigned, the observed values generally fall within 1 pH unit of the expected values. A summary of some of these data is given in Table 2-3. However, there are some dramatic exceptions where much larger pK_a shifts are seen. These are usually associated with hysteresis in the titration curve. That is, an acid titration does not superimpose with an alkaline titration. These effects arise from coupling of the titration to conformational changes.

Suppose a side chain participates directly in an interaction that stabilizes secondary or tertiary structure. This interaction may be impossible to maintain if a change in ionization occurs. For example, consider a tyrosine hydrogen-bonded to a carboxyl of another residue. The titration must be written

$$OH^- + \text{Tyr—O—H} \cdots {}^-\text{O—CO—R} \rightleftarrows \text{Tyr—O}^- + {}^-\text{O—CO—R} + H_2O$$

Even if this interaction is uncoupled to the rest of the structure, it will have a large effect on the observed pK_a of the tyrosine because the free energy of the hydrogen

Table 2-2
Ionizable groups in amino acids: pK_a values

Amino acid	α-COOH	$α\text{-}NH_3^+$	Side chain	pK_a in amino acid	Expected pK_a in a protein
Alanine	2.3	9.9	—	—	—
Arginine	1.8	9.0	$—NH—C(=NH_2^+)NH_2$	12.5	⩾12
Asparagine	2.0	8.8	—	—	—
Aspartic acid	2.0	10.0	—COOH	3.9	4.4–4.6
Cysteine	1.8	10.8	—SH	8.3	8.5–8.8
Glutamic acid	2.2	9.7	—COOH	4.3	4.4–4.6
Glutamine	2.2	9.1	—	—	—
Glycine	2.4	9.8	—	—	—
Histidine	1.8	9.2	imidazolium ring (N—H, NH^+)	6.0	6.5–7.0
Isoleucine	2.4	9.7	—	—	—
Leucine	2.4	9.6	—	—	—
Lysine	2.2	9.2	$—NH_3^+$	10.8	10.0–10.2
Methionine	2.3	9.2	—	—	—
Phenylalanine	1.8	9.1	—	—	—
Proline	2.0	10.6	—	—	—
Serine	2.1	9.2	—	—	—
Threonine	2.6	10.4	—	—	—
Tryptophan	2.4	9.4	—	—	—
Tyrosine	2.2	9.1	—(phenyl ring)—OH	10.9	9.6–10.0
Valine	2.3	9.6	—	—	—

NOTE: The pK_a values in most cases are at 25°C. The expected pK_a values in proteins are determined from model compounds in which titration of side chains is decoupled from charge effects of α-substituents.

bond interaction is probably a few kilocalories. But what if the Tyr and COO^- cannot physically move apart unless other residues move also? Then there will be an additional energy cost to the ionization, either from the repulsion between Tyr—O^- and ^-O—CO—R if these groups remain near each other, or from perturbing the conformation of other residues to let these groups move apart. The total

Table 2-3
Intrinsic pK_a observed for ionizable groups in some proteins

Group	Protein	pK_a
Aspartic and glutamic acids	Lysozyme (3 of ~16 groups)	D
	β-Lactoglobulin (49 of 51 groups)	4.8
	(2 of 51 groups)	7.3
	Serum albumin	4.0
	Insulin	4.7
Histidine	Myoglobin (6 of 12 groups)	D
	(6 of 12 groups)	6.6
	β-Lactoglobulin	7.4
	Lysozyme	6.8
	Insulin	6.4
	Serum albumin	6.9
Tyrosine	Chymotrypsinogen (1 of 4 groups)	9.7
	(1 of 4 groups)	10.4
	(2 of 4 groups)	P
	Insulin	9.6
	Serum albumin	10.4
	Ribonuclease (3 of 6 groups)	9.6
Lysine	Serum albumin	9.8
	Lysozyme	10.4
	Chymotrypsinogen (3 of 13 groups)	P
Arginine	Insulin	11.9
α-Carboxyl	Insulin	3.6

D = always deprotonated at all pH values where native protein structure can be maintained.
P = always protonated at all pH values where native protein structure can be maintained.
SOURCE: Adapted from R. H. Haschemeyer and A. E. V. Haschemeyer, *Proteins* (New York: Wiley, 1974).

free energy change for the ionization can be written as

$$\Delta G_{\text{Tot}} = \Delta G_{\text{ioniz}} + \Delta G_{\text{c}} \quad (2\text{-}1)$$

where ΔG_{ioniz} is the free energy change for ionizing an isolated tyrosine, and ΔG_c is a coupling free energy that includes the contributions from hydrogen-bond loss, electrostatic repulsion, and conformational changes. As shown in Box 2-1, the effect of ΔG_c is to shift the apparent pK_a. Because $\Delta G_c > 0$, the pK_a will be shifted to higher values than those expected for the titration of an isolated tyrosine.

If the conformational change that is coupled to the ionization is reversible, the same coupling still exists when a back titration is performed. Frequently, however, the reformation of a native conformation is extremely slow or even impossible. In this case, when a reverse titration is carried out, the tyrosine will titrate normally because the coupling to the conformational change is now broken. A detailed experimental example of this is shown in Chapter 7 for the titration of tyrosines in ribonuclease. In that case, the specific ionization of these residues can be monitored optically.

Given a total titration curve (such as that shown in Fig. 2-2), the best one usually can do is fit it to the known amino acid composition. Different regions of the curve are assigned in proportion to known numbers of ionizing residues, assuming that they titrate normally. If this cannot account for the total number of protons taken up or lost, one then assumes that some residues are not titrating at all because they are strongly involved in stabilizing the conformation. A more sophisticated analysis requires the computation of shifts in the normal pK_a values due to electrostatic interactions. This will be possible in a reliable fashion only if the tertiary structure is known. As shown in Figure 2-2, in a favorable case, the observed titration behavior can be accounted for with fairly good quantitative accuracy.

Polarity of amino acid side chains

The preceding discussion shows that it is possible to make a fairly clear-cut distinction between charged and uncharged amino acid side chains at any pH. It would be useful if there were other unequivocal ways of grouping the amino acids into classes with distinct properties. One can try to do this on the basis of intuitive chemical feelings about polarity. Side chains that are nonpolar will tend to have low solubility in water. Clearly included in this category are the aliphatic amino acid side chains of Ala, Val, Ile, and Leu. It is also safe to include Phe, Trp, and Met.

At the opposite end of the polarity scale are the charged side chains of Glu, Asp, Arg, and Lys. Asn and Gln, which have amide side chains, are also very polar. The hydroxylic amino acids Ser and Thr are also polar because of the large dipole and hydrogen-bonding capabilities of the hydroxyl. All of these groups are expected to interact strongly with water and to have high solubility.

The polarity of the five remaining amino acids is more ambiguous. Cys and His have pK_a values close enough to 7 that they may actually be charged in many proteins under physiological conditions. If so then they are polar, but when uncharged they are nowhere near as polar as some of the other amino acids in this category. In tyrosine, the aromatic ring more than compensates for the hydroxyl, and Tyr usually behaves as a nonpolar residue. Glycine and proline are special. The basic structure of each differs from that of the other 18 amino acids. Glycine has no optically active α-carbon, and this means that it can exist in conformations inaccessible to the other amino acids. It has no side chain but, from the way it behaves in proteins, Gly should be put in the nonpolar category. Proline is actually an imino rather than an amino acid. Consequently its conformational properties are different from those of all the others. A polarity assignment is difficult but, based on its locations in proteins, Pro behaves as a polar residue more often than not.

As a check on the semiintuitive arguments about side chain properties, the solubilities of the amino acids in water at neutral pH are given in Table 2-4. Using the principle "like dissolves like" and the fact that water is polar, most of these results are qualitatively consistent with the assignments we have made. However, the results in Table 2-4 must not be regarded in quantitative detail. These are the solubilities of zwitterionic amino acids, whereas the properties we need to understand in proteins are those of the side chain as part of a peptide.

There are more fundamental limitations to solubility data. Solubility of a solid sample in water is an equilibrium process. It is affected by the stability of both states: the affinity of solute for water in the solution, and the attraction of solute for solute

Table 2-4
Classification of amino acid properties

Residue	Type	Zwitterion solubility, 25°C (moles/kg)	Side chain transfer ΔG_t, EtOH → H_2O (kcal/mole)
Trp	Nonpolar	0.07	3.00
Ile	Nonpolar	0.26	2.95
Tyr	Nonpolar‡	<0.00	2.85
Phe	Nonpolar	0.17	2.65
Leu	Nonpolar	0.16	2.40
Val	Nonpolar	0.50	1.70
Met	Nonpolar	0.38	1.30
Cys*	Nonpolar	<0.00	1.00
Ala	Nonpolar	1.86	0.75
Gly	Nonpolar‡	3.33	0.00
His	Ambiguous	ND	ND
Pro	Polar‡	14.1	2.60
Ser	Polar	4.02	ND
Thr	Polar	ND	0.45
Asn	Polar	0.19	ND
Glu	Polar	0.29	ND
Asp	Charged	0.04†	ND
Glu	Charged	0.06†	ND
Lys	Charged	3.95§	1.50
Arg	Charged	4.06§	0.75

* Data shown are actually 1/2 cystine.
† With side chain protonated; without a proton, the value is much higher.
‡ Usually.
§ HCl salts.
ND = No data.

in the crystal. Here, we are interested only in the first interaction, but there is no easy way to ensure that the second is not having a major influence on the solubility. In principle, it is possible to assemble data that are more useful than those given in Table 2-4. One wants to know the solubility of gaseous amino acids or side chains in water. There is no practical way to measure these values directly. However, in principle, such information can be derived from thermodynamic cycles. For example, if the thermodynamics of vaporization of the crystal were known, combining this information with the thermodynamics of solution of the solid would yield information about the desired process: equilibrium between vapor and solution.

Charles Tanford developed an alternative approach that may be used in the absence of vapor phase data to provide a simple thermodynamic characterization of the environmental preferences of amino acids. He compared the solubility of amino acids in water (as a typical polar solvent) and in ethanol (as a model for a nonpolar solvent still capable of dissolving amino acids to a sufficient degree to allow convenient measurement). Because the same crystalline reference state exists for both measurements, the free energies of solution can be combined to compute the ΔG for transfer of a mole of residue from ethanol to water at constant concentration. This will be a function of solvent interactions of the charged amino and carbonyl groups as well as the side chains. Tanford chose to use glycine as a standard to represent a side-chain-less amino acid, and he attributed the differences between the ΔG of glycine and those of the other amino acids to the energy contribution of the side chain. This is reasonable because transfer free energies should roughly be additive functions of molecular structure.

A much more extensive discussion of transfer free energies is given in Chapter 5. Here we need only a few of the results. Amino acid solubilities are much less in ethanol than in water. For glycine, ΔG is -4.63 kcal/mole, showing that—if a direct experiment could somehow be done—glycine would transfer spontaneously from ethanol to water. For phenylalanine, ΔG is only -1.98 kcal/mole. The difference between these values, $+2.65$ kcal/mole, is assigned as the free energy of transfer (ΔG_t) of the phenylalanine side chain. Note that this is positive, as it is for most of the amino acid side chains, as shown in Table 2-4. Thus, an isolated side chain, unencumbered by the rest of the amino acid, would spontaneously transfer from water to ethanol. Clearly, such side chains will prefer to be located in the interior of a folded protein, or at least will face inward.

The ΔG_t results are one manifestation of the hydrophobic forces that are an important factor in the determination of the tertiary structures of proteins. What is known about the origin of these forces is discussed in detail in Chapter 5. One can use the ΔG_t values to divide amino acids into hydrophobic and hydrophilic classes. As you can see from Table 2-4, these categories agree moderately well with simple intuitive ideas about polarity. It may seem arbitrary to choose ethanol as a reference nonpolar solvent. However, it turns out that choosing other solvents instead does not seriously alter the general conclusions.

2-2 COMPOSITION OF PROTEINS

Proteins are copolymers of the twenty normal amino acids. Sometimes these are the only substances present when a protein is isolated and analyzed. Quite frequently, other substances are found—either covalently incorporated into the protein or else bound so tightly to it that they copurify. Here, we shall try to summarize what is found in typical proteins.

Amino acid composition

First, consider just the amino acids. Automated methods of analyzing the amino acid composition of a protein have been perfected to a high degree. As a result, a vast body of data exists. A few general features are immediately apparent. The frequency of occurrence of individual amino acids varies enormously from protein to protein. However, in general, certain amino acids (such as Met and Trp) are relatively rare. Others (such as Ala and Leu) usually tend to be among the most common. The amino acid compositions of a number of proteins with very different biological properties are shown in Figure 2-3. These have been chosen to illustrate certain features.

Histones are small, very basic proteins whose role is to complex with DNA in chromatin. They have a relatively large fraction of positively charged amino acids that clearly assist in function by providing some electrostatic stabilization through interaction with the negative phosphates of DNA.

The purple membrane protein exists in vivo bound to membranes, and it is believed to be located largely inside the lipid bilayer. Therefore, it is reasonable that it should have relatively few charged amino acids and be rich in nonpolar residues instead.

Comparison of the other examples shown in Figure 2-3 reveals that small globular or fibrous proteins have a slight tendency toward more polar compositions than those of large globular proteins. This can be rationalized by the surface-to-volume ratio. A larger ratio should lead to more polar residues because, as you will see later, these tend to be located preferentially near the outside of a folded protein tertiary structure. It is difficult to construct a meaningful reference point for an average protein composition. However, the total amino acid composition of a whole organism or tissue falls between the extremes shown by very polar or very nonpolar proteins. Keep in mind that the examples in Figure 2-3 exaggerate observed differences. For any class of proteins, amino acid compositions vary sufficiently to obscure any simple-minded division into polar and nonpolar proteins.

Amino acid compositions represent a large body of easily accessible data that is not clearly related in any simple way to useful structural variables. Numerous attempts have been made to compute average amino acid compositions in the hope that such averages will correlate to some extent with rather general properties of proteins. Sometimes the amino acid composition of a protein is one of the few pieces

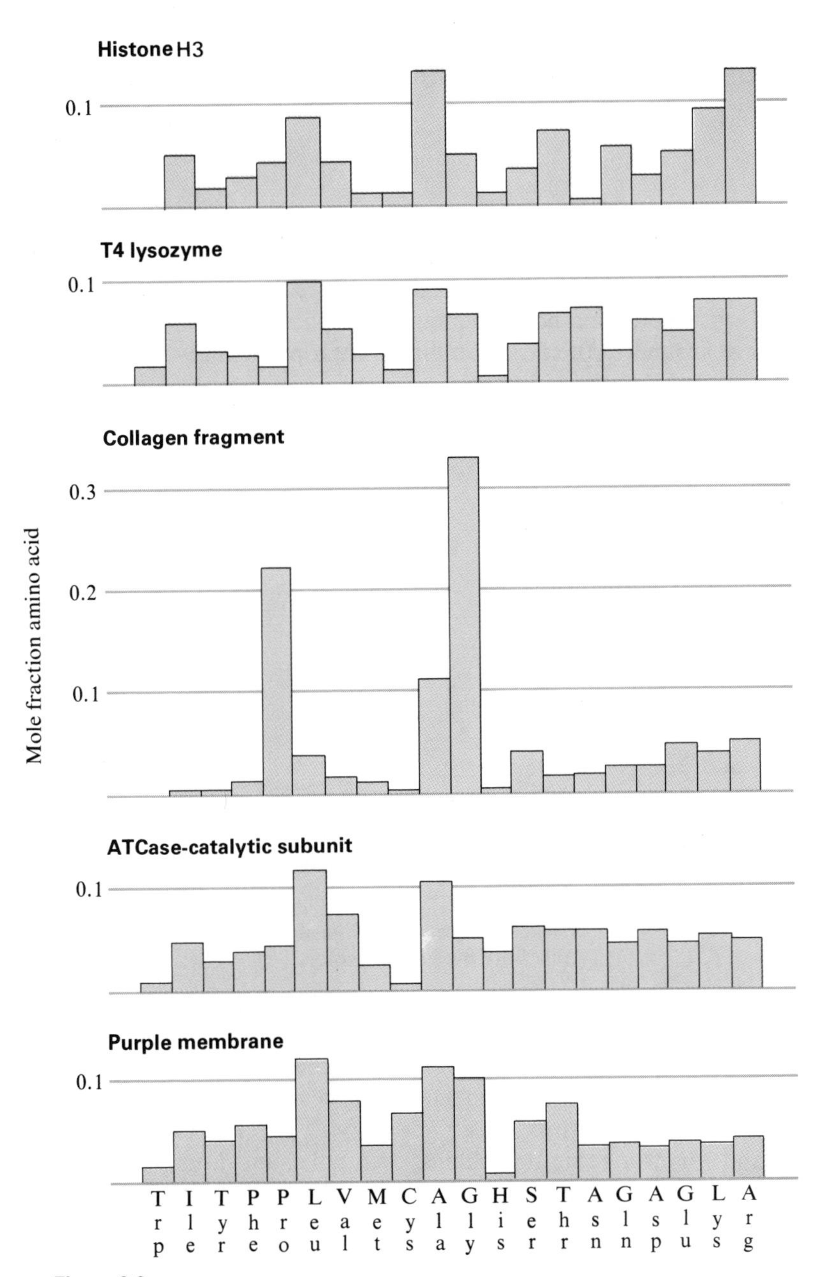

Figure 2-3

Amino acid composition of five proteins. The mole fraction of each amino acid type is plotted in approximate order of decreasing hydrophobicity.

of available data. One would like to be able to use this as a lead for planning further studies.

The availability of ΔG_t values as a measure of hydrophobicity in Table 2-4 suggests that one can set up a quantitative scale of average hydrophobicity, $\bar{H}_\phi$:

$$\bar{H}_\phi = \sum_i \Delta G_{ti}\chi_i \tag{2-2}$$

where ΔG_{ti} is the transfer free energy for the ith type of residue, and χ_i is its mole fraction in a given protein. The values of $\bar{H}_\phi$ for 150 proteins and peptides that were examined ranged from 0.44 to 2.02 kcal (mole residue)$^{-1}$. However, as shown in Figure 2-4, most proteins have average hydrophobicity values that cluster in a narrow range around 1.00. Therefore, this is not a particularly useful parameter for making fine distinctions.

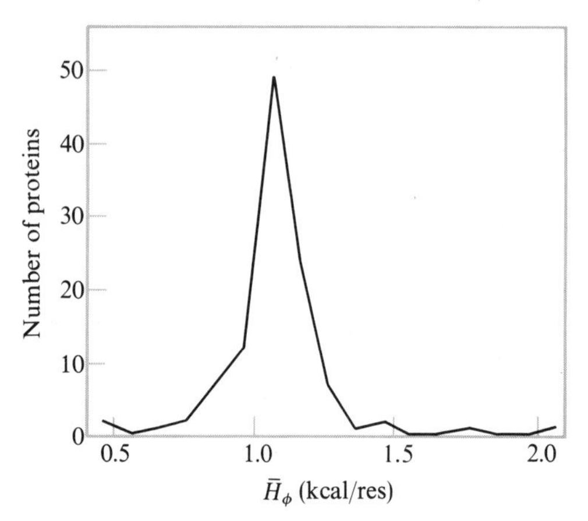

Figure 2-4

Calculated average hydrophobicities of 150 *proteins.* $\bar{H}_\phi$ is defined by Equation 2-2. [After C. C. Bigelow, *J. Theoret. Biol.* 16:187 (1967).]

The $\bar{H}_\phi$ values necessarily stress the fractional composition of some of the nonpolar side chains. A number of alternative scales have been constructed simply by comparing the ratio (R) of the frequencies of occurrence (χ) of whatever particular side chains one wishes to stress—e.g., $R = \sum_k \chi_k / \sum_j \chi_j$. For example, k can be hydrophilic and j hydrophobic side chains, or k polar and j nonpolar, and so forth. One feature of protein structures that is fairly reliable is the tendency of charged or very polar residues to be external. A particular ratio scale (R_3) that weights this factor very heavily is constructed by letting k = Arg, Lys, His, Gly, Glu, Asp, Asn, and His, and letting j = Ilu, Tyr, Phe, Leu, Val, and Met. The particular choice of residues used to construct R_3 is somewhat arbitrary. However, this ratio is convenient because it spreads out different proteins over a wide scale. R_3 was found to vary from 0.36 to 2.03 for 13 proteins whose $\bar{H}_\phi$ values varied only from 0.86 to 1.29 kcal mole^{-1}. It turns out that R_3 and $\bar{H}_\phi$ show some anticorrelation, but they are by no means purely inverse measures. Thus, by using both, we obtain a measure with more information about the system. To decide how to combine both scales, one must choose an objective and use statistics to find the optimal combination.

Predicting properties from amino acid compositions

One objective that should be attainable from amino acid composition is predicting whether a protein associated with membranes will be internal (and thus should have a large number of nonpolar residues and few charged residues) or external (in which case it should resemble a typical soluble nonmembrane protein). Francisco Barrantes recently compared the R_3 values of a large number of membrane proteins with their $\bar{H}_\phi$ values. In his calculations of $\bar{H}_\phi$, he chose to ignore tryptophan. Table 2-5 shows his results. We see that R_3 allows a fairly clear distinction between internal and external membrane proteins, whereas $\bar{H}_\phi$ does not.

Table 2-5
Comparison of R_3 and $\bar{H}_\phi$ values for a sample of proteins

Average variable	205 Nonmembrane proteins	24 Membrane proteins	
		Believed internal	Believed external
R_3	1.26 ± 0.42	0.59 ± 0.18	1.37 ± 0.35
$\bar{H}_\phi$	0.996 ± 0.098	1.197 ± 0.097	0.986 ± 0.075

SOURCE: Data from F. J. Barrantes, *Biochem. Biophys. Res. Commun.* 62:407 (1975).

One can then ask what linear function of R_3 and $\bar{H}_\phi$ will maximize the discrimination between internal and external membrane proteins. This scale, Z, is called a discriminant function:

$$Z = -0.345R_3 + 0.60\bar{H}_\phi \tag{2-3}$$

The average values of Z for internal and external membrane proteins are 0.52 ± 0.11 and 0.12 ± 0.16, respectively. For nonmembrane proteins, an average Z is 0.16 ± 0.17. One can conclude from a statistical analysis of the results that there is only an 8% chance of misclassifying an internal protein or external protein whose Z falls within the given standard deviations.

Table 2-6
Amino acid composition of some membrane proteins

Protein	$\bar{H}_\phi$ (kcal)	Z	Schematic picture
Acetylcholine receptor	1.184	0.37	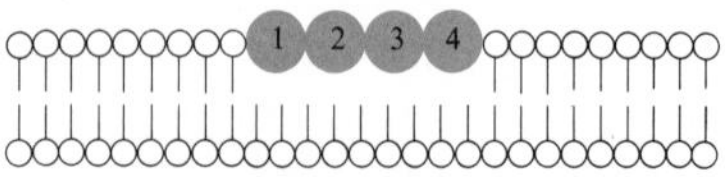
Subunit 1	1.182	0.38	
Subunit 2	1.119	0.29	
Subunit 3	1.140	0.31	
Subunit 4	1.140	0.31	
Acetylcholinesterase	1.059	0.25	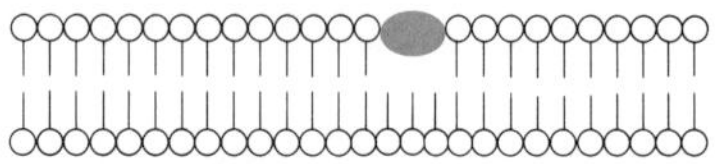
Cytochrome oxidase	1.185	0.45	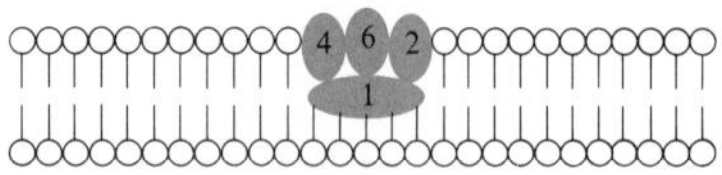
Subunit 1	1.185	0.51	
Subunit 2	1.011	0.25	
Subunit 4	1.033	0.16	
Subunit 6	1.029	0.14	
Na^+—K^+ ATPase	1.272	0.42	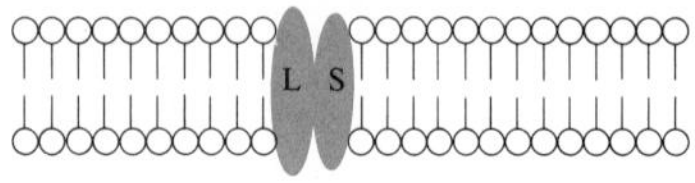
Large subunit	1.248	0.42	
Small subunit	1.309	0.44	
Rhodopsin (bovine)	1.208	0.51	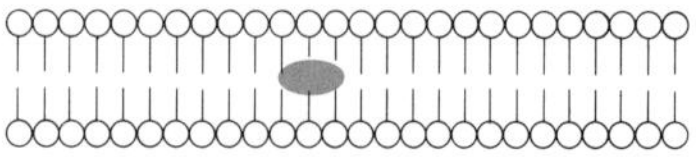
Purple membrane protein	1.247	0.56	

NOTE: $\bar{H}_\phi$ is the average hydrophobicity, as defined by Eqn. 2–2. Z is the discriminant function (a linear combination of average hydrophobicity and the ratio of charged to hydrophobic amino acids), as defined by Eqn. 2–3.
SOURCE: Adapted from M. A. Raftery, R. L. Vandlen, K. L. Reed, and T. Lee, *Cold Spring Harbor Symp. Quant. Biol.* 40:193 (1976).

As an example of the usefulness of the Z scale in suggesting features of the organization of a structure, consider the results shown in Table 2-6. $\bar{H}_\phi$ and Z values are given for several membrane proteins, including two known to be composed of several subunits. One of the first things one would like to know is whether individual subunits are likely to be located internally or externally. All the subunits of the acetylcholine receptor protein and the four subunits of cytochrome oxidase for which data are available have similar $\bar{H}_\phi$ values. However, Z values tell a different story. They suggest that one cytochrome oxidase subunit is buried and two others are external, as shown schematically in Table 2-6. However, the receptor-protein subunits show no differences in Z values. This may indicate that all have similar placement in the membrane. It is risky to take a concept as oversimplified as Z values to be firm structural data. But clearly, as a rough indication of the organization of subunits in complex structures, it appears quite useful.

Many proteins contain more than the twenty normal amino acids

If we want to rationalize some features of protein structure and function from the chemical composition, we must consider not only amino acids, but also any other moieties that may be present. Just about the simplest modification after a protein chain is synthesized is cleavage at one or more internal peptide bonds, to produce a multichain protein. This cleavage has the potential to alter structure and function. In an extreme case, with only noncovalent forces joining them, individual chains may dissociate. However, much more frequently, slight alterations occur in the structure. At pH 7, chain cleavage changes an unchanged peptide bond into charged NH_3^+ and COO^- chain termini. These will have quite different properties and interactions. For example, the protein trypsinogen is a precursor of trypsin. Breakage of a single peptide bond results in the appearance of greatly enhanced catalytic activity. One can see from the three-dimensional structures of the enzyme and precursor that the chain breakage permits several residues near the active site to readjust their position. The NH_3^+ generated by the cleavage appears to move substantially, as illustrated in Figure 2-5. Many other residues also move as a result of the cleavage. These motions open the active site and allow the binding of substrates.

Another common posttranslational modification is the alteration of amino acids. A few examples are shown in Figure 2-6. In some cases, the functional or structural significance of these changes is unknown. However, a change such as adding charge to serine by phosphorylation could have substantial consequences for the local structure. Some of these modifications are simply the first step in more elaborate additions to the protein. For example, many proteins contain covalently attached carbohydrate residues, varying from just a single sugar or two located at one place in the structure to the incorporation of many large complex oligosaccharides. Some of the known modes of attachment are shown in Figure 2-6. Hydroxylated amino acids are a favorite site. The structural consequences of adding carbohydrates, and thus forming a glycoprotein, are not yet well understood. Oligosaccharides are extremely hydrophilic, and this is bound to have some influence. Many membrane proteins contain appreciable numbers of attached sugars. It is a reasonable guess that these tenaciously keep to the aqueous phase and thus ensure the extension of the carbohydrate-containing part of the protein out of the lipid bilayer.

Carbohydrate structures are quite rigid in comparison to peptides. It is not known whether their rigidity has any role in function. It is clear that at least some glycoproteins are quite resistant to irreversible thermal denaturation, but whether the sugars play any role in this is unknown. Somewhat more knowledge is available about the functional role of the sugar residues in proteins. They are of central importance in determining some immunological properties, and are probably involved in cell–cell recognition and in numerous cell-surface phenomena. The sugar moieties of nonmembrane proteins may be like train tickets that regulate secretion from the cell, existence in the serum (or elsewhere), and ultimate departure from the system by degradation.

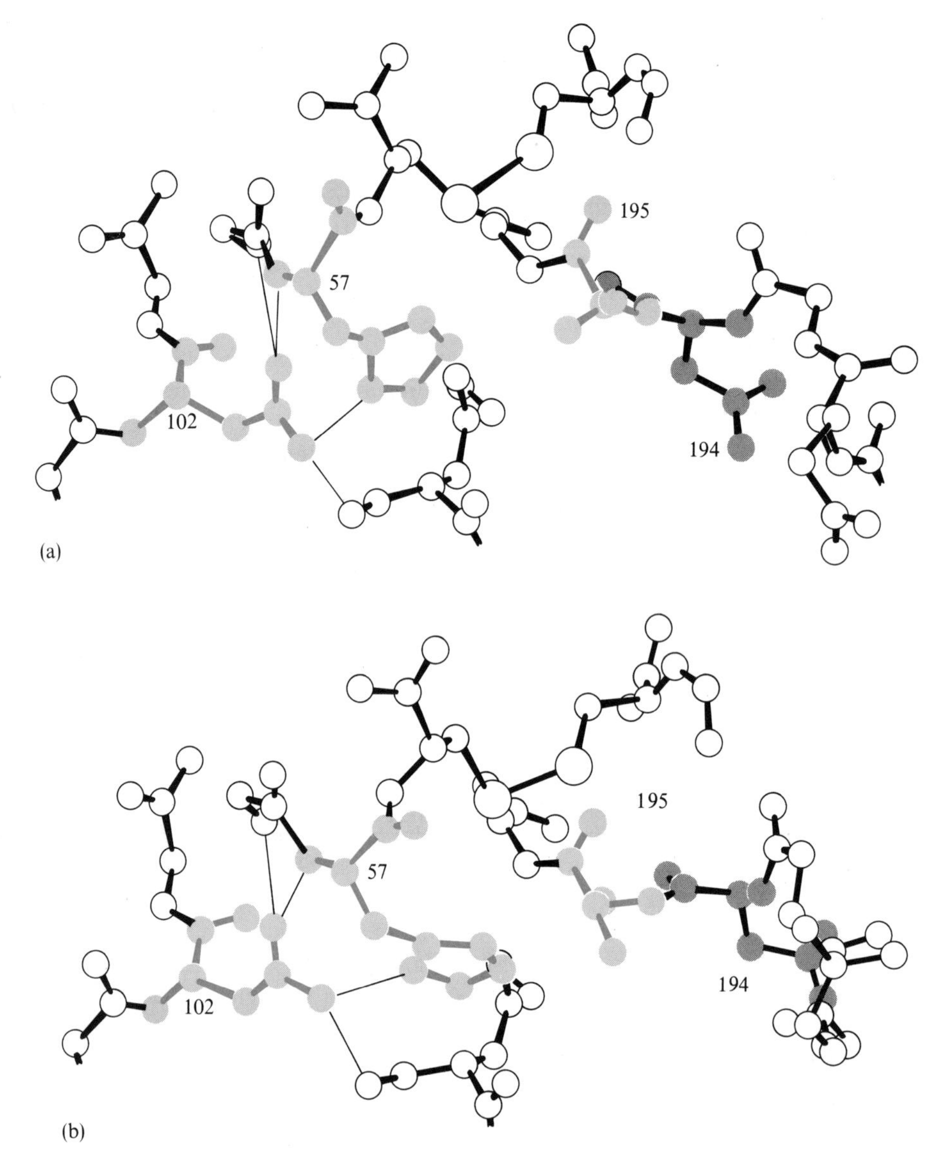

Figure 2-5

Differences in structure near the active site of bovine trypsinogen (**a**) *and trypsin* (**b**). Residues involved in catalysis are His^{57}, Asp^{102}, and Ser^{195}. When the peptide chain of trypsinogen is cleaved to yield trypsin, the Lys^{15}–Ile^{16} bond is hydrolyzed to produce a charged N-terminus in Ile^{16}. This moves to form a salt bridge with Asp^{194}. When it does so, some changes in orientation of the catalytic residues result. Also, significant movements occur among residues to the right of the active site. This movement opens up the binding pocket for the substrate. [Drawing based on material provided by Robert Stroud, and adapted with permission from A. A. Kossiakoff et al., *Biochemistry* 16:654 (1977). Copyright © by the American Chemical Society.]

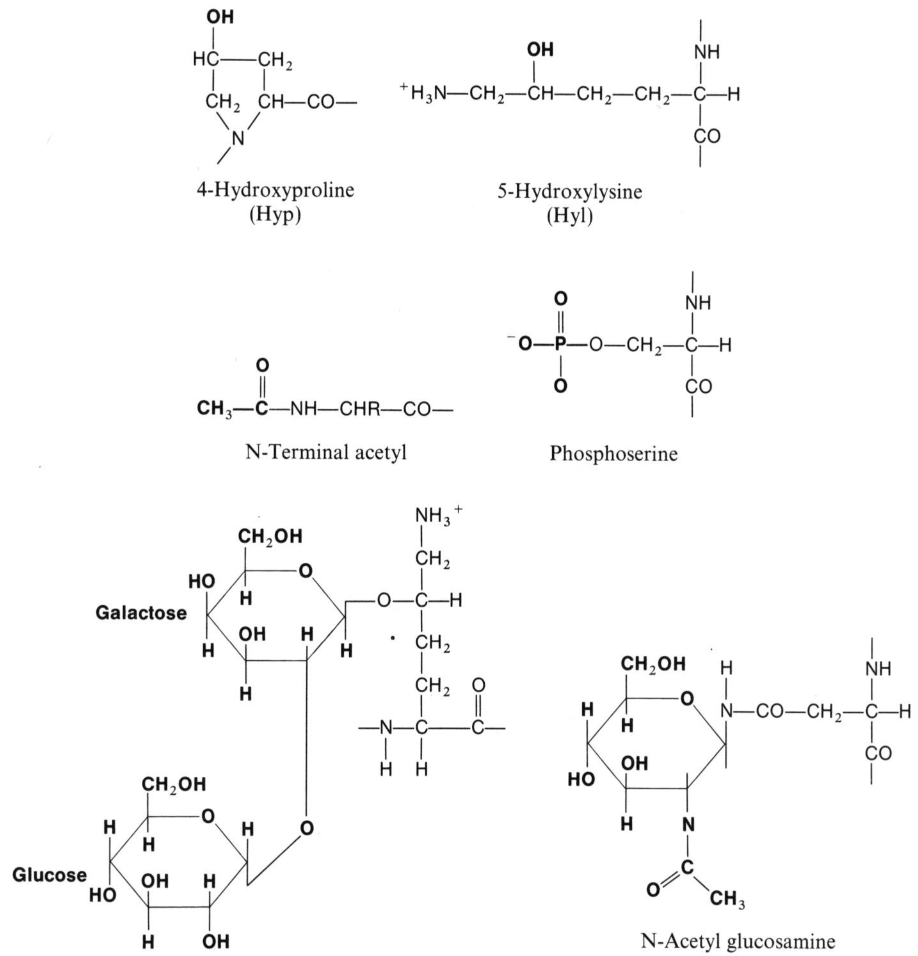

4-Hydroxyproline
(Hyp)

5-Hydroxylysine
(Hyl)

N-Terminal acetyl

Phosphoserine

Galactose

Glucose

N-Acetyl glucosamine

Figure 2-6
Structures of some modified amino acid side chains. All of these modifications take place after the corresponding normal amino acid has been incorporated into a peptide chain on the ribosome.

Metals and other prosthetic groups

Quite a few proteins are found to contain non–amino acid prosthetic groups that play a clear role in function and structure. For example, there is a large class of metalloproteins containing one or more metal ions required for biological activity. Many proteins appear to require noncovalent alkali or divalent alkaline cations for full activity. An example of a Ca^{2+} binding site is shown in Figure 2-7a. Because these ions are present in large quantities in cells, it is easy to see how proteins have adapted to take advantage of them. Their role in most cases probably is simply structural.

Transition metals pose more of a problem. As free cations, they are quite rare in biological systems, and this is fortunate because many proteins are irreversibly damaged when they come into contact with heavy cations. Therefore the challenge for proteins was to find ways to tame and exploit the chemistry and complex bonding of ions such as Fe^{2+}, Zn^{2+}, and Cu^{2+}. Two general approaches have been used. Some proteins contain sites composed solely of normal amino acids that allow tight specific bonding of particular cations. Examples are Zn^{2+} and Cu^{2+} binding sites in a number of enzymes. See Figure 2-7b for an illustrative structure of one of these complexes.

Other proteins are prepared to take on a metal plus one or more associated ligands. Sometimes the ligand dwarfs the metal. Heme proteins are the most common case. Here the ligand, a porphyrin ring, is tetravalent. With some metals, a porphyrin complex has two additional ligand binding sites above and below the porphyrin plane. In hemoglobin and myoglobin, one of these is filled by a histidine of the protein, while the other is left free for bonding to a ligand like oxygen, as shown in Figure 2-7d. More complicated cases exist where proteins bind more than one different kind of metal.

Figure 2-7

Some typical metal binding sites in proteins. All are shown to illustrate the actual three-dimensional structure of the metal complexes. (Metals are shown in color.) **(a)** A Ca^{2+} binding site in thermolysin. [Provided by Brian Matthews.] **(b)** A Zn^{2+} binding site at the active site of carboxypeptidase A. Not shown is a water molecule coordinated vertically. [After D. M. Blow and T. A. Steitz, *Ann. Rev. Biochem.* 39:78. Copyright © 1970 by Annual Reviews, Inc. All rights reserved.] **(c)** Chromatium high-potential iron protein, which contains four irons, four labile sulfur atoms, and four cysteines. Bacterial ferredoxins have a very similar group—compare Figure 2-28. [After L. H. Jensen, *Ann. Rev. Biochem.* 43:461. Copyright © 1974 by Annual Reviews, Inc. All rights reserved.] **(d)** Myoglobin. An iron-porphyrin complex is coordinated to the protein through a histidine. The sixth ligand site is unoccupied as shown; this is the site that binds oxygen. A virtually identical structure exists in hemoglobin. [After R. E. Dickerson and I. Geis, *The Structure and Action of Proteins* (New York: Harper & Row, 1969).]

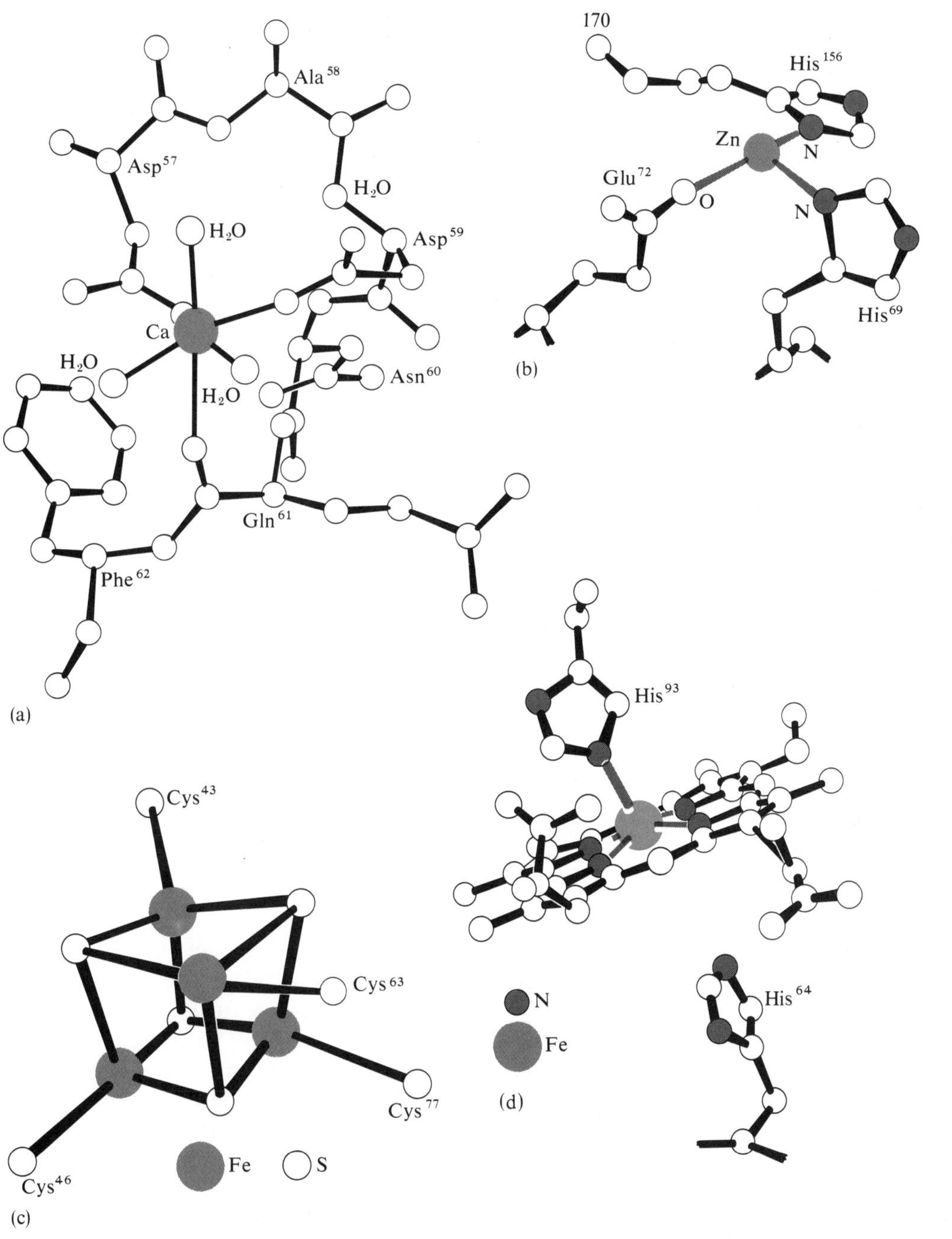
Ala 58
Asp 57
H2O
H2O
Asp 59
Ca
H2O
H2O
Asn 60
Gln 61
Phe 62
(a)
170
His 156
Zn
N
Glu 72
O
N
His 69
(b)
His 93
Cys 43
Cys 63
Cys 77
Cys 46
Fe
S
(c)
N
Fe
His 64
(d)

How much influence does the metal (and any associated ligands) have on the structure of the protein? One way to examine this question is to compare the structure of the holoprotein (containing the metal) with that of the apoprotein (with no prosthetic groups present). For heme-containing proteins like myoglobin and hemoglobin, these structures are very different. This is not surprising because many amino acid residues participate in contacts with the heme in the native structures. Heme removal changes the environment of too large a fraction of the protein for the structure to remain constant.

Sherman Beychok has compared a number of the properties of heme-containing subunits of hemoglobin with those of globin chains from which the heme has been removed. The association of subunits is dramatically altered. The following are some of the equilibria among hemoglobin subunits. The subscript h refers to heme-containing chains, whereas o denotes chains without heme.

Hemoglobin: $2\alpha_h\beta_h \rightleftarrows (\alpha_h\beta_h)_2$

Isolated β subunit: $2(\beta_h)_2 \rightleftarrows (\beta_h)_4$ $\qquad 2(\beta_o)_2 \rightleftarrows (\beta_o)_4$

Isolated α subunit: $2\alpha_h \rightleftarrows (\alpha_h)_2$ $\qquad 8\alpha_o \rightleftarrows 4(\alpha_o)_2 \rightleftarrows (\alpha_o)_8$

Reassociation at 4° C: $\alpha_o + \beta_o \rightarrow \alpha_o\beta_o$

$$2\alpha_o + 2\beta_h \rightarrow (\alpha_o)_2(\beta_h)_2$$

$$\alpha_h + \beta_o \rightarrow \alpha_h\beta_o$$

The overall pattern of results is quite complex. For example, the presence of heme promotes self-association of β subunits and inhibits self-association of α subunits. Moreover, it is clear that the heme can affect quaternary equilibria even though it probably never participates directly in any of these reactions as a component of the subunit interface. Its major effect appears to be to promote a reorganization of the 2° and 3° structure of each subunit.

In contrast to hemoglobin, many other proteins that simply bind an isolated cation show little accompanying structural change. Also, it is frequently possible to substitute one cation for another in many metalloproteins. Replacing iron or zinc with cobalt is particularly useful because the different physical and chemical properties of Co^{2+} let one inquire whether any cation will serve for function, or whether the electronic or bonding properties of the cation are used in more sophisticated ways. One of the reasons why metalloproteins have received so much attention is that the metal often serves as a very informative spectroscopic probe placed right at the functional site. This greatly facilitates a wide variety of physical studies without fear of perturbations caused by insertion of unnatural probes.

Many proteins use small organic molecules rather than metals to assist in their functions. The majority of these molecules seem to be attached to lysine. A number of examples are shown in Figure 2-8. All of these protein adducts bring to a protein a chemical capability not found in normal amino acids. For example, the long

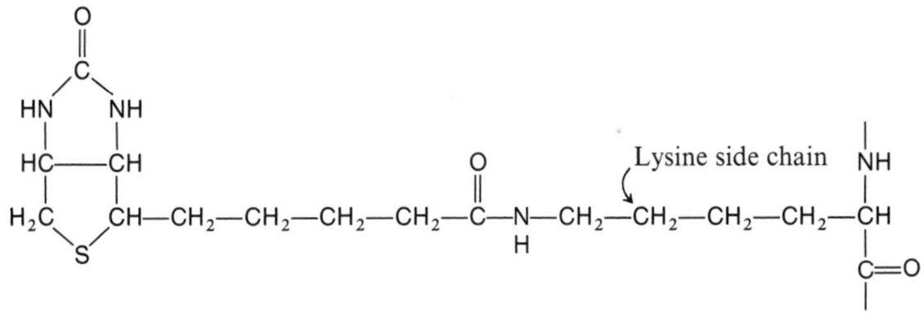

O
C
HN NH
HC—CH
H_2C CH—CH_2—CH_2—CH_2—CH_2—C—N—CH_2—CH_2—CH_2—CH_2—CH
S
H
Lysine side chain
NH
C=O

Biotin

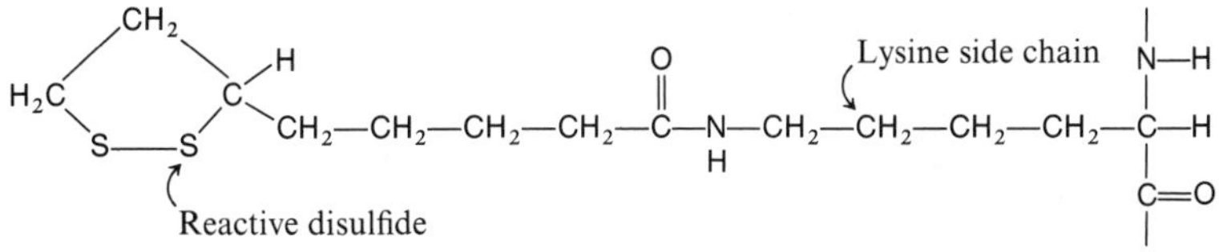

CH_2
H_2C C—H
S—S
CH_2—CH_2—CH_2—CH_2—C—N—CH_2—CH_2—CH_2—CH_2—C—H
O
H
Lysine side chain
N—H
C=O
Reactive disulfide

Lipoic acid

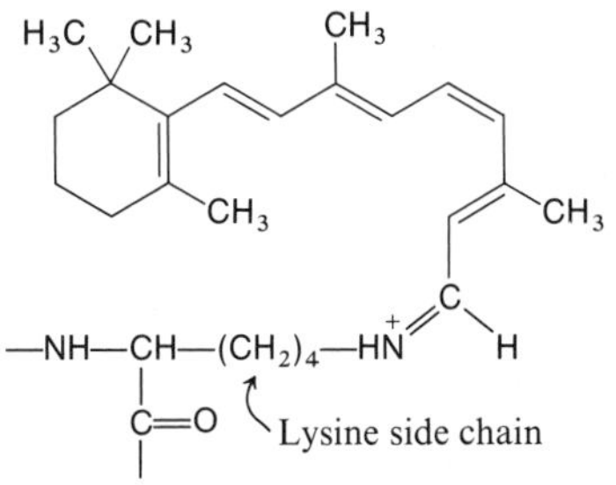

H_3C CH_3
CH_3
CH_3
CH_3
C
—NH—CH—$(CH_2)_4$—HN^+ H
C=O
Lysine side chain

Retinal protonated Schiff's base

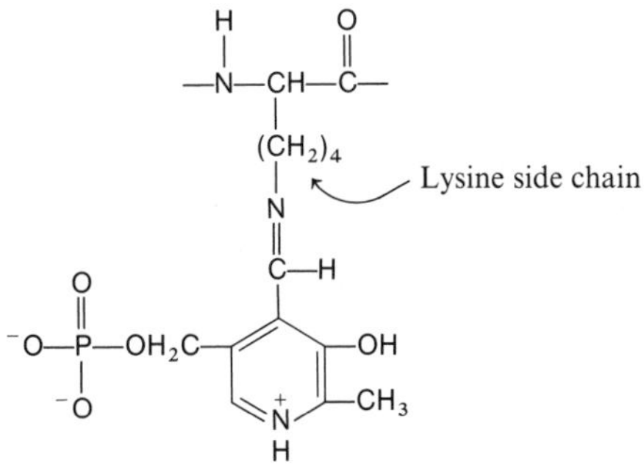

H O
—N—CH—C—
$(CH_2)_4$
Lysine side chain
N
C—H
O
$^-$O—P—OH_2C— —OH
$^-$O
N^+ —CH_3
H

Pyridoxal phosphate Schiff's base

Figure 2-8

Structures of some protein prosthetic groups. All these structures represent ways of introducing chemical or photochemical capabilities not found in the normal amino acid side chains.

flexible chain in biotin and in lipoic acid may allow these coenzymes to move a substrate that has combined with them from one site to another on an enzyme. We do not yet know in molecular detail how any of these organic molecules are situated in proteins, or how the local structure accommodates itself to their presence. However, pyridoxal coenzymes and retinal are useful spectroscopic probes, permitting information about their environment in proteins to be obtained.

2-3 PRIMARY STRUCTURE

The first things to learn about the covalent primary structure of a protein are the number and types of polypeptide chains it contains and the sizes of each of these. In more cases than not, an isolated single protein subunit will consist of a single chain. However, multichain proteins are common, although these may have arisen by posttranslational cleavage of an original single chain. Different chains can usually be separated according to their charge or molecular weight by methods described in Chapters 11 and 12. Approximate sizes are revealed by these same techniques. More precise sizes can sometimes be obtained by chemical analysis of the ratio of of end residues to the total or from the amino acid composition, recognizing that the least prevalent amino acids must exist in integral numbers per chain. The smallest known protein chains occur in a few peptide hormones such as secretin or glucagon and in a number of tiny proteins such as ferredoxin. These have between 25 and 100 amino acids. A more typical single peptide chain will have 100 to 500 amino acids. Some of the largest known chains have upwards of 3,000 residues, but such structures are quite rare except in fibrous proteins.

Determination of the amino acid sequence of the peptide chains of a protein is a well-established technique. Automated analytical procedures exist for many of the stages, especially for the stepwise terminal degradation of chains starting from the amino terminus. In favorable cases, up to 60 or more residues can be sequenced with little human intervention. As a result, amino acid sequence determination is almost a routine procedure for many smaller proteins. It can be more of a challenge for larger proteins or for those in which the solubility properties are unfavorable. However, as a general rule, if you need to know the sequence of a protein and are willing to perform the labor involved, final success is almost guaranteed. The relative ease (or at least certainty) of sequence determination immediately suggests two questions. What can you learn directly from the sequence, and what other experimental approaches are facilitated by knowing the sequence?

Disulfides and other cross-links

The first step after determining a complete sequence, or even partial sequence information, usually is to find out whether any cross-links exist in the structure. The most common of these are disulfide bonds formed by the oxidization of pairs of cysteines. The disulfide product is called the amino acid cystine because it frequently is analyzed as a separate unit. As shown in Figure 2-9, the C—S—S—C bonding system in

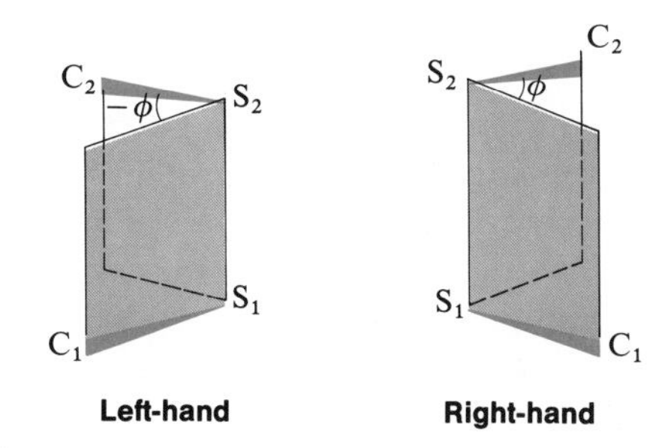

Figure 2-9

Structure of cystine disulfide cross-links. Note that the two methylene groups are not in the same plane; ϕ defines the dihedral angle between them. Two optical isomers are shown for one particular value of $|\phi|$. [After a drawing provided by S. Beychok.]

cystine is distinctly nonlinear, and furthermore the disulfide prefers nonplanar conformations. Therefore, the presence of a cystine imposes a strong constraint in a protein structure. It demands the close approach of two regions of the peptide chain, and furthermore it tends to demand particular angles of approach. This means that, everything else being equal, it should be more difficult to disrupt the structures of proteins containing disulfide bonds. Indeed, some small proteins that contain 5 to 7 disulfides in less than 100 amino acid residues are among the most stable proteins known. They can resist extremes of conditions, such as boiling or strong acid, that will irreversibly denature most proteins quite rapidly.

In general, most proteins can be placed into three classes with respect to SH groups. Quite a few proteins contain neither cysteine nor cystine. Many others contain from one to many cysteines, but no cystine. Then, there are proteins with one or (more typically) several cystines. It is, however, very rare to find a protein that contains both cysteine and cystine. There are good reasons for this. The structure will be unstable with respect to disulfide exchange reactions such as

$$\text{—S—S—Cys}_2 + \text{Cys}_3\text{—SH} \rightleftarrows \text{Cys}_1\text{—SH} + \text{Cys}_2\text{—S—S—Cys}_3$$

These can be either intramolecular or intermolecular. There will only be a narrow range of environmental conditions where it is likely that distinct disulfides and SH groups can be maintained simultaneously in a protein, even if they cannot reach each other sterically. One exception to the rule is serum albumins. These proteins typically have 17 disulfide bonds and one free SH. However, serum albumins are anomalous proteins in other respects as well.

A knowledge of the network of disulfide bonds in a protein can sometimes provide useful hints about its three-dimensional structure. When several chains are present, there can be intrachain and interchain disulfides, and a rather complicated topology can develop for the covalent structure. For example, Figure 2-10 shows a schematic illustration of the primary structure of immunoglobulin G. There are four peptide chains and a total of fifteen disulfide bonds. Of these three are interchain and the rest are intrachain. The topological pattern shown in this figure led to the original suggestion that IgG consists of regions, each constrained independently by intrachain disulfides. As more definitive structural data on IgG accumulated, it became clear that the protein is organized into separated folded domains.

Besides disulfides, there are a few other kinds of cross-links that have been identified in protein. These are, however, much more rare. Some fibrous proteins, especially common in connective tissue, acquire a number of interchain cross-links. These serve to increase rigidity and also probably insure that, if the structure is deformed, it returns readily to the original state. Most of these cross-links originate when the ε-amino of lysine is converted into an aldehyde by lysyl oxidase. This reaction generates a unique functionality that can dimerize through the aldol condensation to form a cross-link between the residues or can react with an unmodified lysine. Either of these reactions can initiate a chain of events that can result in the combination of either three or four amino acid side chains. Some of the known lysine cross-links in collagen and elastin are illustrated in Figure 2-11. Also shown in that figure are cross-links between glutamine and lysine, which occur in fibrin, a protein involved in blood clotting. These cross-links presumably stabilize the clot.

Primary structure and the analysis of secondary and tertiary structures

Knowledge of the primary structure is critical for many other biochemical and biophysical studies. A paramount example is the determination of the three-dimensional structure of proteins by x-ray crystallography. Rarely are data obtained initially at resolution high enough to see individual atoms. Often the results are not even clear enough to allow the path of the peptide-chain backbone to be followed unequivocally through the structure. However, most of the backbone usually is visible at the resolutions typically obtained, and some of the larger side chains stand out clearly. Then what is done is to use the amino acid sequence and the crystallographic electron density map side by side to try to find a unique fit to the structure. This process is discussed in more detail in Chapter 13, but the critical thing to remember is that most x-ray data are not good enough to see the complete sequence of amino acids. However, the sequence data themselves frequently have errors. When used hand in hand, both techniques are measurably strengthened.

A second vital role for primary structure data is the analysis of any techniques that involve chemical modification or the introduction of a probe. If you know only

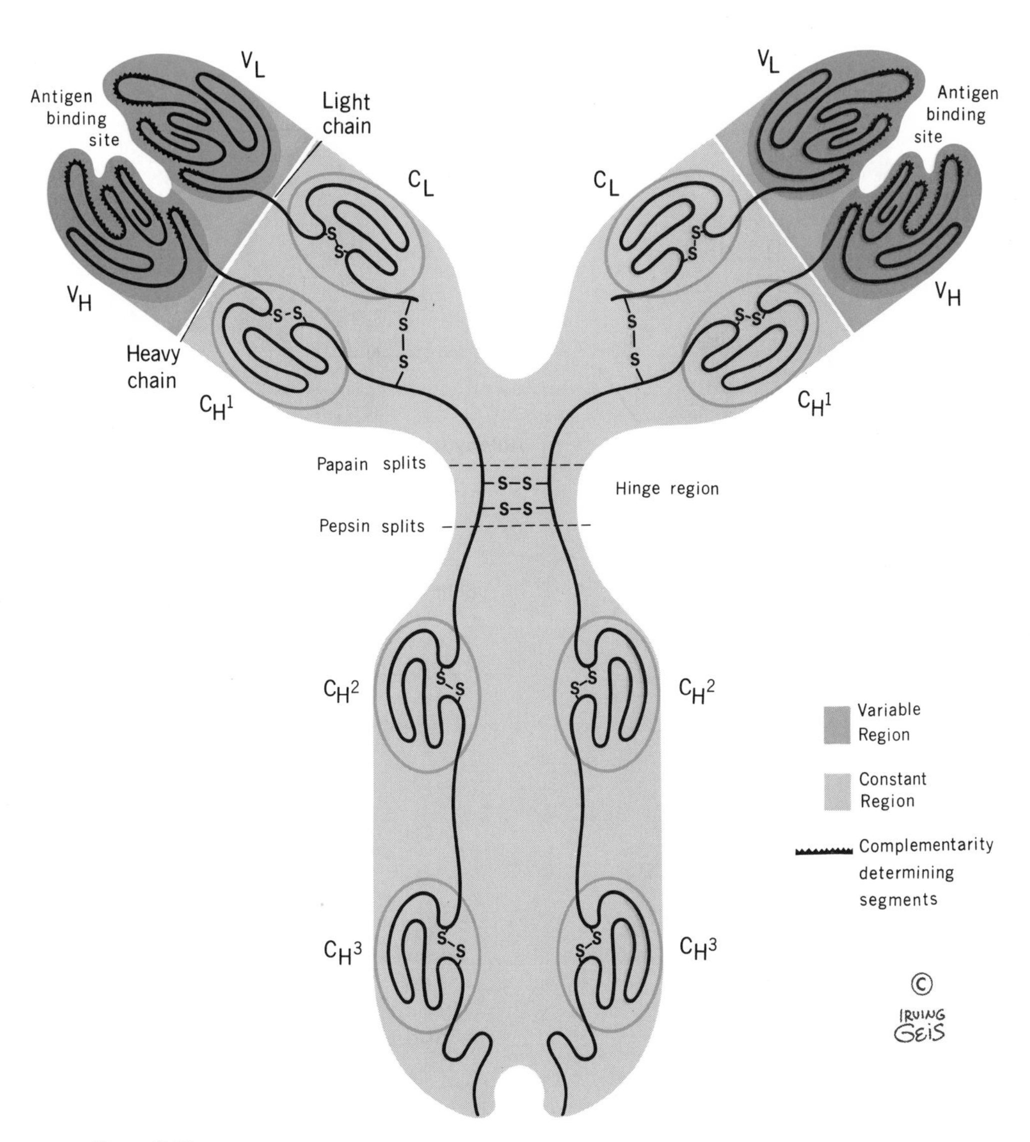

Figure 2-10

The pattern of disulfide cross-links in immunoglobulin G. The subscripts L and H refer to light and heavy chains, respectively; C and V refer to regions of the sequence that are relatively constant or quite variable, respectively, in different IgG species. Each block of each sequence folds into an independent tertiary structure domain, so that the final structure resembles a protein with twelve subunits, even though it is a single covalent entity. [Drawing by Irving Geis.]

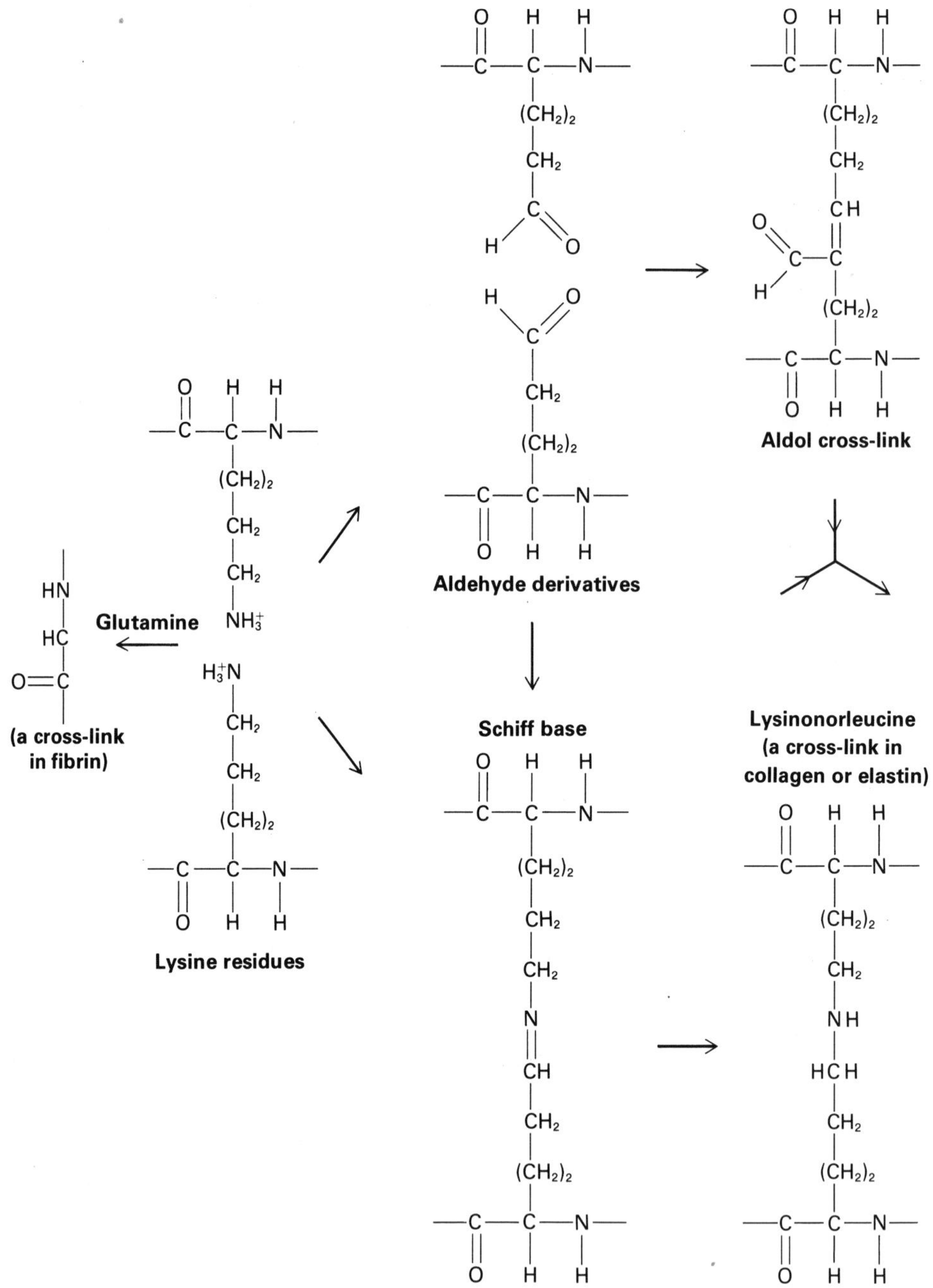

Aldol cross-link

Aldehyde derivatives

Glutamine

(a cross-link in fibrin)

Lysine residues

Schiff base

Lysinonorleucine (a cross-link in collagen or elastin)

(a)

Figure 2-11

Some pathways for the formation of lysine cross-links in collagen, elastin, and fibrin. (**a**) Lysines can combine with each other, with histidine, or with glutamine to yield dimer, trimer, and tetramer adducts. (**b**) Schematic illustration of how intramolecular and intermolecular cross-links can reinforce the array of collagen molecules in a collagen fiber.

Histidine

```
             H   O
             |   ||
        —N—C—C—
         H   |
             CH2
             |
             C———N
             ||    ||
            HC    CH
               \  /
    |           N              |
   HN           |       H      NH
    |           |       |      |
H—C—(CH2)2—CH2—C—C—(CH2)2—C—H—
    |           H       |      |
O=C                    C      C=O
    |                 //  \    |
                     O     H
```

Histidine-aldol cross-link

Desmosine
(a cross-link in elastin)

```
               H   H   O
               |   |   ||
             —N—C—C—
                   |
                 (CH2)2
    |              |                      |
H—N               CH2                    N—H
    |              |                      |
H—C—CH2—CH2       |        CH2—CH2—C—H
    |          \   ||  /                  |
O=C            (pyridinium ring)          C=O
    |               N+                    |
                    |
                   CH2
                    |
                  (CH2)3
                    |
              —C—C—N—
               ||  |   |
               O   H   H
```

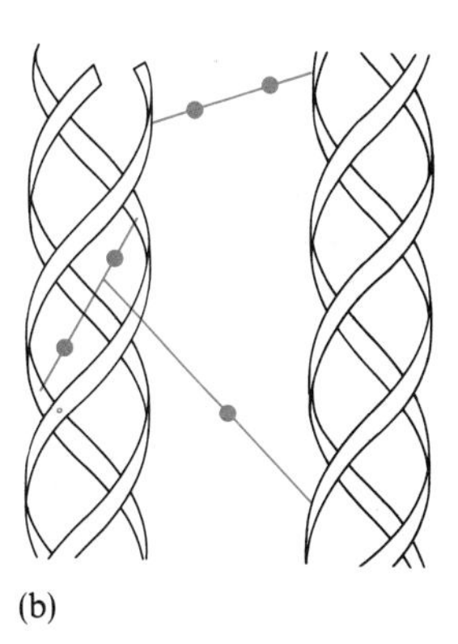

(b)

that a cross-linker has reacted with lysines or that a spectroscopic probe is attached to lysine, you really don't know very much. Is there one unique product? Does it suggest anything about the folding of the peptide chain in the tertiary structure? With a known sequence, the specific sites of chemical modification can be determined by mapping peptide fragments and searching for the incorporated group.

For example, Figure 2-12 shows the backbone of bovine ribonuclease A. Chemical studies on the protein show that 1,5-difluoro-2,4-dinitrobenzene will cross-link lysines 7 and 41. Unexpected sensitivity of the protein to iodoacetate led to the curious finding that histidines 12 and 119 react in an either/or fashion, suggesting that these two positions must be adjacent in the active site of the enzyme. From the sequence and both sets of chemical data, it is immediately clear that a planar arrangement of the peptide chain is ruled out. A much more complex pattern of chain folding in three dimensions must be occurring, and in fact this is what is observed crystallographically.

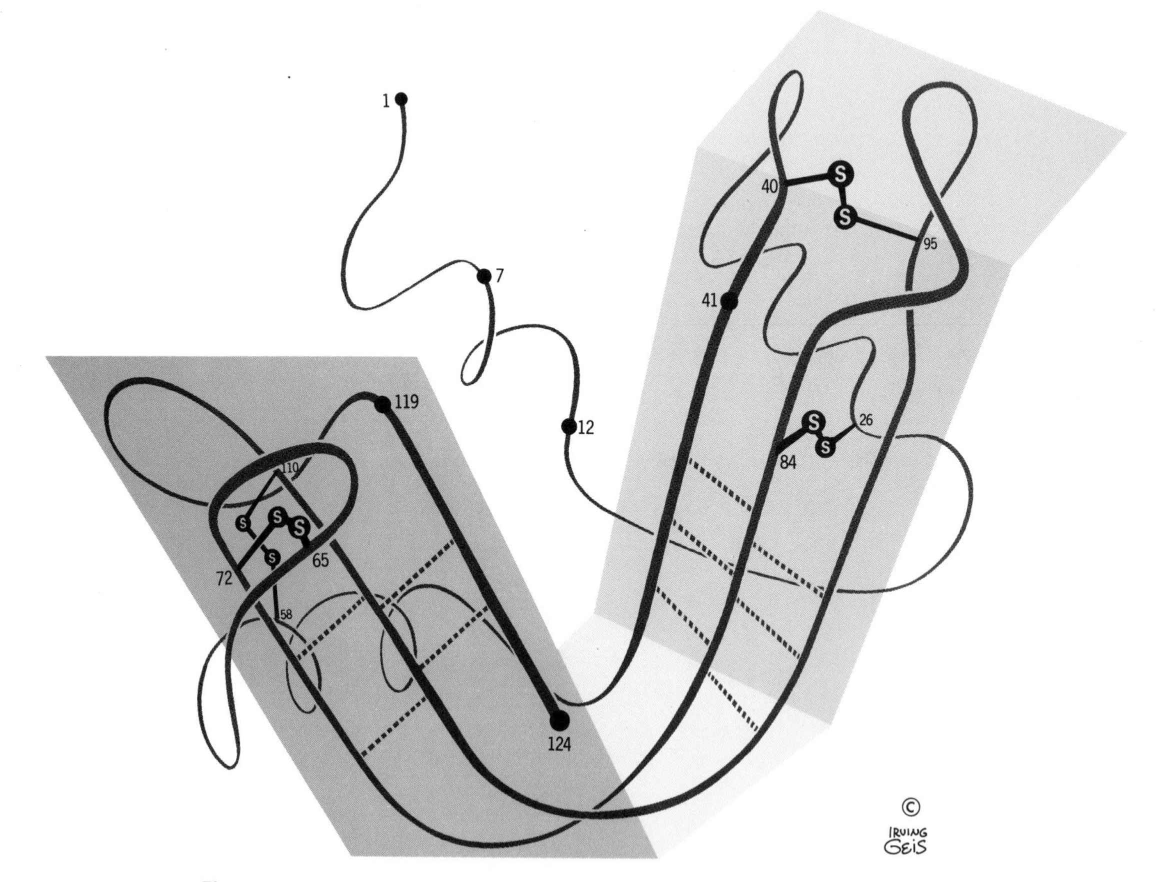

Figure 2-12

Peptide backbone of bovine pancreatic ribonuclease. Also shown are two pairs of residues that must be close in the tertiary structure, either because they have been cross-linked (Lys^{7} and Lys^{41}) or because they are directly implicated at the active site (His^{12} and His^{119}). [Drawing by Irving Geis.]

Primary structure and the prediction of secondary and tertiary structures

Sometimes inspection of the sequence can lead to more decisive predictions about the structure. Moderately reliable schemes now exist for predicting the location of secondary structure regions along the sequence. These are discussed in detail in Chapter 5. One example of the current state of the art is shown in Figure 2-13, which compares the results of a number of different schemes, applied to the protein adenylate kinase. The challenge was to predict, given only the sequence, which amino acids would be involved in three types of secondary structure: α helices, β sheets, and β bends. The predictions were compared with the true three-dimensional structure as deduced by x-ray crystallography. As shown in Figure 2-13, the location of α helices and β bends are rather well predicted by a number of different schemes, but the accuracy of β sheet prediction still leaves much to be desired.

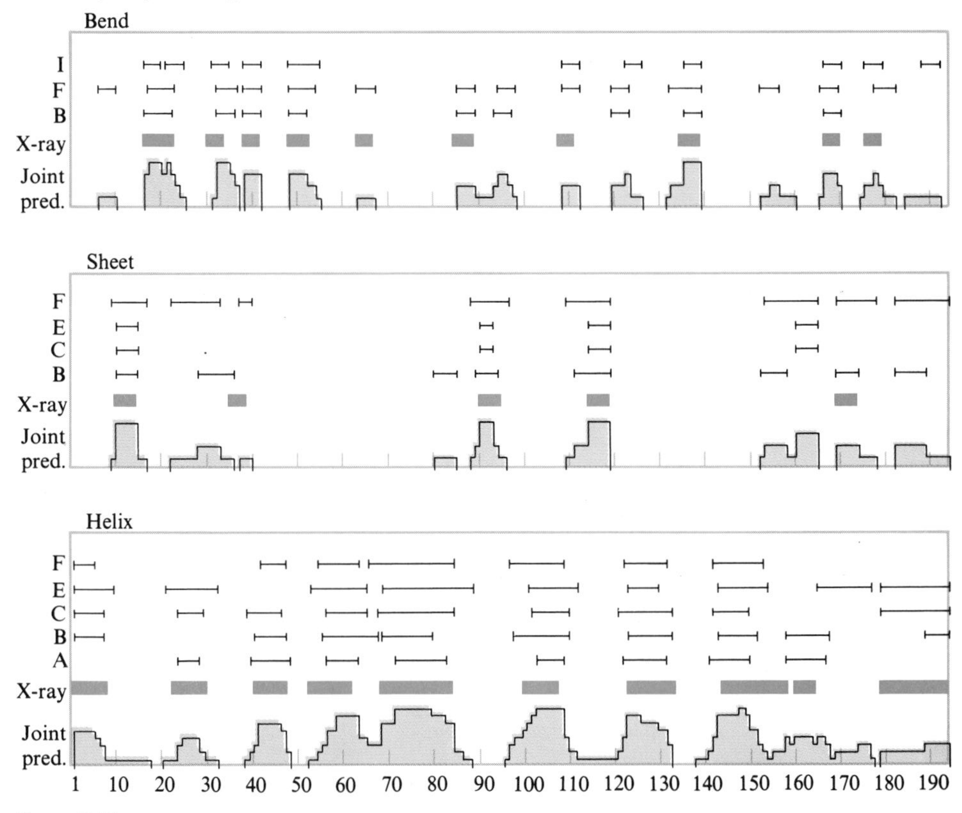

Figure 2-13

Distribution of secondary structure types in adenylate kinase. Regions predicted by several different theoretical methods to be in β bends, β sheets, or α helices are shown. Actual secondary structure regions found by x-ray crystallography are indicated, as is the average predicted distribution considering the various theories jointly. The predictions were made before the crystal structure was known. [After G. E. Schultz et al., *Nature* 250:140 (1974).]

The most ambitious goals for using sequence information are attempts to predict the entire tertiary structure and, from this, perhaps even the function. Some of the principles and simplifications involved in this are given in Chapter 5. Here we show one result that demonstrates the current state of the art. Figure 2-14 compares the three-dimensional structure of bovine pancreatic trypsin inhibitor as measured by x-ray crystallography and as computed starting from the sequence, using general information about the thermodynamic interactions between amino acid residues. The sequence was originally placed in an extended conformation. Forces between residues were calculated from their interaction energies and these were used to compute the expected folding pathway. The agreement is not bad qualitatively. The calculations do indeed predict that a globular structure should form, and furthermore the computed arrangement of the peptide backbone is quite similar in some parts to the true structure. The prediction is not so successful that one can dispense with the laborious efforts involved in protein crystallography, but the computations are close enough to a realistic protein to raise optimism about future prospects.

Some amino acid sequences appear to contain repeats. These can be long range or short range, as shown by the examples in Figure 2-15. This should immediately

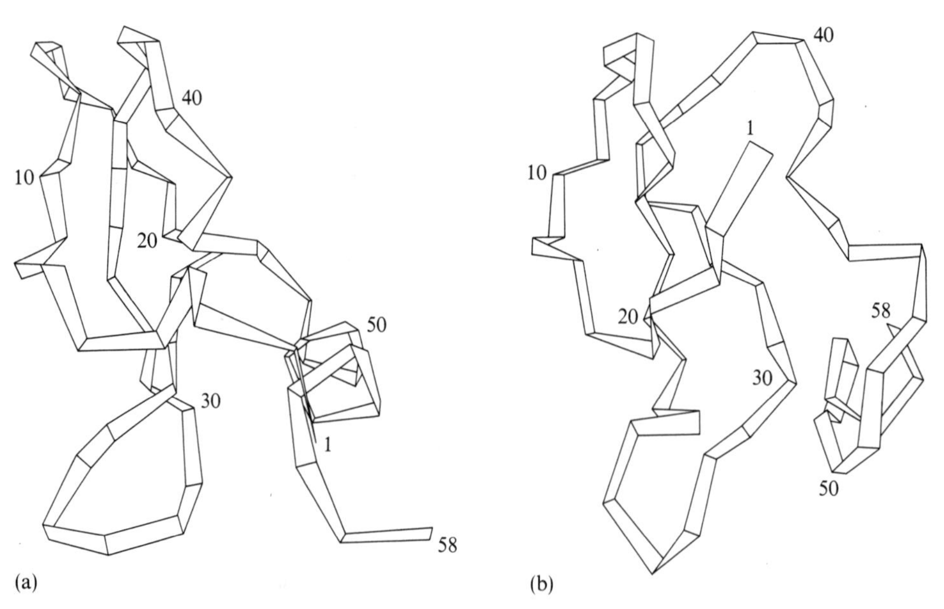

Figure 2-14

Schematic drawing of the peptide backbone of pancreatic trypsin inhibitor. (**a**) As seen in the structure of a crystal. (**b**) As calculated by starting with an unfolded chain (except for a C-terminal helix) and attempting to minimize the conformational energy of the molecule. [After M. Levitt and A. Warshel *Nature* 293 : 693 (1975).]

10 20

NH_3^+–Ala–Tyr–Val–Ile – Asn–Asp–Ser–Cys–Ile–Ala– Cys–Gly–Ala–Cys–Lys–Pro–Glu–Cys–Pro–Val–Asn–Ile – Gln–Gln–Gly–Ser–Ile—
Tyr–Ala–Ile – Asp–Ala– Asp–Ser–Cys–Ile–Asp– Cys–Gly–Ser– Cys–Ala–Ser–Val– Cys–Pro–Val–Gly – Ala–Pro–Asn–Pro–Glu–Asp–COO^-

30 40

(a)

					5					10					15
1	Gly	Tyr	Asp	Glu	Lys	Ser	Ala	Gly	Val	Ser	Val	Pro	Gly	Pro	Met
16	Gly	Pro	Ser	Gly	Pro	Arg	Gly	Leu	Pro	Gly	Pro	Pro	Gly	Ala	Pro
31	Gly	Pro	Gln	Gly	Phe	Gln	Gly	Pro	Pro	Gly	Glu	Pro	Gly	Glu	Pro
46	Gly	Ala	Ser	Gly	Pro	Met	Gly	Pro	Arg	Gly	Pro	Pro	Gly	Pro	Pro
61	Gly	Lys	Asn	Gly	Asp	Asp	Gly	Glu	Ala	Gly	Lys	Pro	Gly	Arg	Pro
76	Gly	Gln	Arg	Gly	Pro	Pro	Gly	Pro	Gln	Gly	Ala	Arg	Gly	Leu	Pro
91	Gly	Thr	Ala	Gly	Leu	Pro	Gly	Met	Lys	Gly	His	Arg	Gly	Phe	Ser
106	Gly	Leu	Asp	Gly	Ala	Lys	Gly	Asn	Thr	Gly	Pro	Ala	Gly	Pro	Lys

Figure 2-15

Amino acid sequence of two proteins showing repeated sequences. **(a)** *P. aerogenese* ferredoxin arranged to show the remarkable similarity of the first and last 27 amino acids. The eight cysteines are involved in coordination to iron atoms, as shown in Figures 2-7 and 2-28. **(b)** A fragment of collagen. With only three exceptions, near the N-terminus, every third residue is a glycine. The large number of prolines prevent α or β structures from forming and dictate that the chain take on a polyproline-type helix. The periodic glycines allow three such helices to be brought together in a staggered triple-strand array to form the collagen triple helix.

suggest that there is symmetry or pseudosymmetry in the three-dimensional structure Long repeats can lead to protein either folding to give a symmetrical site or folding to give multiple domains that might begin to resemble similar subunits of a multi-subunit protein. Short repeating sequences can indicate helical or superhelical structures such as the coiled coils found in collagen and other fibrous proteins.

Sequences and the analysis of function

If something is known about the function of the protein, particularly if it interacts with other large components, the sequence can be extremely illuminating. Figure 2-16 shows two examples of this. Histones bind tightly to DNA. We have already indicated that the high density of positively charged amino acids in histones presumbly acts to stabilize interactions with such a polyanion. As you can see in Figure 2-16a, there is an unusual cluster of positive amino acids in the N-terminal part of the sequence. It is tempting to speculate that this region has an extended structure that coils along an extensive region of DNA. The remaining sections of the protein appear more normal, and perhaps they fold into a more typical globular protein structure.

Glycophorin is a membrane protein isolated from red blood cells. Chemical evidence suggests that it is a transmembrane protein. This means that part of it must be on the outside of the cell, part on the inside, and part must be able to pass through the membrane. It is known that the sugar residues of membrane glycoproteins are almost inevitably on the exterior cell surface. Because the sugar residues are all on the N-terminal half of the glycophorin sequence, this region must be external. The part that passes through the membrane is expected to be very nonpolar (hydrophobic) in order to produce stable interactions with lipids. Inspection of the sequence quickly shows a most unusual stretch of 20 continuous residues that should not object too much to a lipid environment. If this were in an elongated conformation, the chain would be long enough to stretch through the entire lipid bilayer. Once this assignment is made, it seems clear that the short C-terminal polar region is likely to be on the inside of the cell. Note that the transition to polar residues on both sides of the nonpolar stretch is rather abrupt. One can speculate that the molecule was designed to be positioned rather precisely in the bilayer, and that any fluctuations in position toward the exterior or interior of the cell are likely to be very expensive energetically.

Sequence comparisons among closely related proteins

We suggested in Chapter 1 that sequence information gains in power and interest when sequences from a variety of structurally or functionally related molecules are available. Here we give a few examples of this for proteins. One of the largest bodies of sequence data exists for hemoglobins and such related heme proteins as myoglobins. Sequence comparisons within one species show clearly that there are extensive homologies among the α and β chains and myoglobin. It appears likely that in some ancestral

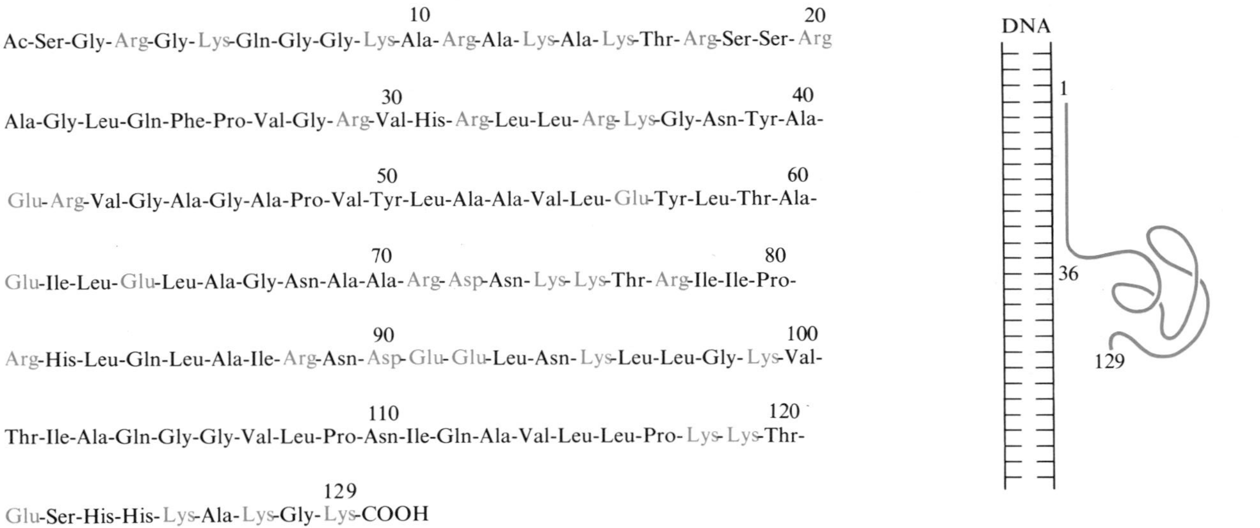

Figure 2-16

Amino acid sequences of two proteins, revealing something about their structure and function. (All charged residues are shown in color.) **(a)** Calf histone 2a. The first 36 residues contain 12 positive charges and no negative charges. Presumably this region interacts with the negative phosphates of a DNA double helix.

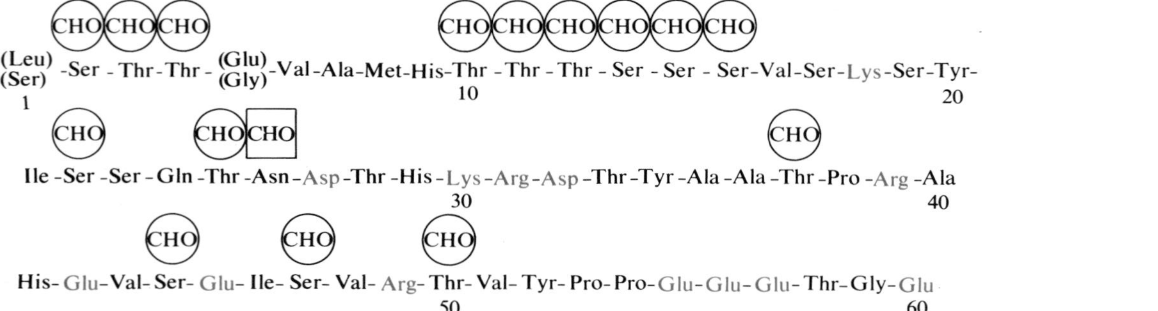

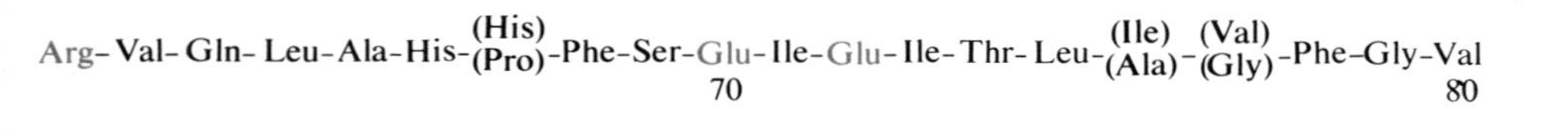

Met- Ala- Gly- Val- Ile- Gly- Thr- Ile- Leu - Leu- Ile- Ser- Tyr- Gly- Ile- Arg- Arg- Leu- Ile- Lys
90
100

Lys- Ser- Pro- Ser- Asp- Val- Lys- Pro- Leu- Pro- Ser- Pro- Asp- Thr- Asp- Val- Pro- Leu- Ser- Ser
110
120

Val- Glu- Ile- Glu- Asn- Pro- Glu- Thr- Ser- Asp- Gln-
130

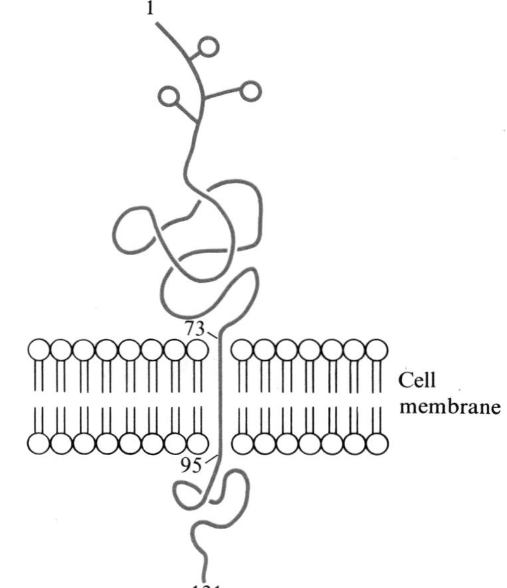

Figure 2-16 (*cont.*)
(**b**) Human erythrocyte glycophorin. CHO denotes positions at which oligosaccharides are attached by an O-glycoside linkage (*circle*) or an N-glycoside linkage (*square*). Glycophorin is a transmembrane protein. The sugars must be on the exterior face of the cell. The large stretch of nonpolar residues from Ile^{73} through Ile^{95} presumably is the region that actually passes through the membrane, and therefore one can guess that the C-terminus is located inside the cell. [After M. Tomita and V. T. Marchesi, *Proc. Natl. Acad. Sci. USA* 72:2914 (1976).]

organism there was only one globin gene that later split into multiple copies, which diverged evolutionarily. By inspection of the sequences, either by visual alignment or by more elaborate computer searches, it is possible to construct a family tree or, in the language of evolution, a phylogenetic tree. Myoglobin is more different from the α and β chains than they are from one another. Therefore, as shown in Figure 2-17, the gene for myoglobin must have separated from that of hemoglobin before α and β chains started to diverge.

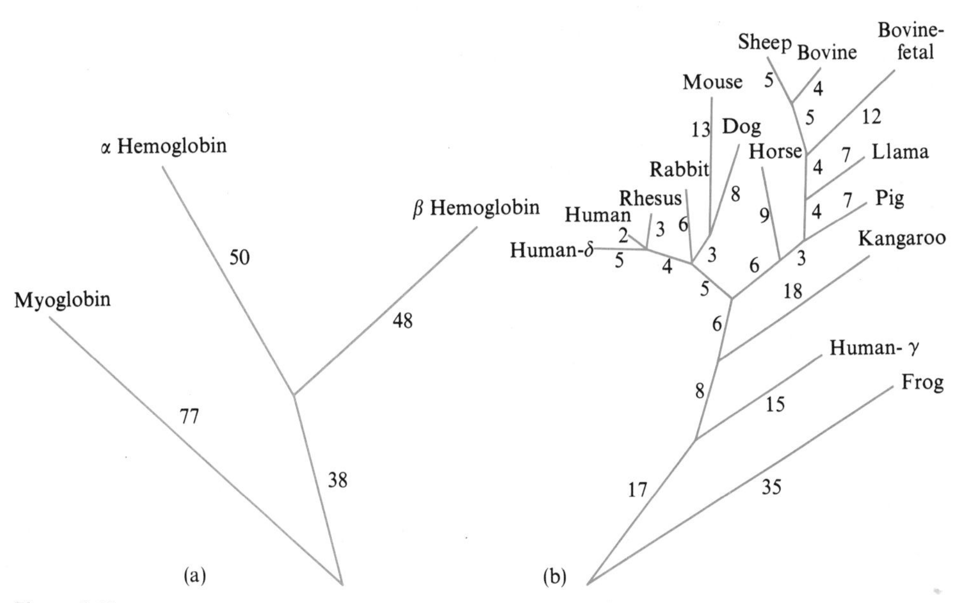

Figure 2-17

A phylogenetic tree constructed by comparing the amino acid sequences of hemoglobins and myoglobin. The length of each branch of the tree is proportional to the number of point mutations that would be needed to pass from one sequence to the next. Branch-points denote the positions of possible common ancestors. The numbers shown by each branch are mutations per 100 residues. **(a)** The whole globin family in the kangaroo. **(b)** An expanded view of β hemoglobin chains and closely related variants in various species. [After M. O. Dayhoff, ed., *Atlas of Protein Sequence and Structure*, vol. 5 (Silver spring, Md.: National Biomedical Research Foundation, 1972).]

Human fetal hemoglobin has 2γ chains instead of β chains. Sequence comparisons indicate that γ is much more similar to the β chain than to the α chain of adult hemoglobin. A variant hemoglobin chain found as a minor component in adult hemoglobin has 2δ chains instead of β chains. These are even closer to β. Thus humans have at least three genes for β-like hemoglobin chains. Because all are functional proteins, it is instructive to compare the sequences. Presumably, residues that differ are not absolutely required in any one form for general functions. The comparison can be extended by contrasting the sequences of human chains to those of other vertebrates.

Figure 2-18

Structure of the β chain of human hemoglobin, drawn to illustrate contacts with neighboring subunits. **(a)** Shows positions that can vary in the β-like chains of normal vertebrate hemoglobins. Note that most of the invariant residues are located in the pocket that binds the heme, or at the α_1–β_2 subunit interface where quaternary structure changes take place during oxygenation. Apparently these regions contain critical structural features requiring a unique choice of residues. In contrast, the α_2–β_2 interface is a tight contact that stays relatively constant during oxygenation. Here, as long as the contact is not broken, quite a few choices of residues appear suitable. **(b)** Shows positions of some of the known human β hemoglobin mutations. Note that these are concentrated in the heme pocket and at the α_1–β_2 interface. This is consistent with the idea that the structure of these regions is critically important for normal function. Quite a few positions in β hemoglobin have never been observed to vary or suffer mutations. Some of this is just statistical variation, but some sites presumably are so critical that any change is lethal. [Drawings by Irving Geis.]

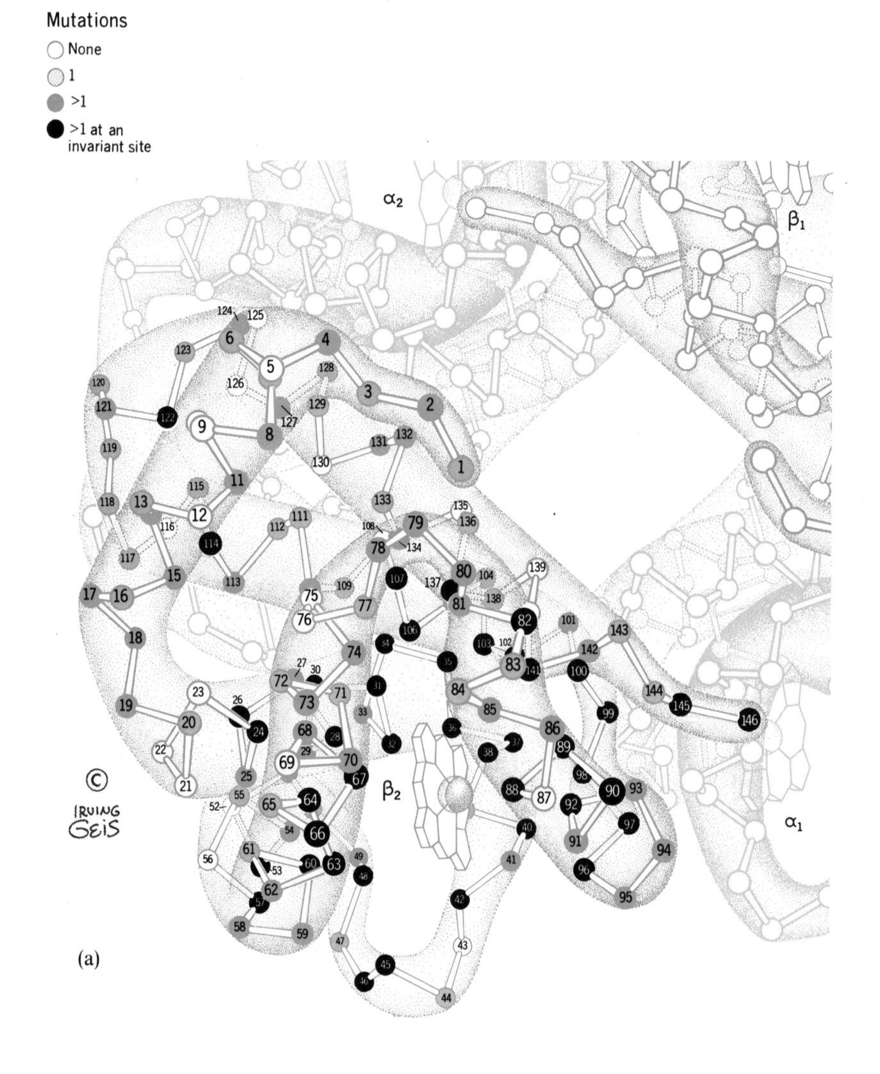

(a)

As expected, with increasing evolutionary distance between the organisms, the hemoglobin sequences diverge more and more. Thus, human hemoglobin is almost identical to that of a gorilla, is fairly different from that of a kangaroo, and is quite different indeed from that of a frog or of a carp.

From the viewpoint of structure and function, there are two ways to look at the overall pattern of sequences. The generalist will say they are all hemoglobins— they all have similar functions. Comparing the sequences of many species, only a fraction of the residues remain identical in all, as shown in Figure 2-18 for the β chain. The points of constancy are likely to be strictly required for function. It is not really quite as simple as this though. Other regions of the structure are presumably equally required for function. However it is likely that, in many cases, pairs (or larger groups) of amino acids have altered in such a way as to maintain the tertiary structure in spite of primary structure variations.

The generalist viewpoint can be carried even further by recognizing that α chains and β chains have nearly identical structures. Comparing sequences of both simultaneously reduces the number of identical critical residues. It can be reduced

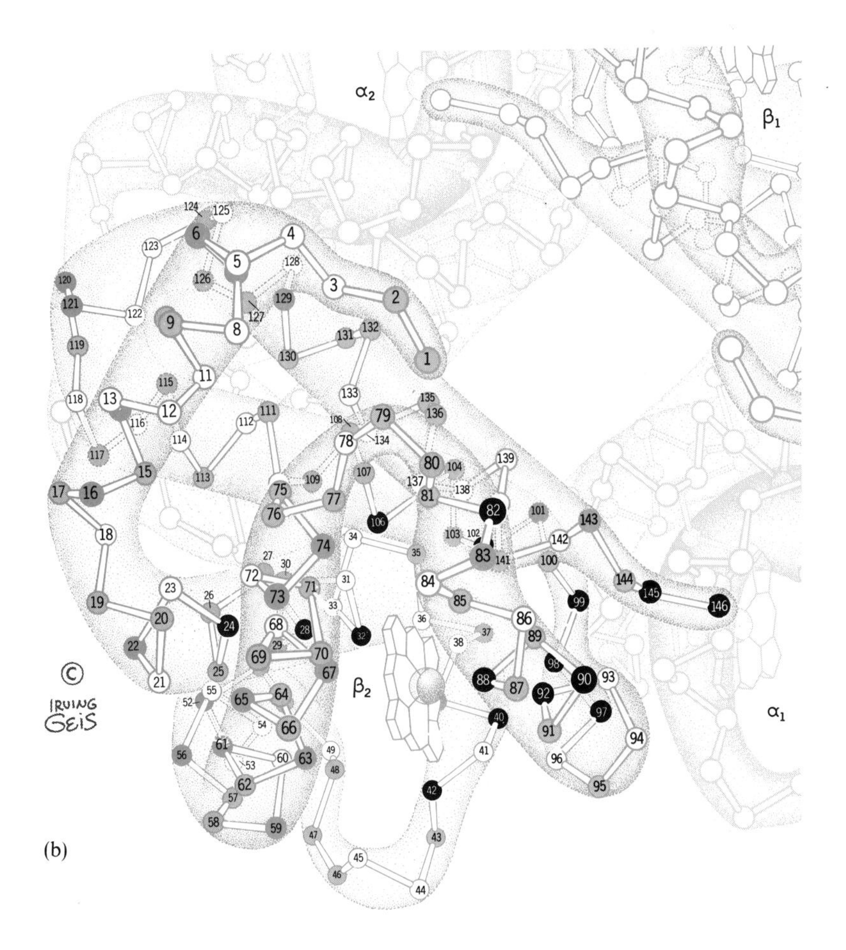

(b)

still further by involving myoglobin sequences too. Here, one has to be a bit careful because hemoglobin functions with associated subunits whereas myoglobin functions alone. Thus the disappearance of constant residues when myoglobin is added to the set of α and β sequences could provide clues about any residues playing a vital role at the subunit interfaces.

Someone interested in the fine details of hemoglobin function treats species variation from a different outlook. Although all these proteins are hemoglobins, the demands different organisms put on the proteins are somewhat different. Cold-blooded and warm-blooded organisms usually operate at different temperatures, high-altitude organisms need different oxygen binding properties, fish need to fill their swim bladders, etc. The sequence should be a gold mine of clues of how the structure of a protein can be fine-tuned to meet these varied individual needs.

Another family of closely interrelated proteins is formed by several serine esterases. These include the proteolytic enzymes, chymotrypsin, trypsin, elastase, and thrombin. Amino acid sequences of these four proteins are compared schematically in Figure 2-19. Not only do the sequences match up very well, but so do the positions of many of the disulfide cross-links and also the location of the hyperreactive serine residue known to be at the active sites of all of these enzymes. It is reasonable to predict that such similar primary structures will lead to similar tertiary structures. This is exactly what is observed, as shown for three of the proteins in Figure 2-20. Note however that—although the sequences and structures are similar enough among these four proteins to class them as related evolutionarily, and also to yield many common features in their mechanism of action—the four proteins are nowhere near identical. From sequence differences and particularly from structure differences, some of the distinctions among the substrate specificities of the four enzymes and subtle differences in their mode of action can be rationalized.

Mutant proteins: hemoglobin

Mutant protein sequences provide an alternative way to try to identify residues that are critically important for function. Because most mutants are discovered as a result of some functional impairment, one can sometimes learn the specific effects of particular amino acid replacements in a protein. However, there are some limitations on the kinds of mutants one can work with. If the mutation is dominant and lethal, the organism (by definition of a lethal mutation) doesn't live long enough to reproduce and may not even develop or live long enough to allow isolation of a sample of its protein. Therefore, some of the most harmful sites for mutation may never actually be observed in practice. With bacteria, there are procedures to circumvent this difficulty—such as the use of temperature-sensitive mutations that are lethal when grown at a restrictive temperature but allow growth at a permissive temperature. Unfortunately, it is much more difficult both practically and ethically to manipulate the genetics of higher organisms.

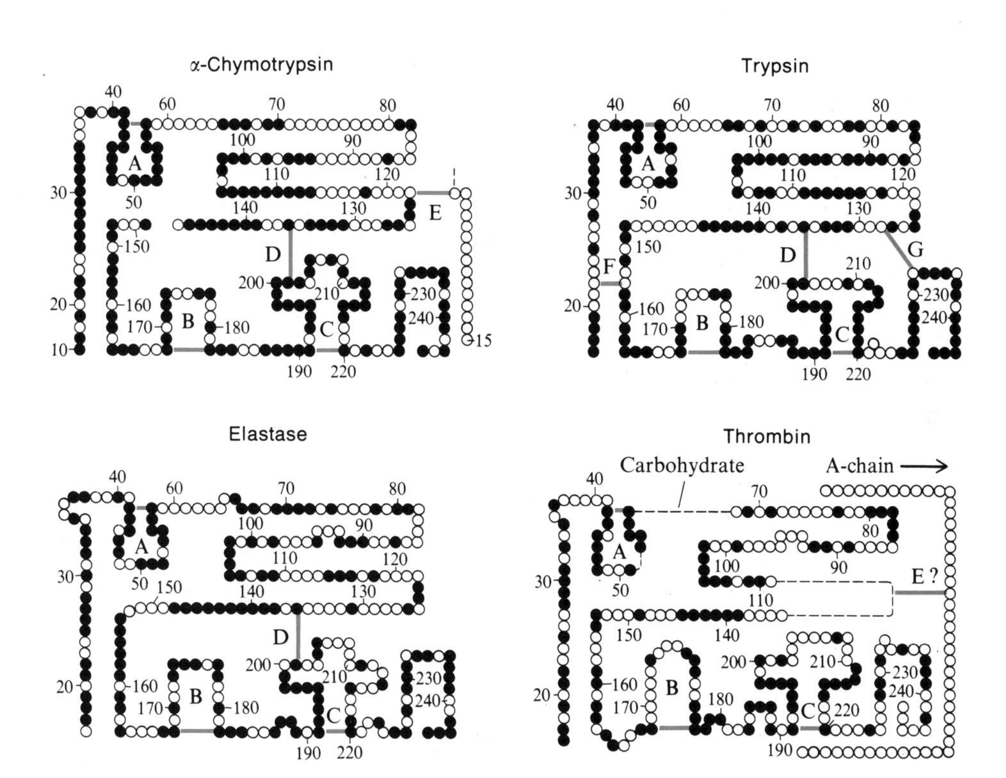

Figure 2-19

Amino acid sequences of four related proteins in the serine esterase family, shown schematically. Black circles show residues that are identical in at least two of the proteins. The letters A through F indicate corresponding disulfide bridges (*colored*). [After B. S. Hartley, *Phil. Trans. Roy. Soc. London Ser. B* 257:77 (1970).]

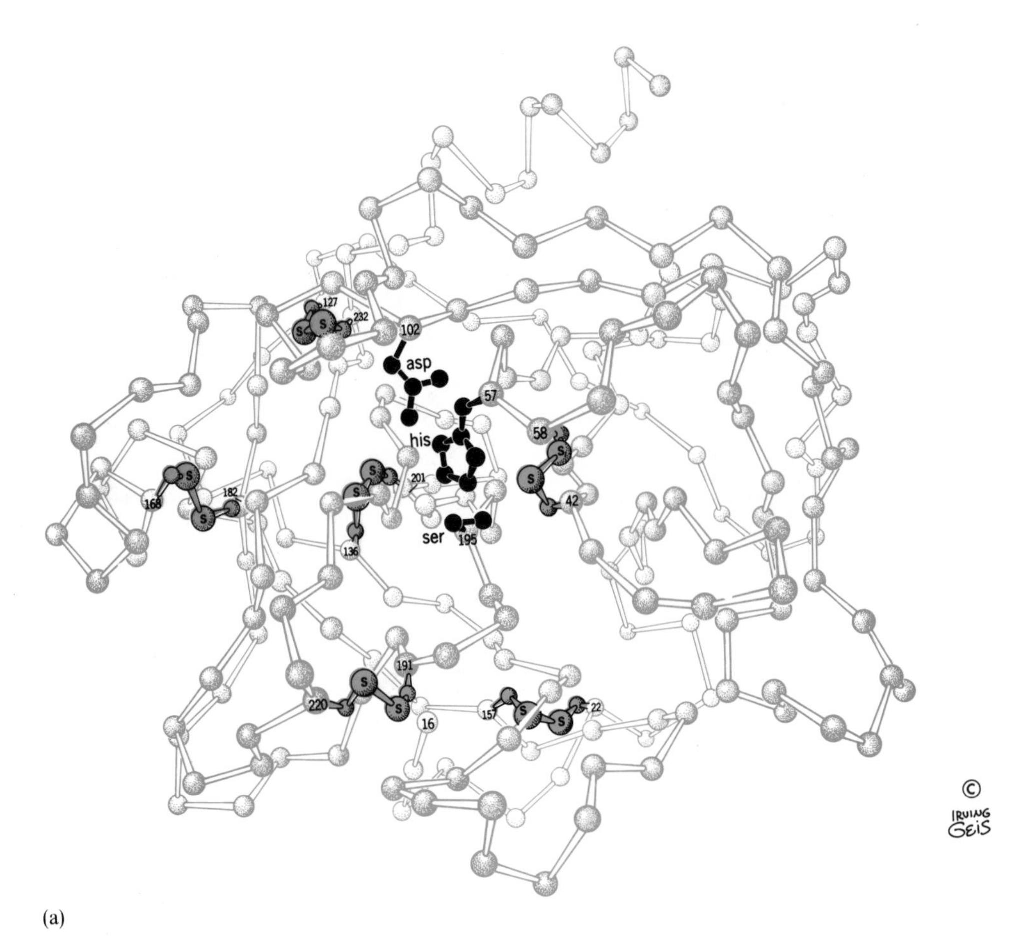

(a)

Figure 2-20

Tertiary structure of three closely related serine proteases. (**a**) Trypsin. (**b**) Chymotrypsin. (**c**) Elastase. Only the positions of α-carbons are shown, except for a few side chains at the catalytic site (*black*) and the disulfides (*colored*). All three proteases cleave peptide bonds in a similar way, but trypsin prefers to cleave after positively charged residues, chymotrypsin after large hydrophobic residues, and elastase after small residues. These different specificities can be understood from differences in the binding pocket adjacent to the catalytic site. [Drawings by Irving Geis.]

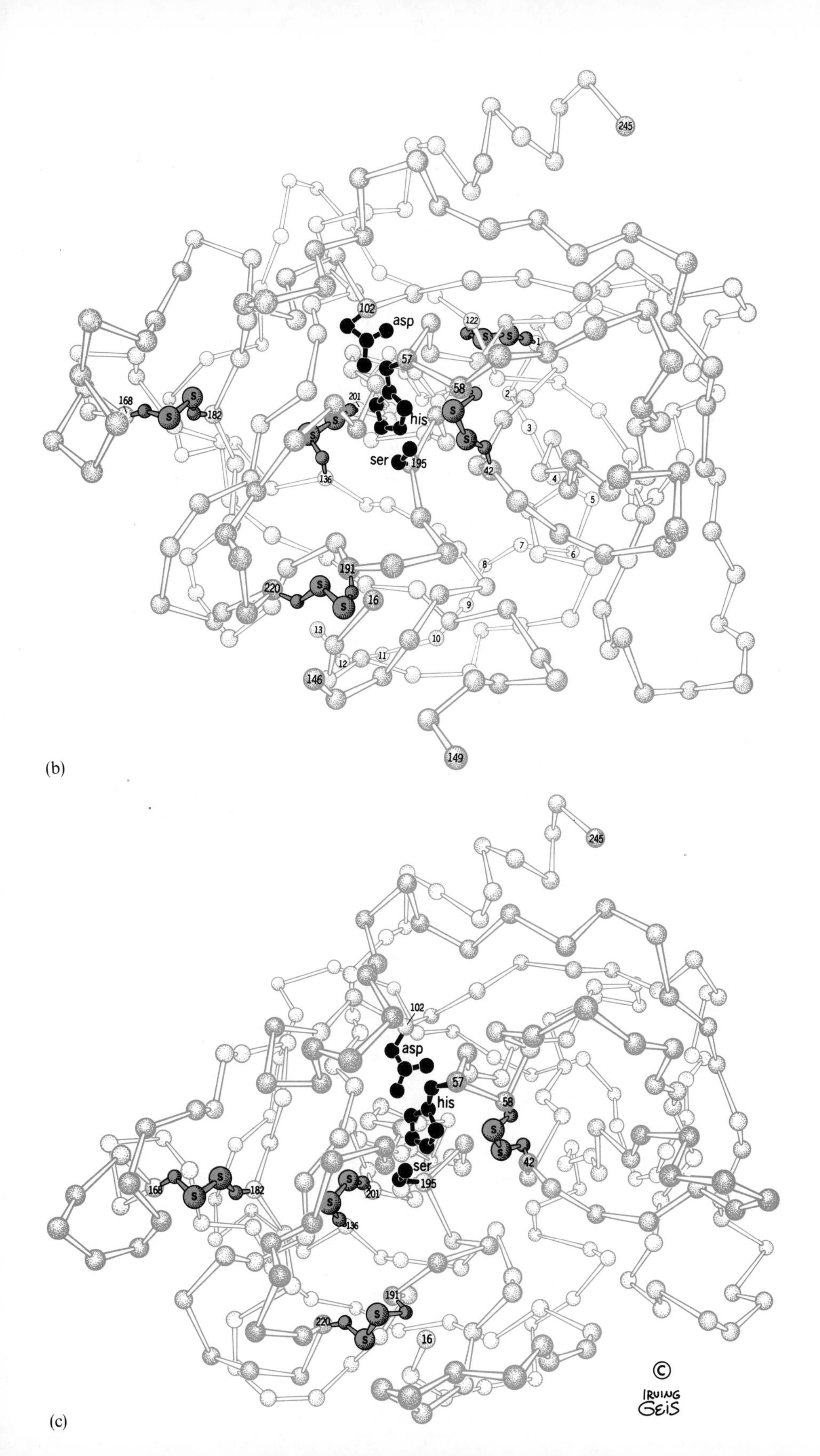
245
102
asp
122
S
S
1
57
58
168
S
S
182
201
S
S
his
2
3
ser
195
S
S
42
136
4
5
7
6
191
8
220
S
S
16
9
13
10
11
12
146
149
(b)
245
102
asp
57
his
58
S
S
42
168
S
S
182
S
201
S
136
ser
195
191
220
S
S
16
(c)
©
IRVING
GEIS

From routine screening, hundreds of human hemoglobin mutants have been found. Figure 2-18 summarizes many of these for the β chain. As you can see by careful inspection of this figure, there is a tendency for mutant variations to show up in regions of the structure that are less susceptible to interspecies variation. This is exactly what one would predict if some particular residues are required for the general functional capability of any hemoglobin.

A discussion of the detailed phenotype of each mutant is a fascinating exercise in the molecular origins of human disease. A detailed description of all these mutants is beyond the scope of this book. However, one that merits special attention is hemoglobin S. This was the first mutant hemoglobin discovered, and it represents a single amino acid change at the sixth position of the β chain (from Glu to Val). It is an important mutant because, unlike most of the others, it is fairly widespread in large segments of the human population with geographic origins in central Africa. Individuals homozygous in the homoglobin S gene have sickle-cell disease. At low oxygen tension, particularly in the capillaries, if an individual participates in vigorous activity, red blood cells change their shape from biconcave disks to elongated, irregular structures. These "sickle" cells become trapped in the capillaries and block circulation, causing serious tissue damage. The sickle cells are also more fragile than normal cells. Their increased rate of destruction leads to sickle cell anemia.

How does a simple amino acid change lead to such a large change in the overall structure of whole cells? Note from Figure 2-18 that the $\beta 6$ position of hemoglobin is not located near the heme groups. Neither is it near the subunit interface of the $\alpha_2\beta_2$ tetramer. Thus, it is reasonable that the oxygen-binding properties of an individual hemoglobin S molecule should not differ very much from those of normal hemoglobin. However, what the mutation does do is to place a nonpolar residue on the outside where there is a negatively charged residue in the normal molecule. Everything else being equal, one might guess that this substitution could increase the likelihood of forming aggregates of hemoglobin molecules. What would have been difficult to predict, *a priori*, is that hemoglobin S in the deoxy conformation readily forms well-ordered tubular aggregates such as those shown in Figure 2-21. These tubes tend to orient in parallel clusters. In order to accommodate such structures, the shape of the red blood cell distorts. The long axis of the sickle cell is parallel to the long axis of the tube clusters it contains. Further details on the molecular mechanism of sickle cell disease are given in Box 2-2.

2-4 SECONDARY STRUCTURE

We have already mentioned that amino acid chains can form a number of well-ordered local structures with helical symmetry or the equivalent. These are important for many reasons. Because of its symmetry, a helix is easy to recognize and describe. Figure 2-22 shows part of the three-dimensional electron density map of myoglobin as determined by x-ray crystallography. To the untrained eye, most of the map is gobbledygook. But the helices stand out immediately as dark circles when viewed

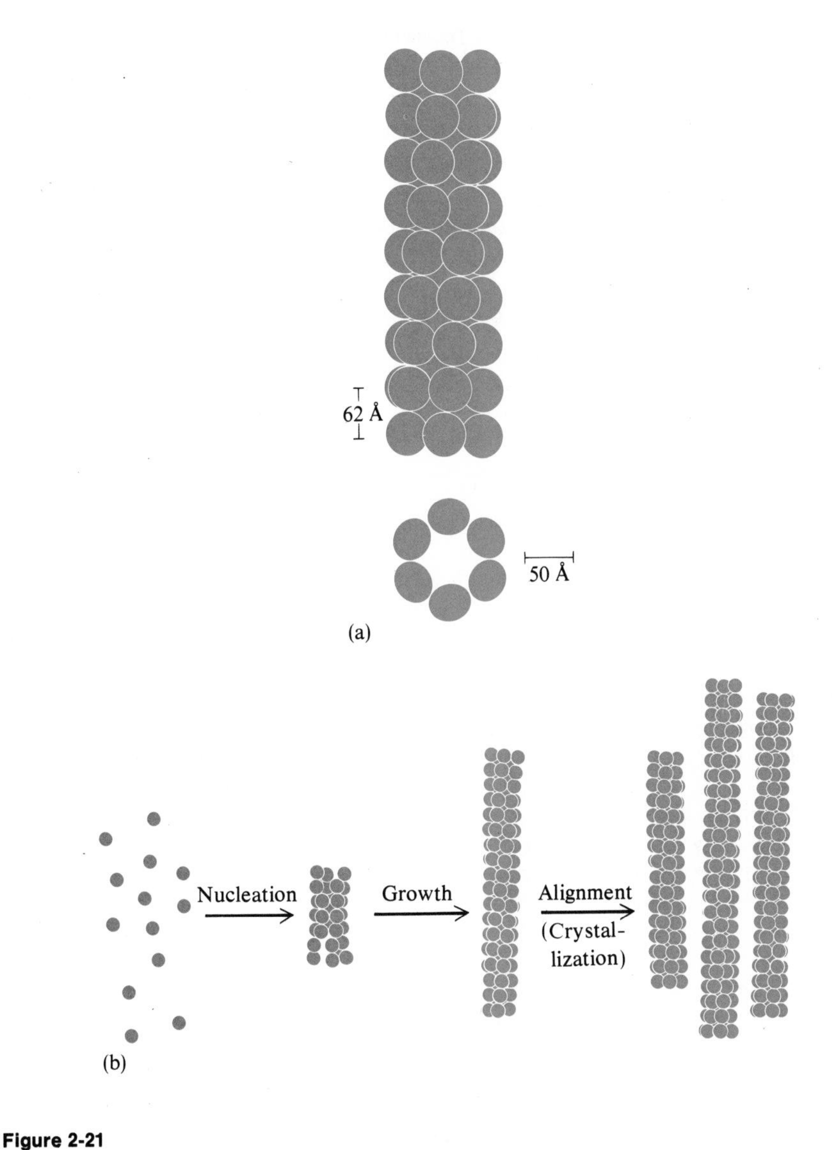

Figure 2-21
Some features of the aggregation of deoxyhemoglobin S, which is responsible for human sickle cell disease. (**a**) Proposed structure of an individual helical fiber of the aggregate. (**b**) A schematic mechanism for the formation of the aggregate stimulated by a decrease in oxygen tension. Nucleation is a very slow step and is rate-limiting. Once it occurs, growth of long fibers occurs rapidly. These fibers then aggregate in parallel arrays that can be accommodated within the erythrocyte only if it distorts its overall shape.

end-on, or as periodically spaced dark regions when viewed sideways (observe the features marked C and E–E′ in Fig. 2-22). For the trained crystallographer, the helices will be apparent at a lower resolution. The observation of helices is often an important clue that the tedious process of x-ray structure analysis is on the right track. Locating the helices precisely can be a stepping stone to focusing in on the current structural solution for the rest of the molecule.

Helices are stablized by neighbor interactions. Thus, unlike most of a protein, they can be dissected out (either literally by enzymatic fragmentation or figuratively through the use of model compounds) and studied independently. This is an important

Box 2-2 STUDIES ON FIBERS FORMED BY DEOXYHEMOGLOBIN S

Three kinds of experimental data have yielded some information on the structure of hemoglobin S (HbS) fibers. Electron microscopy and x-ray fiber diffraction revealed the packing arrangement of hemoglobin tetramers shown in Figure 2-21a. However, the data are not at a sufficiently high resolution to visualize how each individual $\alpha_2\beta_2$ unit is oriented in the fiber. This is frustrating because one would like to know whether there is a direct contact between the mutant $Val^{\beta 6}$ on one tetramer and surface sites on adjacent tetramers in the tubular fiber. If there were, it might be possible to design a selection chemical modification or drug that would interfere specifically with this contact. This potential molecular cure for what is truly a molecular disease would have a significant impact on the health of millions of individuals.

Some additional information on the structure of the fiber is available from optical absorption measurements on oriented samples. The origin of these effects, called linear dichroism, is discussed in Chapter 7. Heme groups will absorb polarized light preferentially when the plane of the heme is oriented parallel to the axis of polarization. Because there are four hemes per tetramer, interpretation of polarization results is not exactly straightforward. However, the results do allow limits to be placed on possible orientations of the tetramer within the tube. Some of the allowed orientations do indeed have the possibility of direct $\beta 6$ intersubunit contacts.

An alternative approach has been to attempt to calculate the energetically most favorable orientation of the tetramers using the known fiber structure. For simplicity, one makes the reasonable assumption that each hemoglobin tetramer is related to the next by a screw axis of symmetry. Thus, only one individual tetramer has to be treated to generate all possible orientations of molecules within the tube. Calculations to date indicate that certain orientations are strongly preferred, but the calculations do not yet go far enough to allow one particular structure to be assigned unequivocally as most stable.

There are some unique features about the physicochemical behavior of HbS in solution. Most aggregation or association phenomena have kinetics quite sensitive to concentration. It is often found that the rate-limiting step in the formation of a tube or fiber is association of several units to form a nucleation center, which then grows rapidly by incorporation of further single units (Fig. 2-21b). For simple mechanisms, the rate of formation of the nucleus, and hence the rate of aggregation or polymerization, will go as c^n, where c is the concentration of step in the overall strategy of "divide and conquer." As discussed in Chapter 5, we understand why certain helical structures are intrinsically stable and, as shown in Chapter 20, we know quite a bit about the factors that regulate their formation and stability.

Helices are a real asset for spectroscopic studies on proteins. Most types of spectra are sensitive to the local structure of individual residues or pairs of residues. There are so many different local conformations in a complex tertiary structure that spectra may become a blur. However, the similarity of the local structure of residues in helices or sheets ensures that their contribution will stand out. Thus, for example,

free units, and n is the number of units in the nucleus. Typically, n is 2 or 3. However, James Hofrichter, Phillip Ross, and William Eaton found that, for the kinetics of formation of hemoglobin S tubes, n appears to be 30 or greater (see figure).

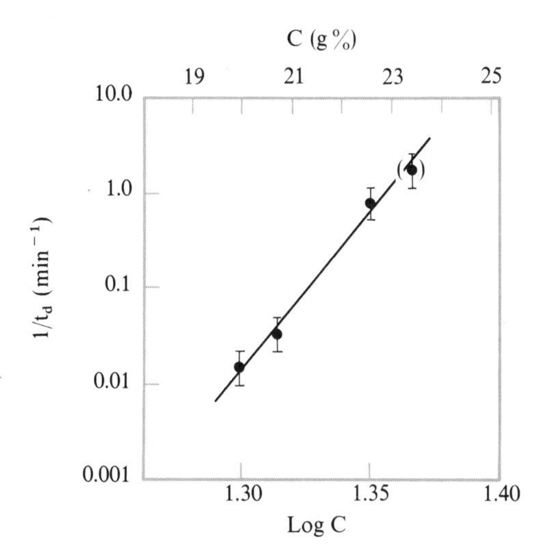

Dependence of the time delay involved in the nucleation step on total concentration (c) of hemoglobin S at 30°C. The slope of the line drawn on this logarithmic plot is 32. Thus the nucleation rate is approximately proportional to c^{32}, indicating that the smallest stable nucleus has at least 32 hemoglobins. [After J. Hofrichter et al., *Proc. Natl. Acad. Sci. USA* 71:4864 (1974).]

This is a staggering result. It is not obvious from the structure of the tube why n is so high. But the implications for possible approaches to the treatment of sickle cell disease are profound. The rate of sickling will be exquisitely sensitive to the concentration of hemoglobin in an erythrocyte, to the volume of the erythrocyte, and to any thermodynamic parameters that control these variables or the strength of individual hemoglobin–hemoglobin contacts. If we can learn how to slow the rate of sickling, it may be possible to improve the treatment of the disease. A red blood cell spends only a short time in the peripheral capillaries. Once it returns to the lungs again and reoxygenates, tube formation is not stable. So, if the sickling rate is slow enough, cells won't have time to sickle as they pass through the capillaries.

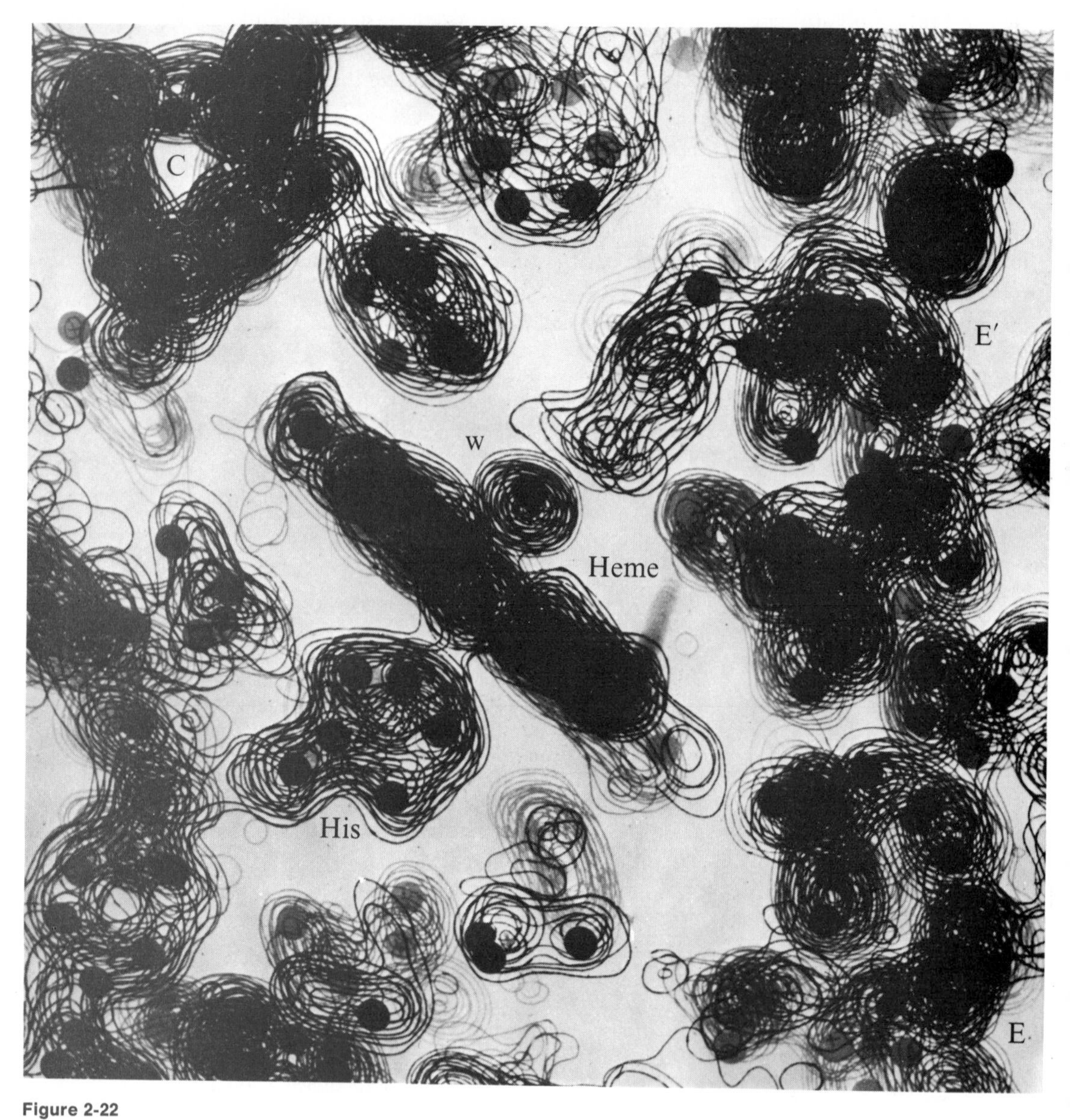

Figure 2-22

Part of the electron density distribution determined for myoglobin by x-ray crystallography. Two α helices are seen end-on at C, and one is viewed sideways between E and E′. Also shown are a side view of the heme, and coordinated histidine and a water molecule (w). (Compare with Fig. 2-7d.) [Courtesy of M. F. Perutz.]

many protein conformational changes are characterized spectroscopically by the accompanying alterations in secondary structure. It is not that this is the only level on which structure is changing, but it is often the most convenient variable to measure and analyze quantitatively.

The α helix

The most famous polypeptide secondary structure is the righthand α helix. Figure 2-23a shows it in a fairly accurate drawing. The basic features are 3.6 residues per turn and a translation along the helix axis of 1.5 Å per residue. This leads to a pitch of $3.6 \times 1.5 = 5.4$ Å. The diameter, neglecting side chains, is about 6 Å. Neglecting end effects, every peptide carbonyl in the α helix serves as a hydrogen bond acceptor for the peptide N–H donor four residues away. Thus, when the α helix is viewed with N-terminus at top and C-terminus at bottom (as shown in Fig. 2-23a), all the carbonyls point up and all the N–H groups point down.

All side chains point out from the helix as long as all the residues are L-amino acids (see also Fig. 1-1). Thus, the same structure can accommodate almost any kind of flexible side chain, so long as it does not get so big as to interfere sterically. However, proline cannot sterically fit into an α helix, as you will see in Chapter 5. Of course, some side chains may not energetically prefer to be in an α helix. For example, poly-L-glutamate at high pH will be a polyanion. Electrostatic repulsion between the side chains greatly destabilizes the helix and instead will promote an extended coil structure at low ionic strength because this maximizes the distances between the negative side chains and thus minimizes the repulsive electrostatic free energy.

In one of the most impressive feats of structural chemistry, Linus Pauling and Robert Corey postulated the exact structure of the α helix (as well as accurate structures for several other ordered polypeptide conformations) six years before an α helix was seen at molecular resolution for the first time in the crystal structure of myoglobin. It is interesting to follow the reasoning they used. Bond lengths and angles were known for a variety of small peptides from crystallographic studies. It was reasonable to expect that these would remain constant in any secondary structure. The peptide-bonded NH and CO in virtually all known structures had essentially identical dimensions and adopted a planar *trans* conformation:

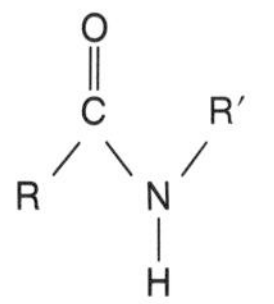

```
      O
      ‖
      C      R′
     / \    /
    R    N
         |
         H
```

Assuming this conformation to be identical in all more-complex structures was an enormous simplification. It immediately suggested that, because each peptide

Figure 2-23

Three of the most common polypeptide secondary structures. Only atoms in the peptide-chain backbone are shown. A parallelogram indicates the plane of each peptide bond. (**a**) The right-handed α helix. (**b**) Two antiparallel β-sheet strands. Note how alternate side chains will be placed on opposite sides of the sheet. (**c**) The left-handed polyproline II helix. [Drawings by Irving Geis.]

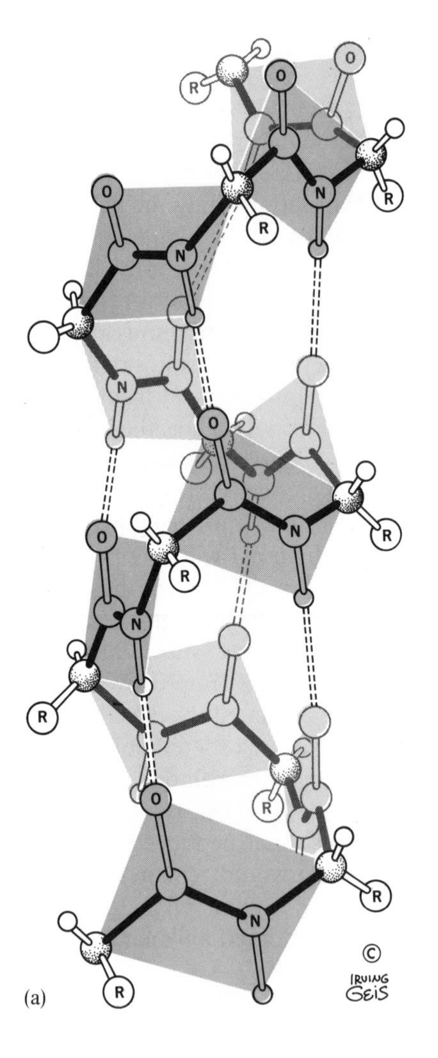

backbone could have the same structure, the overall backbone could have a regular secondary structure: some kind of helix. So, Pauling and Corey searched for polypeptide helices that could accommodate any of the common amino acid side chains. To keep the number of possible models to be examined within bounds, they applied two additional constraints: maximize the number of interpeptide N—H···O=C

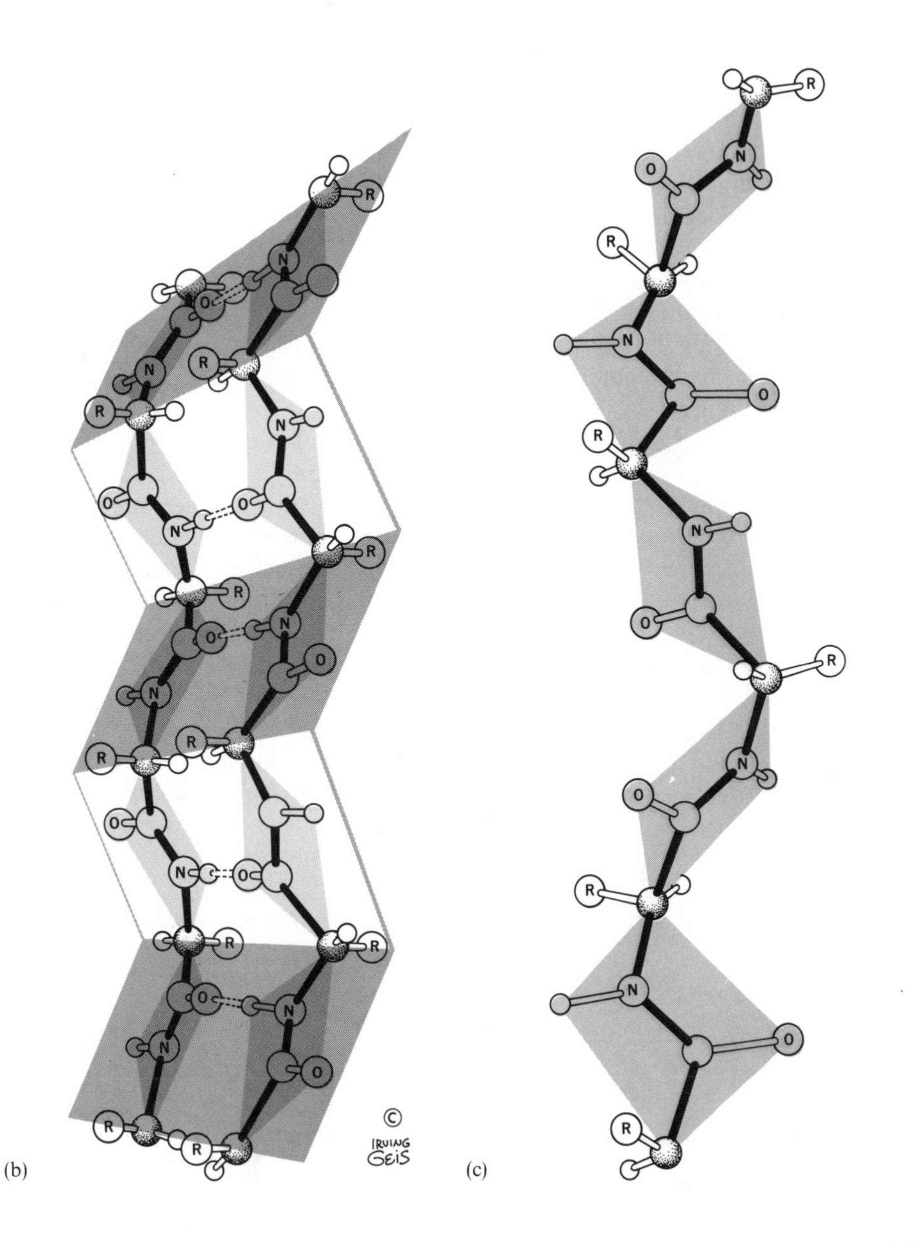

hydrogen bonds, and find conformations in which these are as close to linear as possible. The rationale behind this constraint was that the hydrogen-bond enthalpy in vapor-phase model systems is on the order of 5 to 7 kcal/mole. Conformational energies for rotation about single bonds tend to be much less than this, and so it was assumed that hydrogen-bond energies would completely dominate.

β Sheets and other secondary structures

As is apparent from Figure 2-23, the right-hand α helix nicely meets all the criteria used by Pauling and Corey in their search. A similar-looking left-hand α helix was also postulated, and so were two β-sheet forms. The antiparallel β sheet is illustrated in Figure 2-23b. Each single strand is a 2-fold helix—i.e., a pleated structure. All peptides participate in hydrogen bonding, but these bonds are all interstrand rather than intrastrand. The geometry of the peptide backbone in the β sheets approaches the most extended chain conformation allowed by normal bond lengths and angles; the displacement along the helix axis is 3.47 Å per residue. In many proteins, parallel and antiparallel β chains combine to produce a large structural core. An example of this is shown in Figure 2-24. (See also Box 2-3.)

All four secondary structures have been found to exist, either in proteins or in model polypeptide systems, exactly as Pauling and Corey predicted. However, as you will see in Chapter 5, the criteria originally used to construct them are not quite correct and have been modified to be consistent with our current understanding of forces that stabilize macromolecular conformations. Part of the difficulty has to do with effective hydrogen-bond strengths. But another serious complication is the effect of the side chains. These will determine in large part whether a given sequence will prefer to form no ordered structure, the α or β structures, or a few alternatives that also have been found in proteins.

The 3_{10} helix is occasionally found in short stretches in globular proteins. It is essentially a variant of the α helix and has 3 residues per turn instead of the 3.6 residues per turn of the α helix.[§] A much more common feature is the β bend, illustrated in

[§] Two other variants of hydrogen-bonded single helices occasionally seen in short stretches in proteins are the α_{II} helix and the π helix. For illustrations of these structures, see Dickerson and Geis (1969).

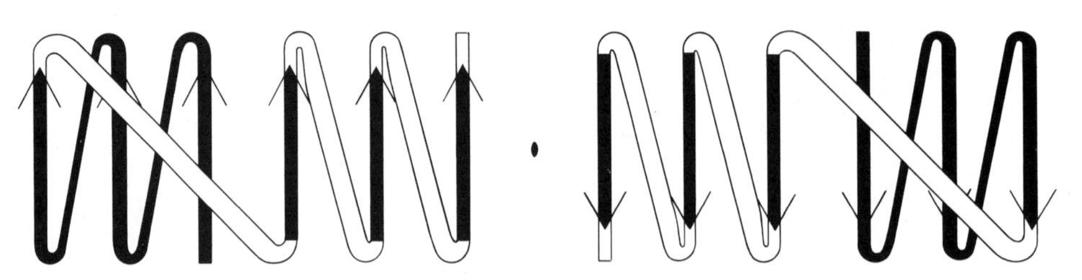

Figure 2-24

The hydrogen-bonding pattern of the β sheet of the two identical subunits of horse-liver alcohol dehydrogenase. Within each subunit there is a six-stranded parallel sheet. Between the two subunits is a two-stranded antiparallel sheet. The location of a C_2 symmetry axis is also shown. [After M. G. Rossman et al., in *The Enzymes*, 3d ed., vol. 9, ed. P. Boyer (New York: Academic Press, 1975), p. 61.]

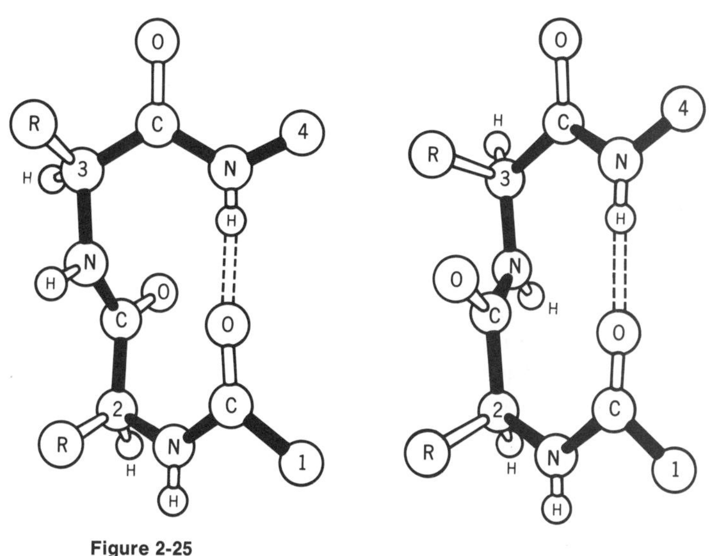

Figure 2-25

Structure of two kinds of β bends. [Drawing by Irving Geis.]

Figure 2-25. In this way, peptide chains can turn corners and still maintain energetically favorable conformational angles and one intrachain hydrogen bond. It is fair to call β bends secondary structure because they are the result of neighbor interactions and could be considered the start of a helix with zero pitch.

Polyproline helices and collagen

The final set of common polypeptide secondary structures comprises the polyproline helices. Proline cannot form normal α or β structure because of the constraints imposed on the backbone by the five-membered ring. However, it can form two single-helical structures that are unique among polypeptide secondary structures in that they contain no hydrogen bonding. Polyproline I is a left-handed helix with 10 residues per 3 turns. It has all *cis* peptide bonds, which is quite a rare occurrence with normal peptide structures. The polyproline II helix is left-handed and has *trans* peptide bonds. Interconversion between the two forms can be caused by changes in solvent. In normal aqueous buffers, helix II is the stable form. It has 3 residues per turn and a displacement along the helix axis of 3.12 Å, compared with only 1.5 Å in the α helix. The structure of polyproline II shown in Figure 2-23 has a much more extended chain than the α helix. In this extended helix, each side chain can get far away from all the others. In fact, the polyproline helices are stabilized in significant part by such steric constraints.

Box 2-3 β-SHEET TOPOLOGIES

Some of the variety of β-sheet structures found in proteins are illustrated schematically in the figure. Each sheet is shown projected into a plane; solid arrows indicate the directions of the individual strands (pointing toward the C-terminus). Connections between strands that occur above the plane are shown as wide double lines, those in the plane as narrow double lines, and those below the plane as single lines. Sheets that appear to form cylindrical barrels are bracketed by hyphens. Only topological connectivity is represented. No attempt is made to show the length or conformation of individual strands or connections.

To gauge the degree of abstraction represented by the topological representations in the figure, compare the results with more realistic drawings of the structures of carbonic anhydrase (Fig. 2-34), lactic dehydrogenase (Fig. 2-39a), lysozyme (Fig. 2-33), superoxide dismutase (Fig. 2-43), and the variable domain of immunoglobulin G (Fig. 2-43).

Jane Richardson has made a number of useful generalizations from the known structures of β sheets, including the following.

1. All crossover connections (except one) are right-handed. Those shown above the sheet slant from upper left to lower right; the ones below the sheet slant in the opposite way. [See J. S. Richardson, *Proc. Natl. Acad. Sci. USA* 73:2619 (1976).]
2. Parallel β structure is seen only in sheets with five or more strands.
3. Pure parallel and antiparallel sheets are favored over mixed sheets.
4. Knotted topologies are absent. None of the sheets shown in the figure, if held at the N- and C-termini and pulled out straight, would end up as a knot.
5. The N-terminal strand is somewhat more likely than the C-terminus to be in the middle of the sheet. This is consistent with the notion that chain folding can commence before synthesis of the peptide chain is completed.

Schematic structures of β sheets arranged from top to bottom by number of strands, and from left to right by the relative amounts of antiparallel and parallel β structure. Abbreviations used for some of the proteins are the following: LDH = lactate dehydrogenase; STI = soybean trypsin inhibitor; Lys = egg-white lysozyme; T4Lys = T4 phage lysozyme; HiPIP = high-potential iron protein; PTI = pancreatic trypsin inhibitor; LADH = liver alcohol dehydrogenase; S. Nucl = *S. aureus* nuclease; RNase = pancreatic ribonuclease; Therm = thermolysin; Rhod = rhodanase; AdenylK = adenylate kinase; cyt.*b*5 = cytochrome b_5; HexoK = hexokinase; Rubr = rubredoxin; PGK = phosphoglycerate kinase; Chym = chymotrypsin; ConA = concanavalin A; PGM = phosphoglucomutase; ImmC = immunoglobulin constant domain; ImmV = immunoglobulin variable domain; Cu,Zn SOD = Cu,Zn superoxide dismutase; TIM = triose phosphate isomerase; GPDH = glyceraldehyde phosphate dehydrogenase; CarbAnh = Carbonic anhydrase. [After J. Richardson, *Nature* 268:495 (1977).]

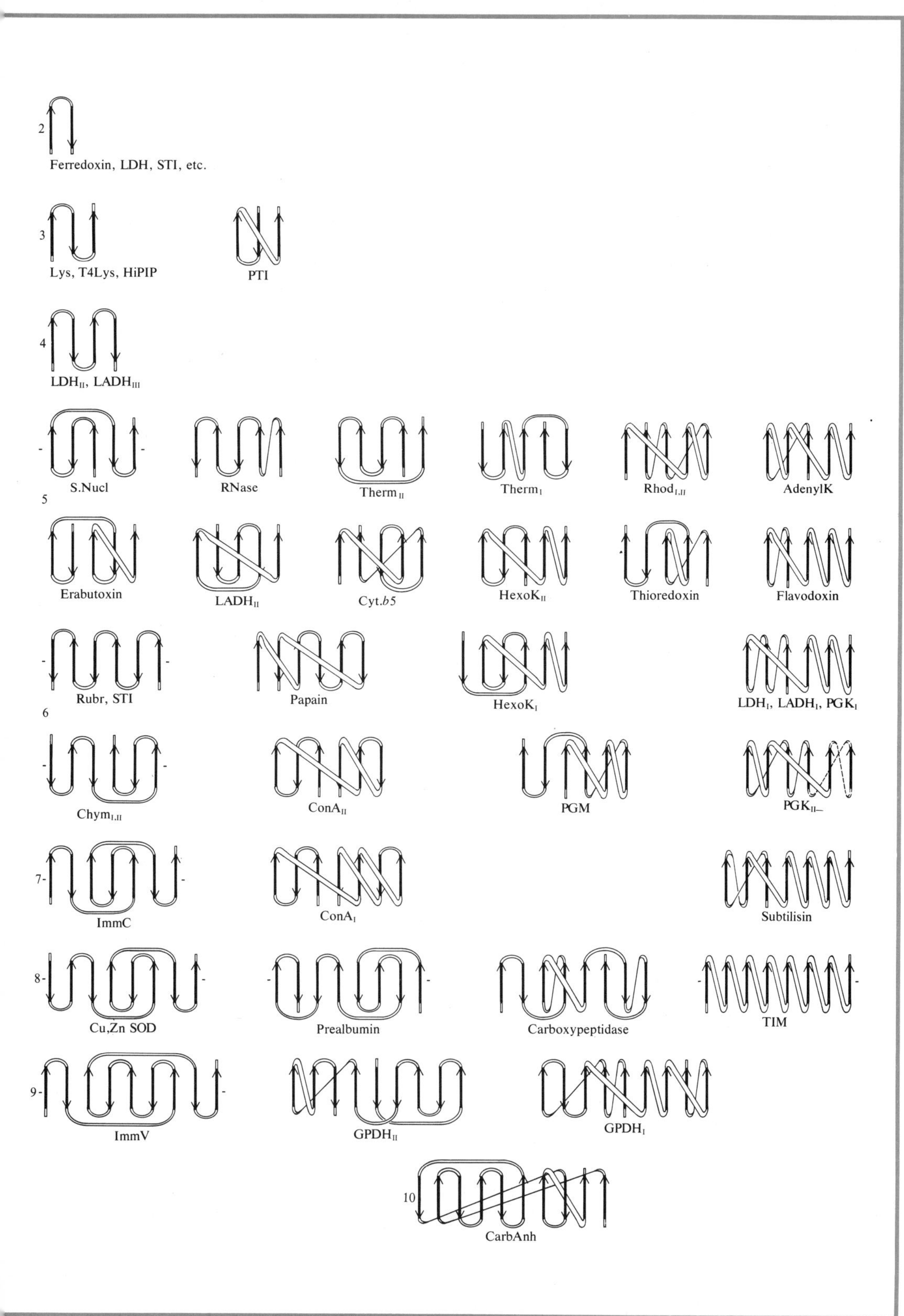

2
Ferredoxin, LDH, STI, etc.
3
Lys, T4Lys, HiPIP
PTI
4
LDH$_{II}$, LADH$_{III}$
S.Nucl
RNase
Therm$_{II}$
Therm$_{I}$
Rhod$_{I,II}$
AdenylK
5
Erabutoxin
LADH$_{II}$
Cyt.*b5*
HexoK$_{II}$
Thioredoxin
Flavodoxin
Rubr, STI
Papain
HexoK$_{I}$
LDH$_{I}$, LADH$_{I}$, PGK$_{I}$
6
Chym$_{I,II}$
ConA$_{II}$
PGM
PGK$_{II}$
7
ImmC
ConA$_{I}$
Subtilisin
8
Cu,Zn SOD
Prealbumin
Carboxypeptidase
TIM
9
ImmV
GPDH$_{II}$
GPDH$_{I}$
10
CarbAnh

Poly-L-proline is a synthetic structure. Runs of consecutive prolines, or even stretches with a large fraction of proline, are exceedingly rare in most proteins and, as a result, only occasionally does one see a secondary structure resembling the poly-L-proline II helix. However, a dramatic exception occurs with proteins in the collagen family. These have a most unusual composition dominated by proline, hydroxyproline, and glycine. Tropocollagen contains three polypeptide chains: two, called α_1, are identical; the other, α_2, is quite similar. Each has a molecular weight of about 95,000. All have a glycine every third position in the sequence, while one (or occasionally two) of the intervening positions is usually filled by proline or hydroxyproline as shown in Figure 2-15. The high proline content pretty well restricts accessible structures for each of the collagen chains to polyproline helices. Tropocollagen, as isolated, is a 285,000 dalton molecule with a rodlike shape 3,000 Å long and only 15 Å in diameter. It is a triple helix formed by twisting three polyproline II helices about each other as shown in Figure 2-26. The triple helix contains interchain

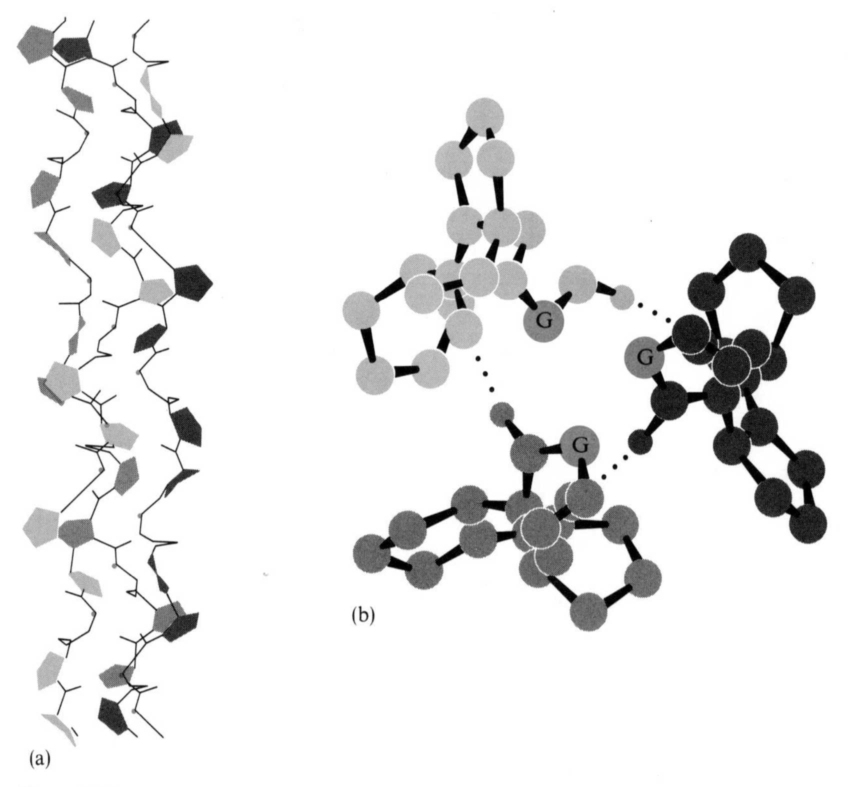

Figure 2-26

The triple-stranded collagen helix, shown (**a**) in a side view and (**b**) in an end view. In the end view, interstrand hydrogen bonds are indicated. The sequence shown is (–Gly–Pro–Pro–)$_n$. Note that the prolines are outside, but the glycine is inside. Every third residue must be glycine because there is no space in the center for a larger residue. All glycine α-carbons are shown in color. [After L. Stryer, *Biochemistry* (San Francisco: W. H. Freeman and Company. Copyright © 1975).]

hydrogen bonds. All of the proline rings point toward the outside. The glycines are all on the inside. Their critical feature is their small size. The three strands are wound so tightly that there would be no room for any larger residue. The locations of the glycines are staggered so that at each level of the triple helix (2.9 Å vertical spacing per residue) there is always a glycine on one strand.

The pseudorepeating sequence of the individual tropocollagen strands raises potential problems for the formation of the triple helix. It would be easy for three strands to come together out of phase by three residues or by any multiple of three residues. Then, instead of a rod, the structure of collagen would have frayed ends that would tend to aggregate intramolecularly in an irregular way. The biological role of collagen is to be packaged into neatly ordered fibers. These rigid, enormously long structures are used for mechanical strength in connective tissue. In the fiber, tropocollagens are arranged in a regular quarter-staggered array as shown in Figure 2-27. Frayed ends would be extremely disruptive. To make sure the three

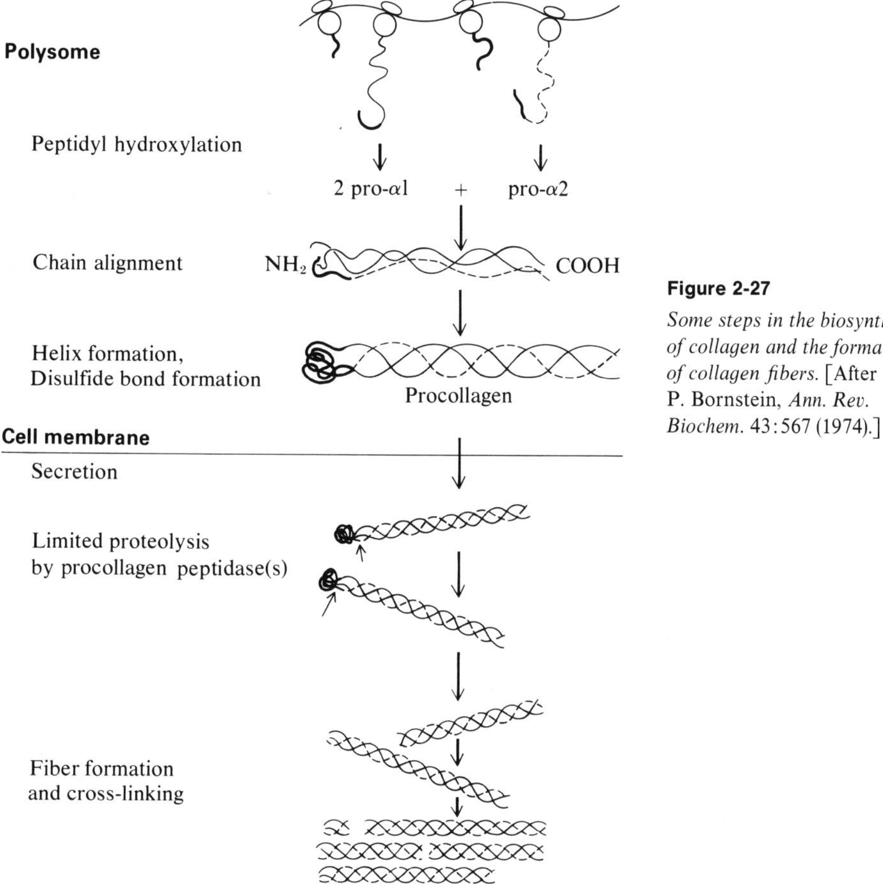

Figure 2-27

Some steps in the biosynthesis of collagen and the formation of collagen fibers. [After P. Bornstein, *Ann. Rev. Biochem.* 43:567 (1974).]

strands of tropocollagen align precisely, nature synthesizes a precursor, procollagen. Its three strands have a molecular weight of 120,000 each. The extra parts, all at the same end, have properties more characteristic of normal globular proteins. They fold together, become cross-linked by disulfides, and provide the structural nucleus that ensures correct alignment in the collagen helix. Then the extra regions are removed proteolytically.

Occurrence of secondary structures in proteins

In a typical protein whose three-dimensional structure has been determined by x-ray crystallography, around 60% of the residues can be classified as participating in one of the above kinds of secondary structures. Table 2-7 summarizes some of the information currently available. You can see that most proteins tend to have appreciable

Table 2-7
Secondary structures in protein crystals

	Helices			β Sheets				β Bends	
Protein	%	No.	Comments	%	No. of sheets	No. of strands	Comments	No. of β(I)	No. of β(II)
Carbonic anhydrase	20	7	α, α_{II}, 3_{10}	37	2	13	P, A, (C)	3	1
Carboxypeptidase A	38	9	α, α_{II}	17	1	8	P, A	10	3
Chymotrypsin (elastase, trypsin)	14	3	3_{10}, α	45	2	12	A, 2C	11	6
Concanavalin A	2	1	—	57	3	18	A, A2A	—	—
Cytochrome b_5	52	6	α, 3_{10}	25	1	5	P, A	1	1
Cytochrome *c*	39	5	—	—	—	—	—	3	3
Ferredoxin	—	—	—	4	1	2	A	—	—
Flavodoxin	30	4	—	30	1	5	—	—	—
Insulin	52	3	α, 3_{10}	6	1	1	A, A2A	—	—
Lactate (or malate) dehydrogenase	45	10	—	20	3	12	P, A	6	4
Lysozyme	40	6	α, α_{II}, 3_{10}	12	2	6	A	4	2
Myogen	57	6	α, 3_{10}	4	1	2	—	5	—
Myoglobin (hemoglobin)	79	8	α, 3_{10}, π	—	—	—	—	6	—
Nuclease	24	3	—	14	1	3	A	—	—
Pancreatic trypsin inhibitor	28	2	α, PP	33	1	3	A	—	—
Papain	28	5	—	15	1	7	A	—	—
Ribonuclease	26	3	α, 3_{10}	35	1	6	(P), A	1	1
Subtilisin	31	8	—	16	2	8	P, A	16	1
Thermolysin	38	7	—	22	1	10	P, A, (C), CL	—	—

NOTE: Percentages given represent the percentages of primary structure involved in helices and in β sheets, respectively. Symbols used are as follows: $\alpha = \alpha$ helix; $\alpha_{II} = \alpha_{II}$ helix; $3_{10} = 3_{10}$ helix; $\pi = \pi$ helix; A = antiparallel β structure; A2A = "across twofold axis"; C = cylindrical β structure; CL = clover-leaf β structure; P = parallel β structure; PP = polyproline structure; parentheses indicate distorted structures.
SOURCE: Adapted from A. Lilias and M. Rossmann, *Ann. Rev. Biochem.* 43:480 (1974).

amounts of both α and β structure. *P. aerogenese* ferredoxin is a striking exception with virtually no recognizable secondary structure. It is easy to rationalize this. The protein has only 54 amino acids. Eight are Cys, as shown in Figure 2-15. These come together four at a time to form ligands at the two iron–sulfur cubic structures of the protein, illustrated in detail in Figure 2-7.

A schematic drawing of the tertiary structure of ferredoxin is shown in Figure 2-28. You can see that the peptide chain can barely manage to struggle from one coordination site to the next. This constraint leaves no room left for formation of secondary structures. Not all proteins with prosthetic groups have this difficulty. Myoglobin and hemoglobin have enormous amounts of α helix, yet they clearly manage to fold this in such a way as to provide a heme-binding site (Figs. 1-4 and 2-18).

The data in Table 2-7 must be viewed with one caution: a highly biased set of proteins is represented. Most are quite small, and all have been successfully crystallized

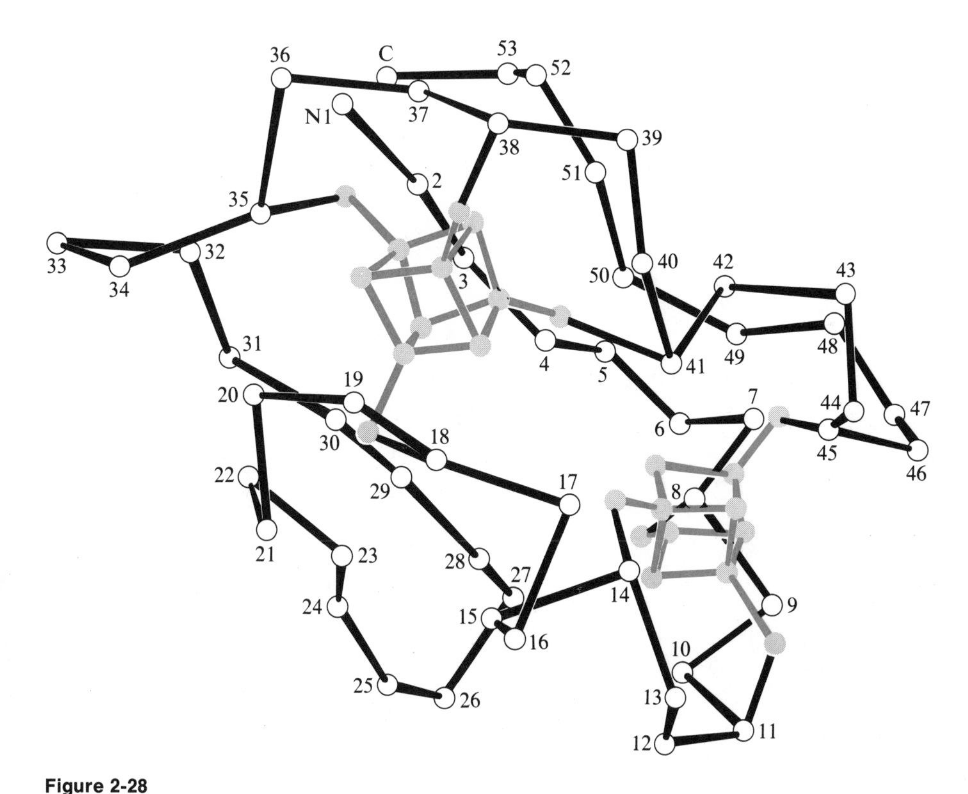

Figure 2-28

Tertiary structure of P. aerogenese ferredoxin, determined by x-ray crystallography. Atoms shown are α-carbons (○), Fe, labile S, and cysteine S (●). Iron sulfur cubes are shown in color. [After L. H. Jensen, *Ann. Rev. Biochem.* 43:461 (1974).]

in a manner well-ordered enough to allow high-resolution difffraction and isomorphous replacement data to be obtained. It is not inconceivable that a high fraction of secondary structure could assist these processes. Actually one can learn fairly accurately that proteins have helices and sheets from solution spectroscopic studies. But to go further—to ask how many helices, how long, how many strands in the sheets, etc.—usually requires a crystal structure. The detailed arrangement of secondary-structure features is a major aspect of tertiary structure, and we consider this next.

2-5 TERTIARY STRUCTURE

A number of tertiary structures have already been illustrated. They are complicated and usually take much study before many important features can be recognized. It is very difficult to show many different aspects of the structure in a single illustration. Some can only be appreciated by handling an actual model. A space-filling model is often not very illuminating. Figure 2-29 shows such a model for ribonuclease. It shows

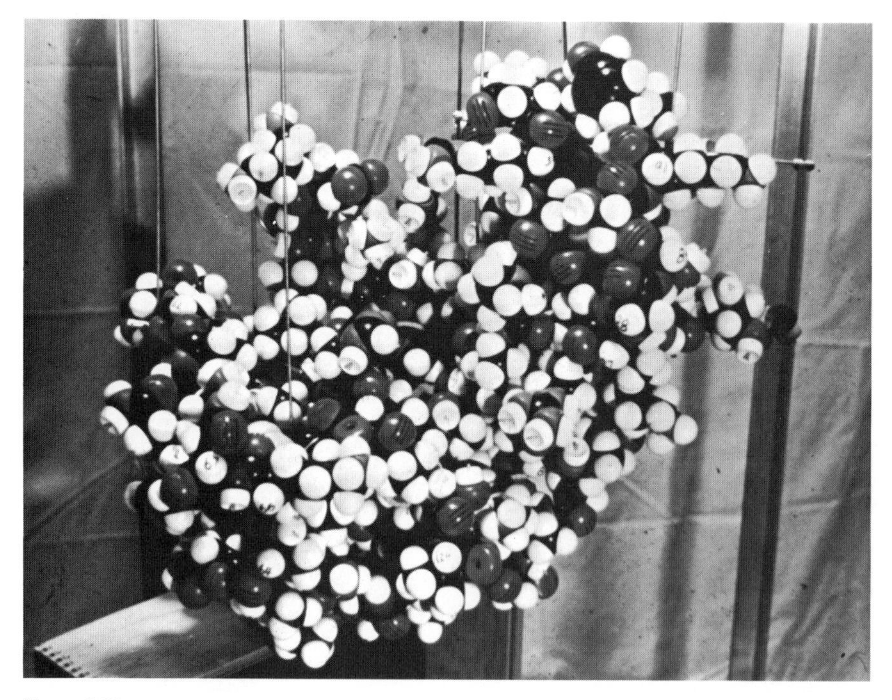

Figure 2-29

A space-filling CPK molecular model of bovine pancreatic ribonuclease S. The large groove running vertically and accentuated at the top of the molecule is the substrate binding site. (Compare with Fig. 16-14.) [Photograph provided by F. M. Richards.]

that proteins are fairly solid, and some details of the true complexity of the outside surface and shape of a binding pocket are visible. All details of the path of the peptide chain and any secondary structure are obscured.

A few more details are visible if models are constructed with all the hydrogen atoms omitted. Figure 2-30 shows such a model for hexokinase A. The drawing was actually prepared by a computer.

A more abstract model of lysozyme is shown in Figures 2-31 through 2-33. Here, lines are used to represent every chemical bond in the backbone and side chains. The artist has cleverly distorted the structure somewhat to improve visibility but, even so, is is hard to deal with such a tangled web. And this is a very small protein. For much larger ones, such drawings will not be very useful.

A fourth level of abstraction is to show only the positions of α-carbons and perhaps indicate where a few critical side chains, prosthetic groups, or disulfides are located. Such a drawing was shown for part of lysozyme in Figure 1-12. Another example, for thermolysin, is given in Figure 2-34. These α-carbon drawings produce an easily visible structure whose apparent openness is completely misleading, as you can see by comparing it with the other kinds of models. A final type of tertiary structure illustration is shown in Figure 2-35 for human carbonic anhydrase. Here certain prominent features such as a twisted β sheet are exaggerated to draw attention to them, whereas others are oversimplified. Such a drawing is really a cartoon of the structure. Throughout the book we use the simplest kind of models needed to illustrate the particular issues at hand. You must keep in mind that what they *don't* show can be as important a part of the structure as what they do show.

General organization of the peptide backbone

The easiest features to recognize in the simpler kinds of models are secondary structures. Table 2-7 summarizes what one should look for. The proteins in this table have between 100 and 300 residues per subunit. The percentage of α helix and the number of helices shown in the table indicate that a typical α helix in a globular protein will be about 10 residues long. Such a helix will have a length of 15 Å, which is roughly equal to the radius of a subunit of these proteins. The β sheets are much fewer in number, but each involves many more residues. Most of the extensive β-sheet networks have from 4 to 6 strands and involve 20 to 40 residues. These can extend over a major portion of the tertiary structure, and they can be more difficult to recognize than are α helices. Antiparallel sheets are slightly more common than parallel sheets, but many of the bigger sheets have both parallel and antiparallel strands. An example is the secondary structure of human carbonic anhydrase, shown in Figure 2-35, as it sits within the tertiary structure. Note that the extensive β sheet is twisted, somewhat resembling a left-handed propeller. This is a general feature of all known extensive β sheets in globular proteins. It can be rationalized from the particular internal rotation angles favored by individual residues in β sheets. See Box 2-3 for more details about β sheets.

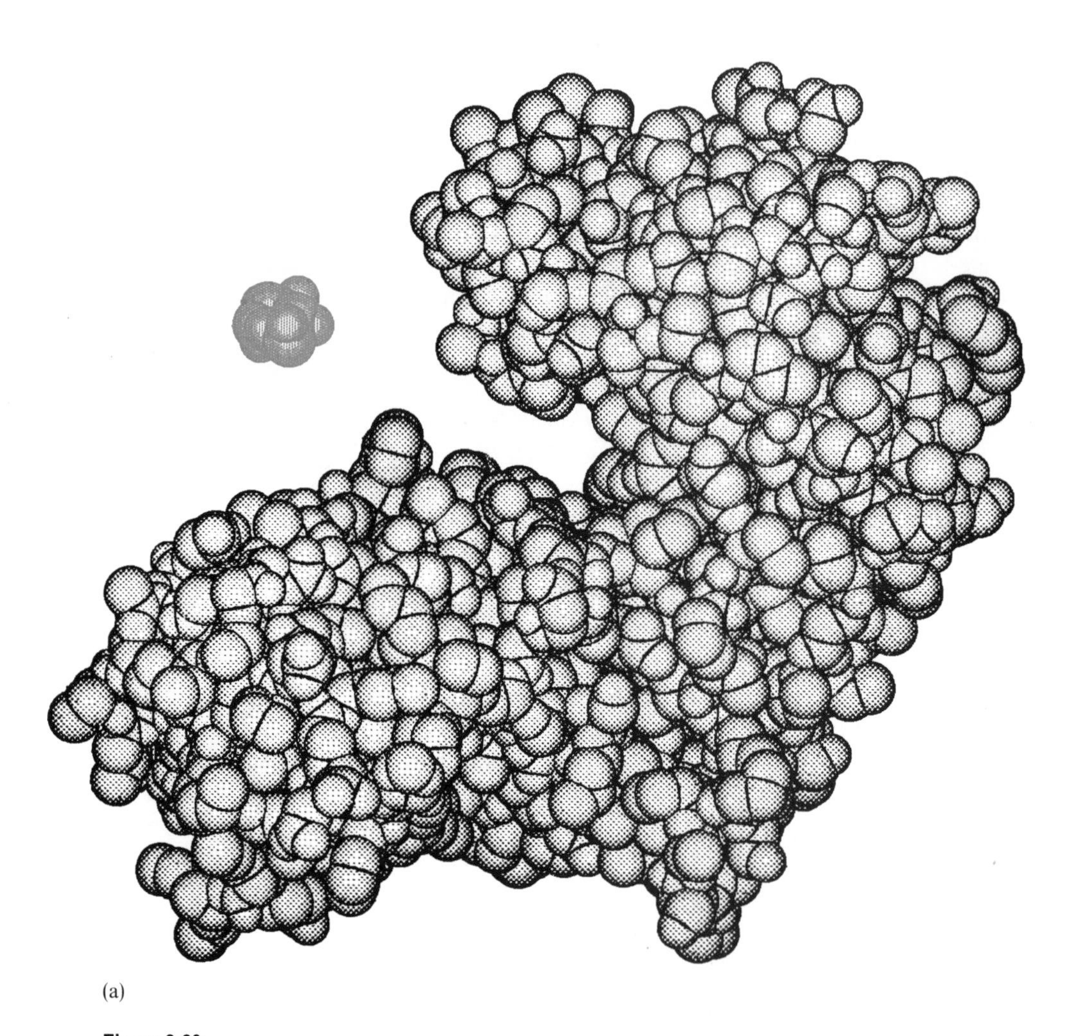

(a)

Figure 2-30

Tertiary structure of yeast hexokinase A, drawn by a computer interfaced to an electrostatic plotter. All atoms except hydrogen are shown separately. The hydrogens are taken into account by increasing the radius of any heavier atom that contains attached hydrogen. (**a**) Free hexokinase and its substrate, glucose (colored). (**b**) Hexokinase complexed with glucose. Note the striking change in enzyme structure that accompanies glucose binding. The enzyme so surrounds the glucose that an enzyme structure change is required to release the bound ligand. [Drawings provided by T. A. Steitz.]

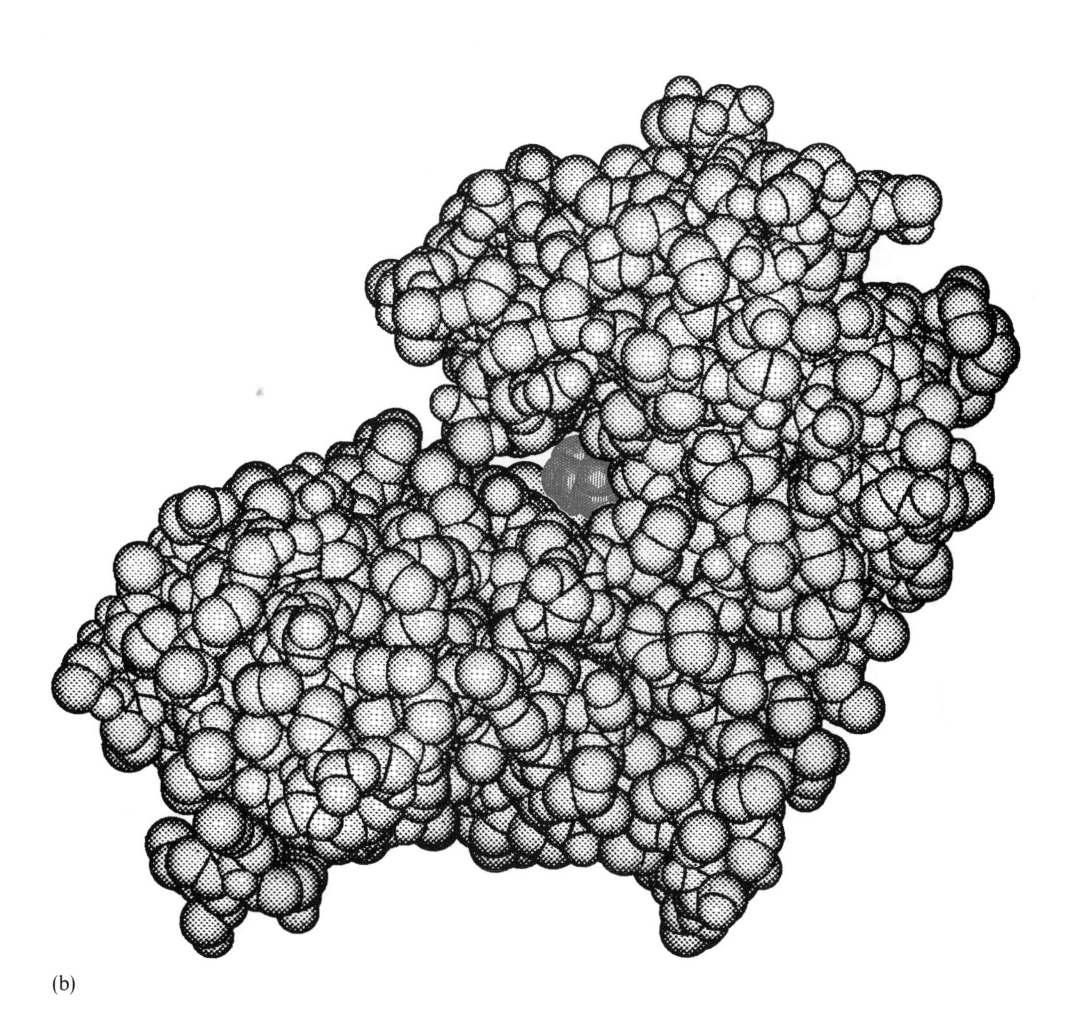

(b)

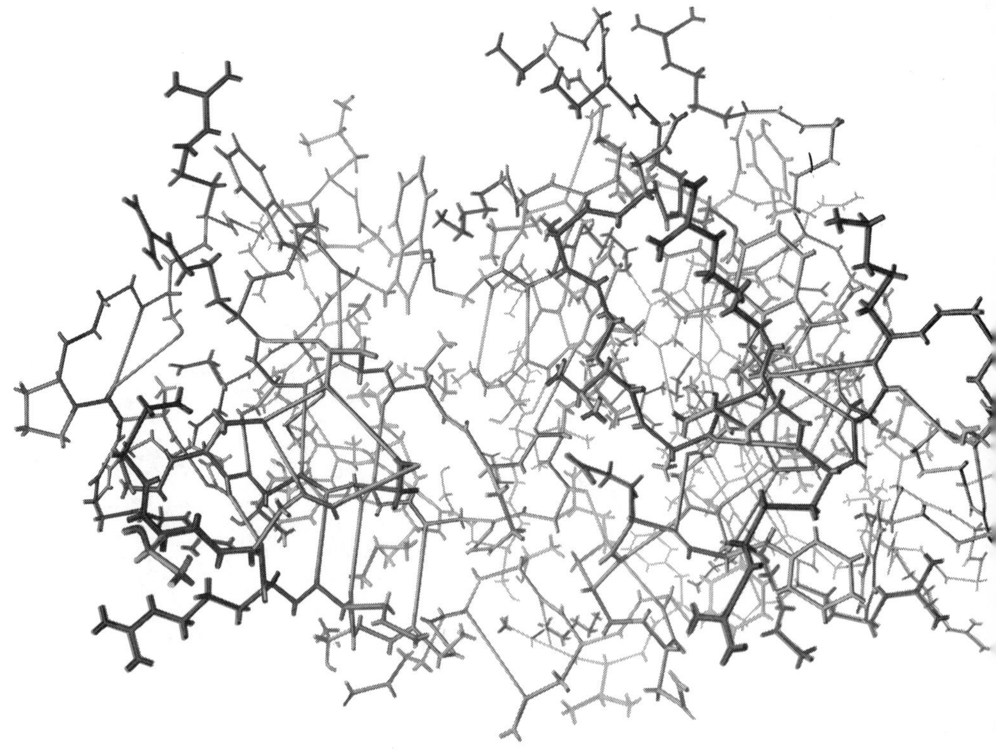

Figure 2-31

Location of charged groups within the tertiary structure of hen egg-white lysozyme. In this and the two following figures, all atoms of the protein are shown. The artist has made some slight distortions in the structure to improve visual clarity. The Lys, Arg, Asp, and Glu side chains are shown in color, as are the N- and C-termini. [Drawing by Irving Geis.]

Figure 2-32 (facing page, top)

Location of hydrophobic side chains in hen egg-white lysozyme. All of the following side chains are shown in color: Trp, Tyr, Phe, Pro, Ile, Leu, Val, Ala, Cys–S–S–Cys, and Met. See caption of Figure 2-31 for further details. Note the large nonpolar cluster of residues on the right-hand side of the molecule and a smaller cluster to the left of the sugar binding cleft. [Drawing by Irving Geis.]

Figure 2-33 (facing page, bottom)

Location of hydrogen bonds between residues in hen egg-white lysozyme. All such hydrogen bonds are shown in color. Some constitute secondary structure regions such as the β sheet at lower left and two α helices that run vertically near and to the right of the center of the molecule. Others, however, are not involved in any recognizable secondary structure form. [Drawing by Irving Geis.]

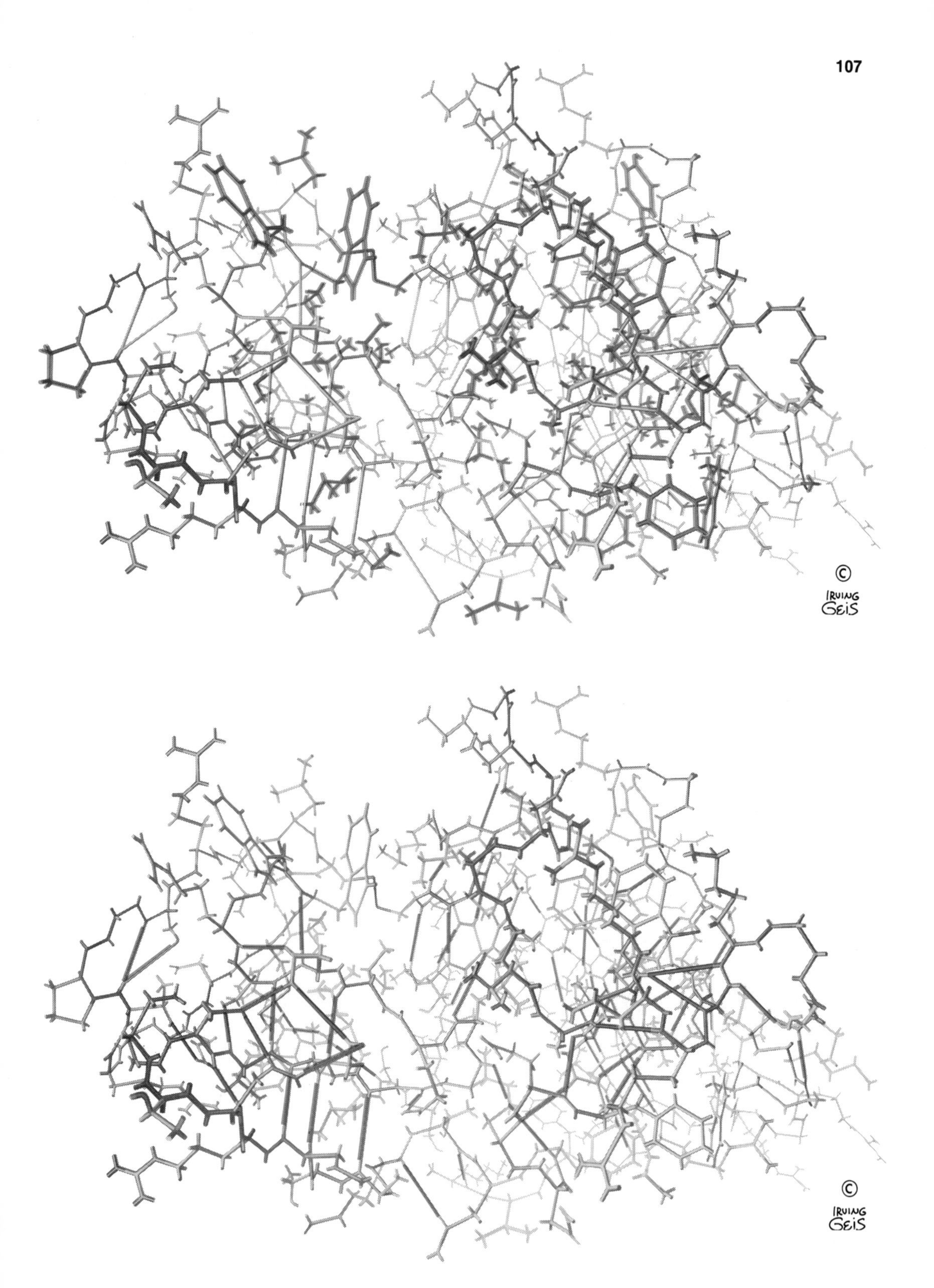
©
IRVING
GEIS
©
IRVING
GEIS

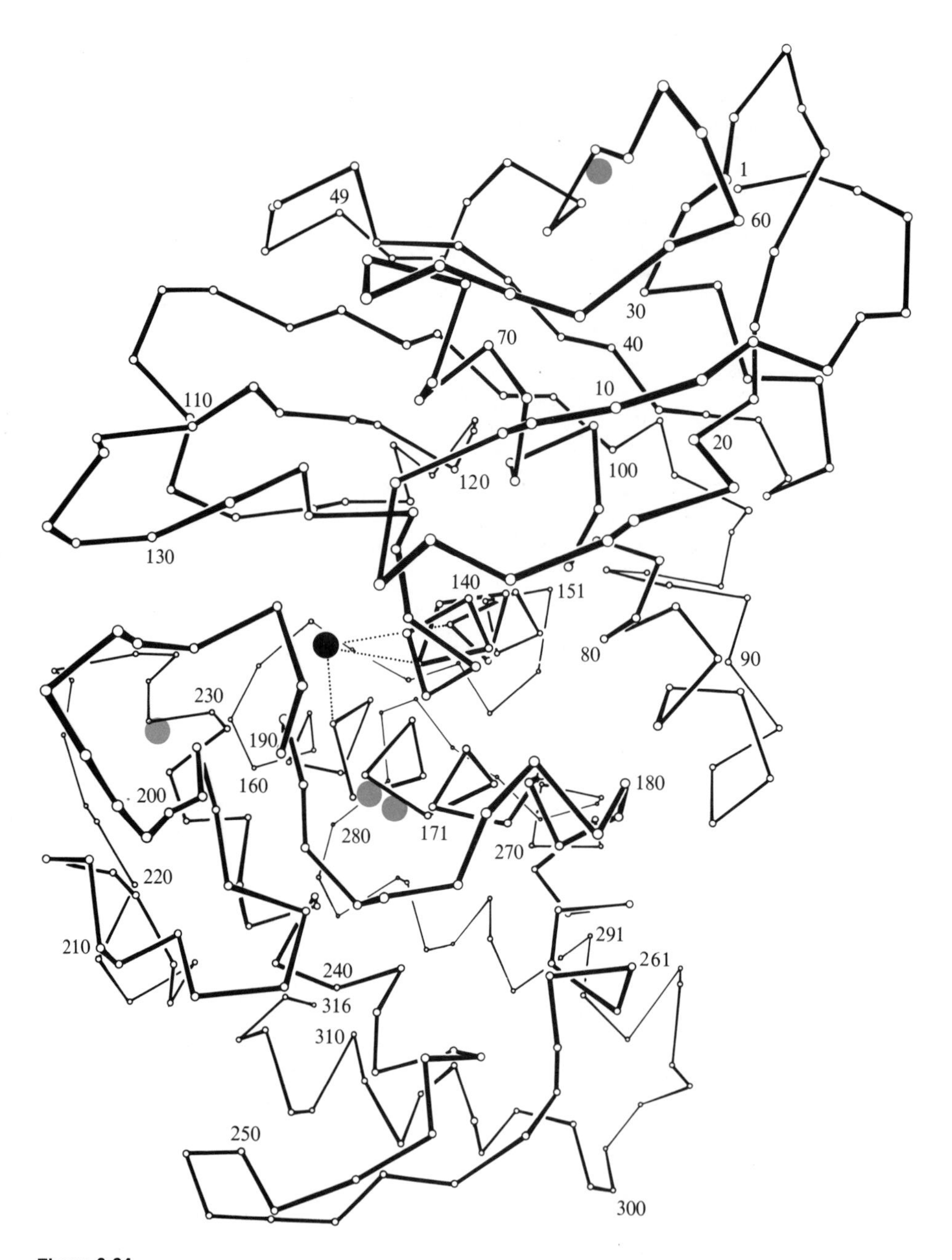

Figure 2-34

Tertiary structure of thermolysin, a protease with unusually high thermostability. Note that the structure is segregated into two quite separate and independent folding domains. The α-carbons are shown as hollow spheres connected together to trace the peptide chain backbone. The zinc atom is shown as a black sphere; four Ca^{2+} ions are indicated as colored spheres. Numerous α helices are visible. [From B. W. Matthews et al., *Nature New Biology* 238:37 (1972). Drawing provided by Brian Matthews.]

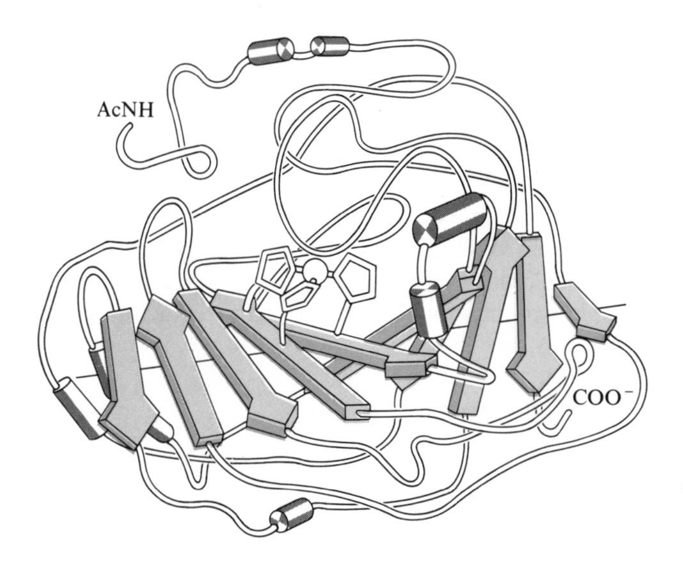

Figure 2-35

An idealized drawing of the tertiary structure of human carbonic anhydrase, as determined by x-ray crystallography. Shown are the peptide chain and the three histidines that coordinate to a zinc ion at the active site. Helices are represented by cylinders; individual β-sheet strands are drawn as arrows from the amino to carboxyl ends. Note the twist of the β sheets. [After K. K. Kannan et al., *Cold Spring Harbor Symp. Quant. Biol.* 36:221 (1971).]

In some proteins, two such twisted sheets are combined to make a barrellike structure. (For example, see the two proteins in Fig. 2-43.) There does not appear to be any striking preference for α and β structures to occur at constant portions of the tertiary structure—such as inside versus outside, or N-terminus versus C-terminus. Nor is there any particular tendency for the well-defined structural units to cluster within the tertiary structure. The α helices are not usually in parallel bundles in folded globular proteins. There may not be many neat ways to pack adjacent secondary structure units, except for cases with very special side-chain sequences.

A major question to ask of any tertiary structure is whether it must be considered as a unit, or whether it appears to be subdivided into domains. Examples of both cases exist. In most of the structures we have shown thus far, the peptide chain wanders extensively back and forth through the structure. However, in thermolysin (Fig. 2-34) the folding pattern can be separated into two fairly independent regions. This makes it worthwhile to study whether or not the individual regions can exist as stable folded structures in the absence of the rest of the protein. If so, it becomes possible to study their reassociation, and therefore to learn something directly about specific interactions and forces involved in the maintenance of tertiary structure.

The presence of domains may reflect how the tertiary structure of a protein can form from an extended peptide chain. The rates of forming sections of folded or organized structure will be especially fast for regions of an extended peptide chain that are near each other within the primary structure. This is because it is more likely in a random configuration that these two regions will be near each other in space. In Chapter 21, we discuss the dynamics of folding in a quantitative framework. Here we just note that, if a protein structure divides into domains, it is likely that these domains are significant intermediates in the folding pathway. It is also likely that the protein may be able to undergo conformational changes in which the relative disposition of the domains changes markedly while their internal structures remain fairly constant. A protein with domains is thus a much better candidate for showing a flexible structure than is one in which various regions are interlocked by crisscrossed chains. (An example is immunoglobulin G, illustrated in Figs. 1-9 and 2-10.)

Environment of individual residues

Protein tertiary structures provide penetrating clues to the forces that must be involved in stabilizing them. Here we state a number of general rules that come from the study of the many available tertiary structures. The independent physical chemical evidence supporting or explaining these rules is treated in Chapter 5. The rules are not absolute, and exceptions often are connected to functional properties.

1. The conformations of individual peptide groups fall in the most stable ranges predicted from model compound studies and semiempirical energy calculations. Formation of secondary and tertiary structure seldom occurs at the expense of the local steric stability of small groups of residues. When it does, of course the properties of these residues will be altered, and sometimes their usefulness in catalysis or other functions is increased. But, in general, it appears that the need for individual residues to be in their most stable conformations goes a long way toward directing what secondary and tertiary structures can occur.

2. Charged residues are on the surface of the protein. (A typical example is shown in Fig. 2-31.) We have already argued how energetically costly it would be to bury charges. The distribution on the surface varies. Sometimes it appears fairly uniform. A staggered array of positive and negative charges would be lower in energy than a clustering of like charges. This latter distribution does often occur however, and such tertiary structures would be expected to be more susceptible to disruption by lowered ionic strength.

   Anisotropic charge distributions will lead to a large overall molecular dipole moment. It is likely that particular charge distributions play significant biological roles, either in assisting the binding of other large or small molecules, or possibly in leading to preferred orientations in the presence of high electric fields such as those present across some membranes. Electrical forces are the most long-range of the strong interactions prevalent in biological systems, and presumably one will eventually find cases in which these forces are specifically exploited to achieve a particular functional objective. In a few cases, charges have been found inside a protein, and these are inevitably on residues directly involved in catalysis or some other function.

3. Nonpolar, hydrophobic residues are found preferentially on the interior of the tertiary structure of a globular water-soluble protein. Note that this rule is stated with several hedges and qualifications. An example of a case where it is obeyed in a clear-cut way is shown in Figure 2-32. Even here, locations are not restricted quite as extensively as those of charged residues. The energy loss of exposing a few nonpolar residues is not all that great. For oligomeric proteins, the intersubunit contacts are often very nonpolar. However, an isolated subunit can be quite stable as an independent unit, yet still have a partially hydrophobic exterior needed to interact with other subunits.

   A total exception to the rule is expected for proteins that interact extensively with lipids. Although we do not yet know the tertiary structure of any membrane protein or lipoprotein at a high enough level of resolution to see individual side chains, it will be a great surprise if such structures, when available, do not show extensive external hydrophobic residues. A few cases are known of proteins completely insoluble in water but readily soluble and functionally active in organic solvents. In these cases, one can guess that the structure may be inside-out compared with a normal protein. Charged groups may be inside and interacting with each other, stabilized by bound ions where necessary. Hydrophobic groups would form a surface around them to interact with the solvent.

4. Virtually all hydrogen-bond donors and acceptors are located where they can potentially be satisfied. Either they are on the surface where they can hydrogen-bond to water, or they are incorporated into ordered secondary structure regions, or they appear to be located where they can form internal tertiary hydrogen bonds. (This is illustrated in Figure 2-33.) The cost of *not* making hydrogen bonds could be excessive if carried out on a large scale. Therefore, polar side chains tend to be on the surface, because it is more likely that they can find hydrogen-bonding partners. Internal polar side chains frequently hydrogen-bond to each other or to nearby portions of the peptide backbone.

   Note, however, that most tertiary structures are known only by interpretation of x-ray structural data. These data are rarely at high-energy resolution to show hydrogen bonds. The interpretation of the data probably tends to result in more hydrogen bonds than may really occur, because from our current state of knowledge we think such bonds *ought* to occur.

Some details of how tertiary structures are organized can be seen by studying the distribution of side-chain types on α helices within the tertiary structure. If a helix is near the surface, because of its basic structure, side chains will have to point inside and outside the protein in a smooth progression as you travel along the chain. Thus one can predict that there will be a periodic distribution of polar and nonpolar side chains for such a helix. This is indeed observed. See, for example, one of the hemoglobin helices described in Figure 2-36. Note also how this helix is bounded on the N-terminal by prolines. Such proline locations are fairly common on proteins. It is almost as if they serve the role of preventing such stable secondary structures as α helices from continuing themselves indefinitely. Sheets on the other hand can bend back on themselves through β turns and therefore are less prone to outrun the structure. Side chains of β sheets point above and below the plane. A sheet near the surface should have a predictable distribution of amino acid side chains.

Packing density of amino acids in folded proteins

How well are things packed inside a protein? Is it really an almost solid structure or are there spaces, either empty or filled with solvent? A few large holes have occasionally been seen in the tertiary structures of proteins. Myoglobin has a space that nicely accommodates a xenon atom, explaining why the protein binds one equivalent of this inert gas. Ribonuclease S has a few holes that appear to be empty; lysozyme and α-chymotrypsin have more holes, apparently filled with solvent. Sixteen internal water molecules have been found in the crystal structure of α-chymotrypsin.[§] However, by visual inspection most proteins look fairly tightly packed, as you can see by the examples shown in Figures 2-29 and 2-30.

It is possible to make a much more quantitative inquiry. One approach is to use the detailed coordinates to calculate a packing density. This is the ratio of the volume enclosed by the Van der Waals envelope of all the atoms in a region to the total volume of the region. The packing density is 0.74 for closest packed spheres, 0.91 for closest packed infinite cylinders, and 1.00 for a continuous solid. Definitions of packing density near the surface are rather ambiguous and, if one chooses too small a set of atoms to look at, it fluctuates quite a bit. But, if one divides up a protein into moderately sized regions and calculates packing density as shown in Figure 2-37, some interesting features emerge.

The average packing density of internal regions of ribonuclease S is about 0.75, which is in the range found for crystals of typical small organic molecules. For example, Gly–Phe–Gly crystals have a packing density of 0.749. However, the packing density fluctuates quite a bit, even when averages are taken over large zones of the protein. It is noteworthy that regions adjacent to the active-site groove show quite a low packing density, whereas regions next to these (farther from the active site) have a much higher than average packing density. This may have some functional significance. Loosely packed regions have the possibility of bulk flexibility, or they may

[§] Do not confuse the internal solvent and holes discussed here with the external solvent and holes that lie *between* protein molecules in a crystal. There is a large amount of such solvent, but this is not considered part of the tertiary structure.

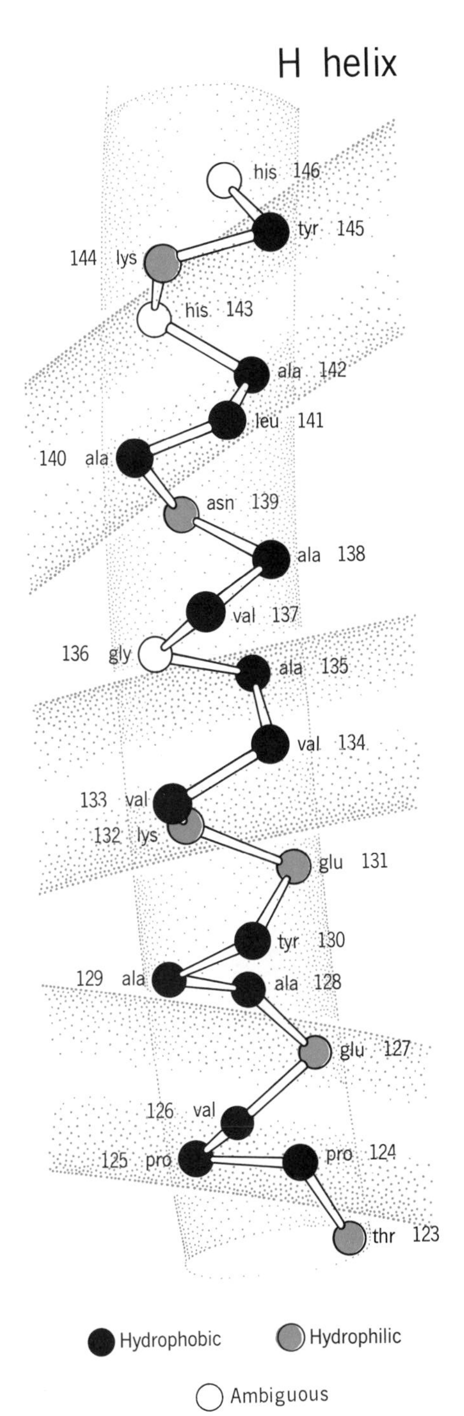

Figure 2-36

The H helix of the human hemoglobin β chain as it sits within the tertiary structure of the β subunit. The hydrophobic side chains are mostly on the near face of the helix as it makes contact with the A, E, and F helices that cross in front of it. [Drawing by Irving Geis.]

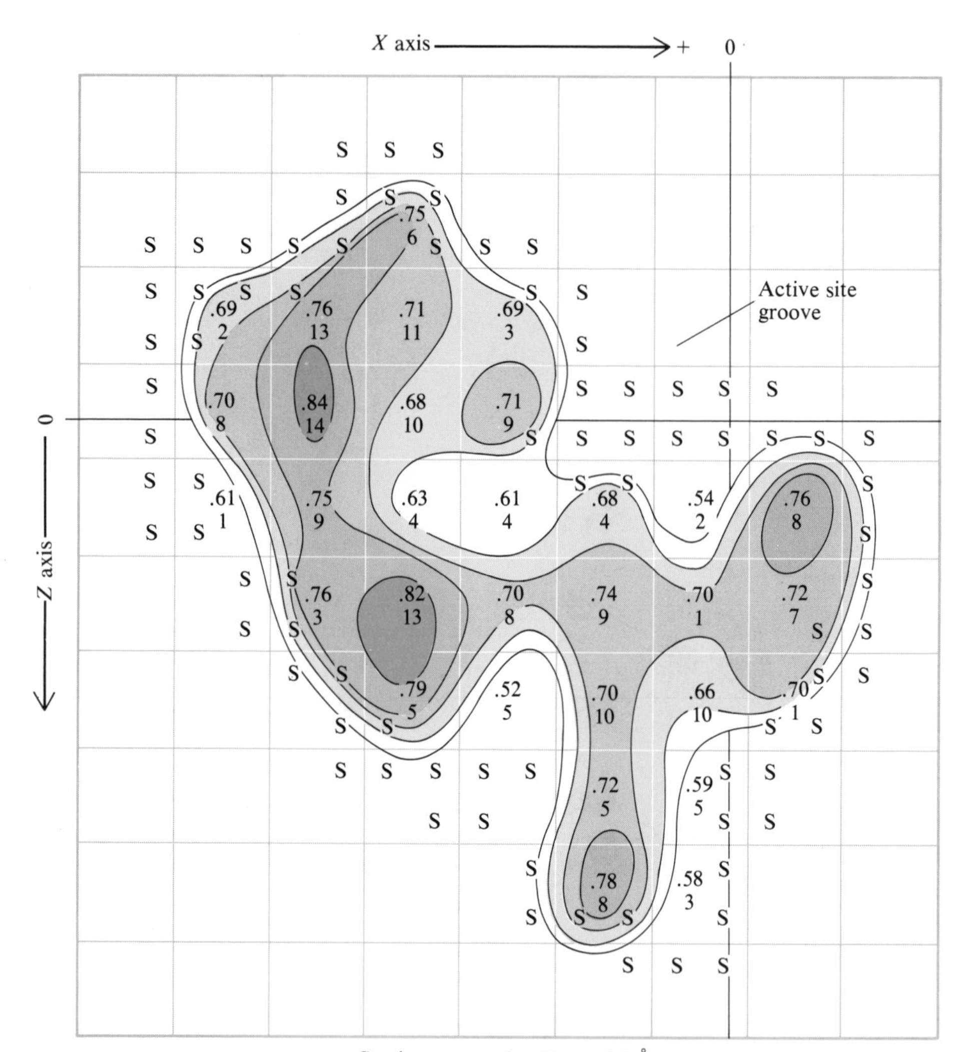

Figure 2-37

A section through the center of bovine pancreatic ribonuclease S (a fragment containing residues 21 through 124). A grid of 5.6 Å cubes is superimposed on the structure. Within each is shown the number of atoms in that cube, and the average packing density for those atoms (fraction of total volume occupied). Also shown (in color) are approximate zones of packing density. A hypothetical layer of solvent molecules denoted by S was used to define the protein surface. [From F. M. Richards, *J. Mol. Biol.* 82:1 (1974).]

allow internal motion, side-chain rotations, etc. The densest regions are likely to be extremely rigid. So, the distribution of packing density shows which kinds of motions are probable for the structure as a whole. It suggests a slightly mobile yet constrained active site, but any large motions would involve the rigid regions. These would move as a unit and thus could transmit any conformational changes over a long distance.

Volumes and mass densities of proteins

By adding up the Van der Waals volumes of all the atoms, one can compute the molecular volume of a protein. This turns out to be 16,900 Å^3 for ribonuclease S, neglecting the residues not well defined in the x-ray structure, and is approximately 18,100 Å^3 for lysozyme. These values are up to 10% higher than apparent volumes measured in solution from the known molecular weights and experimentally determined densities. (See Chapter 10.)

The experimental densities, ρ (in g cm^{-3}) assume that bound water has the same density as bulk water. This might have been a sensible assumption, but the discrepancy with the x-ray results suggests otherwise. The surface of a protein contains many charges, and it is known that charged groups bind water tightly, increasing its density. This electrostriction effect is about 16 Å^3 per charged group. Ribonuclease S has 24 charges that could give an apparent volume decrease in solution of about 400 Å^3 per protein molecule. This effect would show up as a decrease in the apparent volume of the protein but, as you can see, it is not large enough to account for more than a fraction of the discrepancy.

The problem gets worse if the mass density (ρ) is computed from the x-ray coordinates as a function of the distance from the protein center of mass. Results for several proteins are shown in Figure 2-38, where ρ is plotted for spherical shells at increasing radii of 2 Å. At very short distances, the densities fluctuate widely because the actual volumes of the shells (and thus the numbers of atoms they contain) are so small. But then a plateau is reached where ρ is roughly independent of distance. For various small proteins, it is between about 8 and 16 Å from the center of mass. At larger radii, densities fall off gradually. This is because of the irregular detailed shape of each protein. Note that no solvent molecules have been included in these calculations.

In a simple case, modeling a protein as a uniform solid object, the pleateau density should correspond to the actual measured solution density. However, it is low by from 10% to more than 25%. This discrepancy is even worse than those discussed above. There are three possible explanations. Water molecules bound to the outside may be much more dense than bulk water, not just due to electrostriction but perhaps as a more general phenomenon. If one considers a jacket of water one molecule thick, it would have to have a density between 1.35 and 1.60 g/cm^3 to resolve the discrepancies. This seems unreasonably large, considering that bulk water is only about 1.0 g/cm^3.

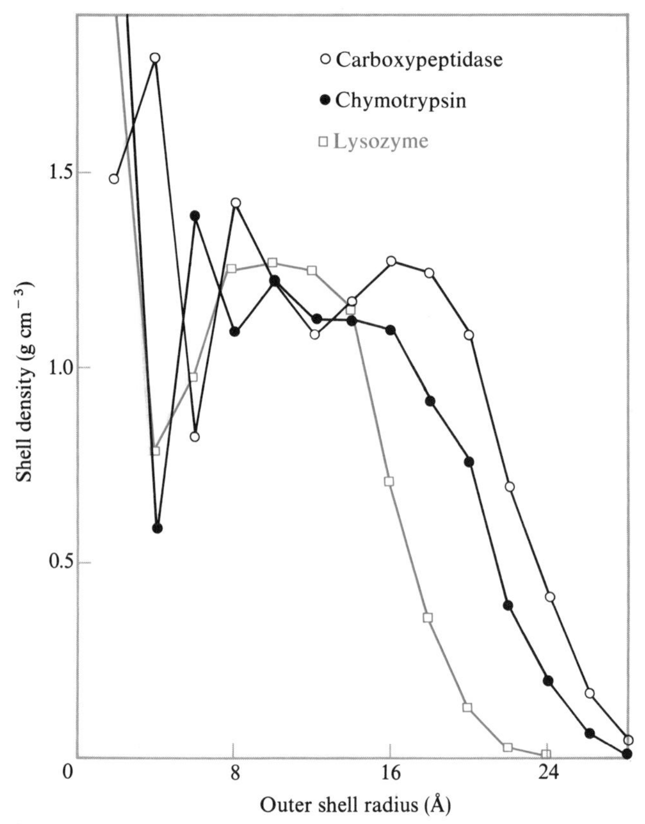

Figure 2-38

Densities of spherical shells for several proteins as a function of the distance of the shell from the center of mass. All atoms except solvent were used to compute the density in g cm^{-3}. [After W. Kauzman, *Nature* 248:447 (1974).]

A second possible explanation is that the outside regions of a protein may be more dense than those inside. The packing densities shown in Figure 2-37 do not show this trend, but remember that there is some tendency for hydrophobic residues to be found internally. These have lower ρ values than do polar residues. A purely hydrocarbonlike interior could easily explain the plateau densities of 1.1 to 1.2 g/cm^3. In fact, one could include some peptide chains and still keep within this value but, in order to make the total agree with the observed protein solution density, the outside regions would have to have densities of 1.50 to 1.55 g/cm^3. This is simply too high when compared with known densities of solid peptides.

A third explanation is that many more water molecules actually exist in the protein interior than can be found by crystallography. To resolve the discrepancy by this mechanism alone requires 100 to 300 water molecules. This seems quite unlikely. One must conclude that all three mechanisms probably contribute to some extent.

This serves to illustrate the real complexities that underlie even relatively simple questions about protein tertiary structures.

Is the tertiary structure static or dynamic?

The density calculations just discussed illustrate the need for information about the flexibility of protein regions and the permeability of a protein to solvent or other small molecules. There are no general answers to these questions, but a few techniques are potentially capable of providing some insight. The motions of aromatic amino acid side chains can be monitored by spectroscopic techniques such as fluorescence polarization. In most cases, such residues show rotational rates expected for the protein as a whole, suggesting that they are rigidly attached to the protein. However, vibrational motions with a small angular amplitude would not necessarily be seen by such measurements. Externally attached probe molecules usually show a similar stiffness, but here it is not clear whether they interact with the protein surface in such a way as to hinder normal motions.

NMR relaxation, particularly ^{13}C, is a powerful method for examining the motion of certain side chains because individual atom types such as ε-carbons on lysine can be monitored independently. In general, results indicate that surface side chains have considerable rotational freedom. Much less freedom, if any, is detectable for internal residues. This result is consistent with x-ray crystallographic results. Many surface side chains appear disordered in protein crystals, implying rotational motions. The bulk of a protein appears rigid, but one must remember that x-ray data on macromolecules is rarely available at a high enough resolution to be certain about this.

Assessing the permeability of a protein or the accessibility of individual residues is a very complicated issue. In general, one asks whether small molecules can enter the interior of a protein and either participate in chemical reactions or change the spectroscopic properties of internal residues. Different small molecules sometimes appear to give different results. For example, oxygen molecules appear able to reach and quench the fluorescence of most or all tryptophans in a globular protein. Iodide ions can reach only a fraction of them. Both are equally effective as quenchers if the tertiary structure of a protein is disrupted by denaturants. It isn't known whether the differences with native structures are due to the charge on the iodide, to its large size, or to more subtle local environmental effects on the quenching reaction.

A variety of covalent chemical modifications of amino acid side chains show that the tertiary structure protects many residues that would normally be reactive in free amino acids or short peptides. If a particular side chain can be modified, it is generally safe to conclude that it is exposed. If the reaction proceeds without a loss in functional activity, one can be fairly confident that the chemical reaction has not seriously perturbed the structure. However, if a residue fails to react, experience has shown that it is not necessarily safe to conclude that it is buried. Local environments, even on the surface of a protein, can differ markedly from normal aqueous solution.

Calculations of the dynamics of protein structures offer a very promising way of examining flexibility and permeability. (See the example discussed in Box 2-4.)

Methods of comparing tertiary structures

The tertiary structures of many proteins have been determined by x-ray crystallography. Fairly detailed comparisons are useful, but the difficulty is finding convenient ways of doing this. When the tertiary structures are extremely similar, visual observation allows them to be placed at the same orientation; then the occasional differences can be seen simply by inspection. Compare, for example, the structures of several serine esterases (shown earlier in this chapter) or the α and β hemoglobin chains and

Box 2-4 MOLECULAR DYNAMICS CALCULATIONS ON PROTEINS

Recently, the dynamics of protein tertiary structures have been examined by Martin Karplus. He used the technique of molecular dynamics to calculate the expected motion of residues in the protein bovine pancreatic trypsin inhibitor. In this technique, the forces on all residues are estimated from a previously calculated potential function (such as those shown in Chapter 5). The force on each atom or residue is

$$\mathbf{F}_i = -\sum_j \nabla V_{ij}$$

where V_{ij} is the potential of interaction between atoms i and j.

For free atoms or molecules of mass m_i, each force produces an acceleration $\mathbf{a}_i$ according to Newton's laws of motion.

$$\mathbf{F}_i = m_i \mathbf{a}_i$$

In a molecular dynamics calculation, one solves these two equations simultaneously for all atoms i at small time increments. Then new positions for each atom are computed, and the calculations are repeated.

Several important features emerge from the results of these calculations, as illustrated in the figures. The root mean square deviation of a residue position from the x-ray structure is about 1 Å. This is very substantial and indicates that considerable breathing must be occurring, which allows small molecules to permeate. Not all residues move equally. Certain regions show much more flexible structure than others. However, few motions are large enough to render the x-ray results a bad representation of the peptide chain.

myoglobin. As the differences become larger, it is easy for them to obscure the similarities. It becomes hard to find orientations of the two molecules that facilitate comparisons. Computers can, by brute-force trial and error, find orientations that maximize similarity, but sometimes at a cost of the loss of any intuitive feeling for the structure. What is needed is some way of abstracting or simply summarizing the structures. At least two approaches to this problem have been proved useful. Both of these approaches place heavy emphasis on secondary structure elements, because these elements are visually relatively easy to identify.

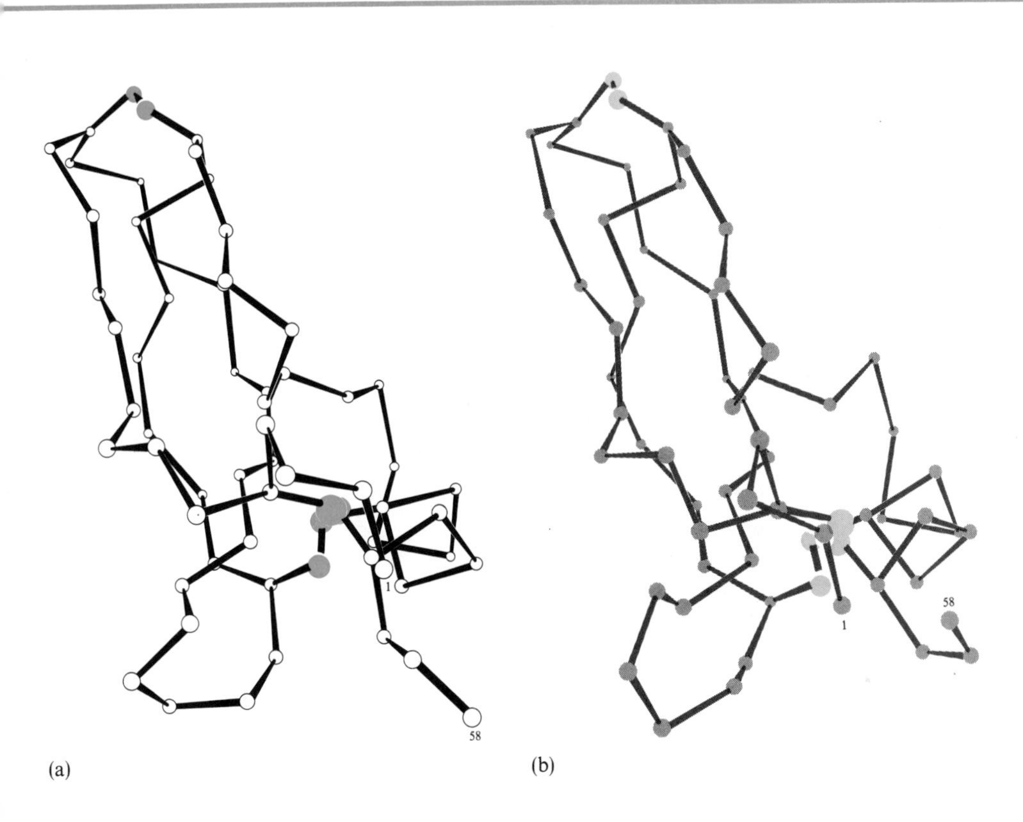

(**a**) The x-ray structure of bovine pancreatic trypsin inhibitor. The α-carbons are illustrated as spheres, sulfur atoms of cystine as colored spheres. (**b**) The molecular structure calculated after molecular motion has been allowed to occur for 3×10^{-12} sec.

The most commonly used method is to prepare some kind of cartoon that expresses the major features of the secondary structure and their connectivity in the simplest possible way. The cartoons, unlike the real structures, are easy to verbalize and remember. Figure 2-39 shows a set of these for a series of proteins with nucleotide binding sites. Structural similarities and differences among this set are readily apparent. They can be used as a basis for a subsequent, more detailed, and more rigorous comparison using actual models or atomic coordinates.

An alternative method for secondary structure comparison is less direct but removes the need to find mutually consistent viewing orientations. The idea is to use

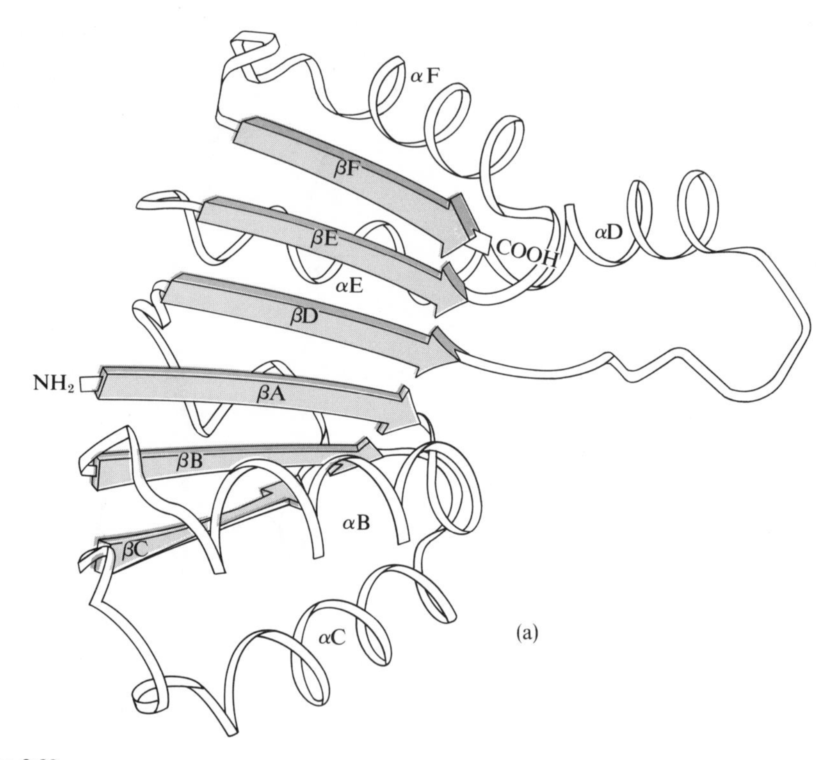

Figure 2-39

Nucleotide binding protein domains and related tertiary structures in a series of proteins. **(a)** Schematic tertiary structure of part of lactate dehydrogenase. Arrows show strands of β sheets drawn from the amino to the carboxyl end. **(b)** Cartoons of similar tertiary structure patterns in lactate dehydrogenase and other proteins. Triangles indicate β sheets viewed with the N-terminus closest to the observer. Coils indicate α helices. A, N, and F indicate positions of the known binding sites for the adenine moiety of NAD or ATP, for the nicotinamide moiety of NAD, and for flavin, respectively. [After M. G. Rossman, in *The Enzymes*, 3d ed., vol. 9, ed. P. Boyer (New York: Academic Press, 1975), p. 61.]

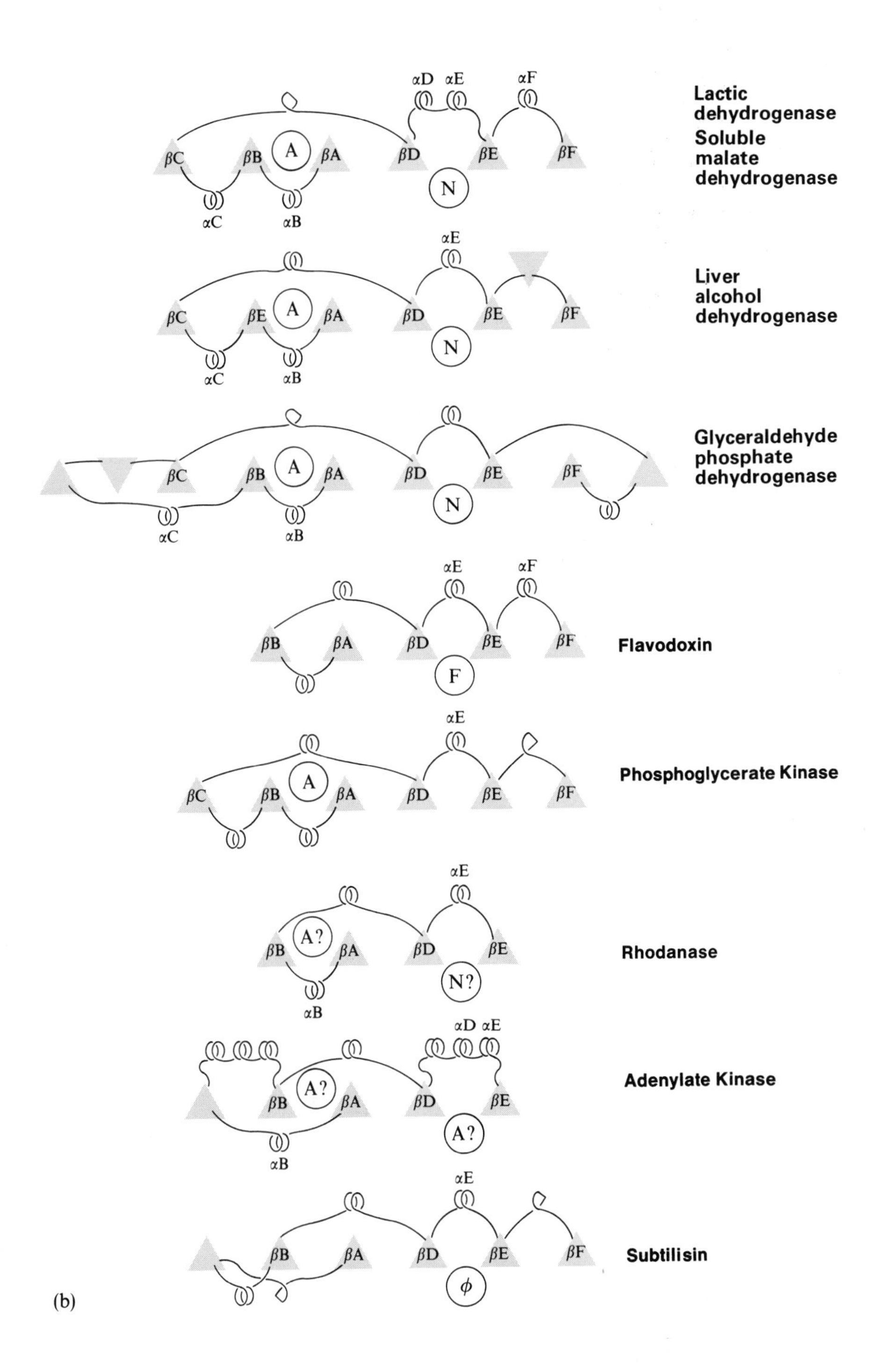
αD αE αF
βC βB A βA βD βE βF
N
αC αB
Lactic dehydrogenase
Soluble malate dehydrogenase
αE
βC βE A βA βD βE βF
N
αC αB
Liver alcohol dehydrogenase
βC βB A βA βD βE βF
N
αC αB
Glyceraldehyde phosphate dehydrogenase
αE αF
βB βA βD βE βF
F
Flavodoxin
αE
βC βB A βA βD βE βF
Phosphoglycerate Kinase
αE
βB A? βA βD βE
N?
αB
Rhodanase
αD αE
βB A? βA βD βE
A?
αB
Adenylate Kinase
αE
βB βA βD βE βF
φ
Subtilisin
(b)

internal distances rather than coordinates. For example, one can compute all inter-α-carbon distances and divide these into two or more categories representing residues near or far in the tertiary structure. This throws away most of the information in the structure, because distances are quite a bit less definitive than coordinates. However, the resulting summary is easily displayed as a two-dimensional map (Fig. 2-40). Here, both the vertical and horizontal axes are amino acid residues numbered from the N-terminus. Residue pairs located near each other in the primary structure will be plotted near the diagonal of such a map. Naturally these residue pairs will tend to be close in the tertiary structure as well.

A separate map is constructed for each protein. Similar secondary structure features usually stand out strikingly. For example, α helices will be straight lines with a slope of -1, displaced three residues from the diagonal because the hydrogen bonds between residues four apart in the α helix make α-carbons three residues apart among the closest pairs in the structure. Any β sheets also will show up as straight lines. Parallel β sheets will have a slope of -1, but displaced farther from the diagonal. Antiparallel sheets will have a slope of $+1$.

The schematic structure maps can help to spot similarities in structural features and ease the way for a comparison of actual coordinates. They are also extremely useful in determining whether the chain folding pattern is segregated into distinct domains. These will appear on the map as clusters of near contacts widely separated from other such clusters. From the map pattern of each domain, it is possible to judge whether these internal structures are similar. For example, lactate dehydrogenase shows four domains, as illustrated in Figure 2-40. The first two have very similar structures and, as it turns out, they also have similar functions. This enzyme binds the dinucleotide NAD. Each of the two similar domains serves to bind one of the mononucleotide moieties of the coenzyme. Schematic structures of these domains are given in Figure 2-39.

Relationship between structural and functional similarity

The simplest case of structural similarity is two or more proteins with rather similar structures and related functions. However, it is especially interesting to compare cases where only the structure or the function, but not both, show obvious gross similarities. A striking example occurs when we compare a whole series of proteins capable of binding nucleotides. The two domains just mentioned, which bind NAD in lactate dehydrogenase, are found also in soluble malate dehydrogenase, glyceraldehyde-3-phosphate dehydrogenase, and liver alcohol dehydrogenase. Figure 2-39 shows a secondary structure cartoon of these regions. These four proteins show no extensive primary structure homology. Furthermore, the location of the dinucleotide binding domains within the primary structure is quite different. In three, it occurs within 150 residues of the N-terminus, but in liver alcohol dehydrogenase it doesn't start until residue 196. The location of the domains within the quaternary structure of the

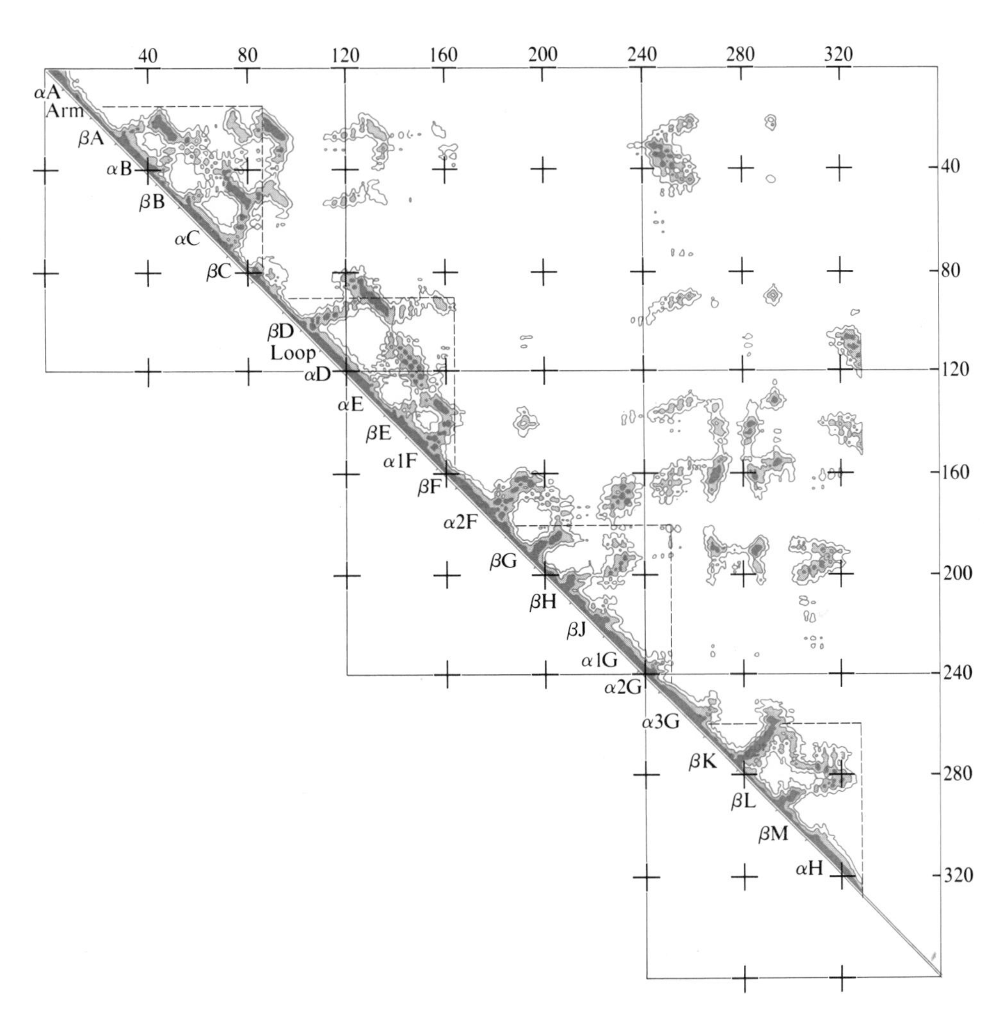

Figure 2-40

A plot of all distances between pairs of α-carbon atoms in lactate dehydrogenase. Contours are shown enclosing pairs of atoms separated by 0, 4, 8, 12, and 16 Å. Horizontal and vertical axes give the residue number. Only half the sequence matrix is shown because the other half is identical ($r_{ij} = r_{ji}$). Particular secondary structure features are named along the diagonal. Shown in dashed boxes are four structural domains. [After A. Liljas and M. G. Rossmann, *Ann. Rev. Biochem.* 43:475 (1974).]

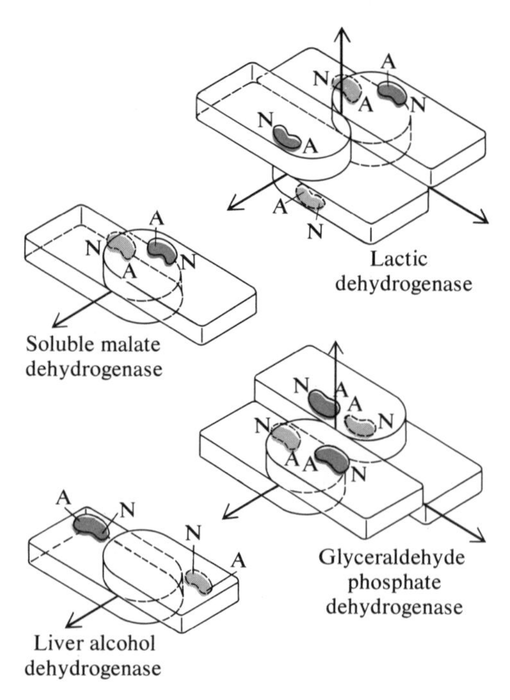

Figure 2-41

Schematic quaternary structures of four proteins, all of which contain similar NAD binding sites. Twofold rotational symmetry axes are shown by arrows. [After M. G. Rossmann et al., in *The Enzymes*, 3d ed., vol. 9, ed. P. Boyer (New York: Academic Press, 1975), p. 61.]

proteins is not consistent either. As shown in Figure 2-41, in glyceraldehyde-3-phosphate dehydrogenase, all four NAD binding sites are clustered in the quaternary structure, whereas in lactate dehydrogenase they are segregated into pairs.

The remarkable thing is that the same or very similar structural domains are found in proteins that do not bind NAD. For example, flavodoxin, a protein that binds FMN, has one domain equivalent to one of the two lactate dehydrogenase domains. Phosphoglycerate kinase has two such domains, one of which is involved in ADP binding. Adenylate kinase also appears to have similar domains, as does rhodanase. Even subtilisin, a proteolytic enzyme with no known nucleotide interactions, also has a lactate dehydrogenase–type tertiary structure domain.

It appears that nature, having found an efficient way to construct a binding site, has repeated the same theme over and over again. This raises a number of interesting questions to which we don't yet have answers. Is this simply due to evolution? Did all these proteins in the distant past derive from some common ancestral binding protein? If so, then subsequent divergence of structure and function has erased all the similarities except for the tertiary structure of these domains. Is the domain similarity an expression of simple structural constraints on proteins? Perhaps there are not many ways a protein chain can fold in a thermodynamically stable conformation to produce binding sites. Or are there some unique physical features of this particular domain—structure, flexibility, rigidity, or whatever—that make it particularly desirable or effective for a functional site? The answers may have to await a more detailed understanding of physical properties of these domains and of the forces that stabilize them.

The family of similar or equivalent nucleotide binding domains was found by Michael Rossmann after an intentional search for similar structures that could underlie related or near-identical functions. Note that it involves only part of the tertiary structure of each protein. Another interesting example involves only a tiny fraction of the total structure. The catalytic sites of the chymotrypsin family of proteases is strikingly similar to the catalytic site of subtilisin (Fig. 2-42). This similarity is generated despite the fact that the overall pattern of folding of the peptide chains is completely different. These site equivalences imply similar catalytic mechanisms.

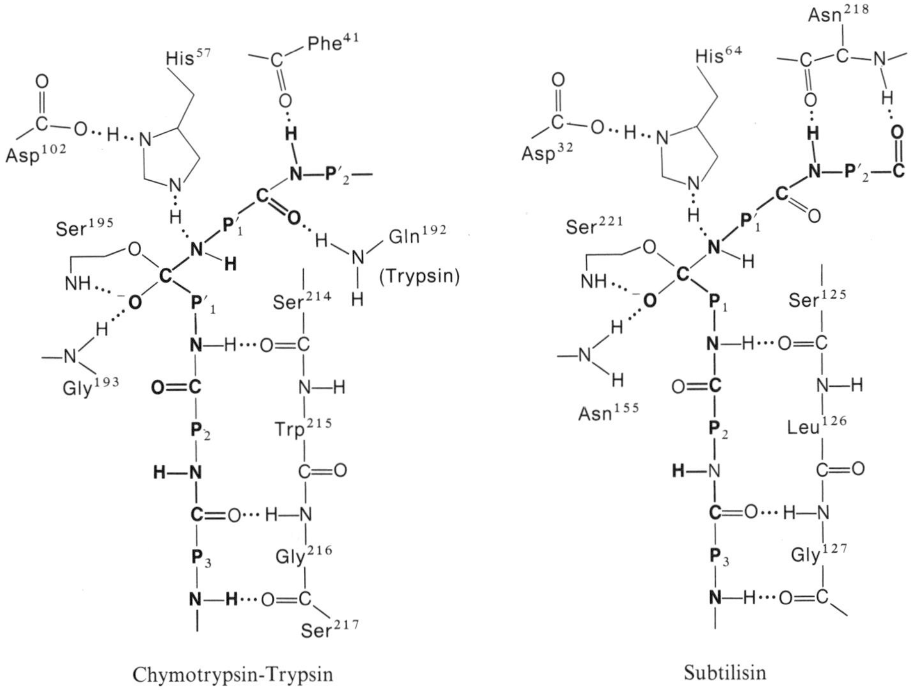

Chymotrypsin-Trypsin

Subtilisin

Figure 2-42

Similarity of substrate binding sites in chymotrypsin and subtilisin. A peptide chain substrate is shown in boldface. Note the similar pattern, not only of residues Asp^{102} or Asp^{32}, Ser^{195} or Ser^{221}, and His^{57} or His^{64}) involved in catalysis, but also of other residues that just assist in the binding of the peptide chain. [After A. Liljas and M. G. Rossmann, *Ann. Rev. Biochem.* 43:475 (1974).]

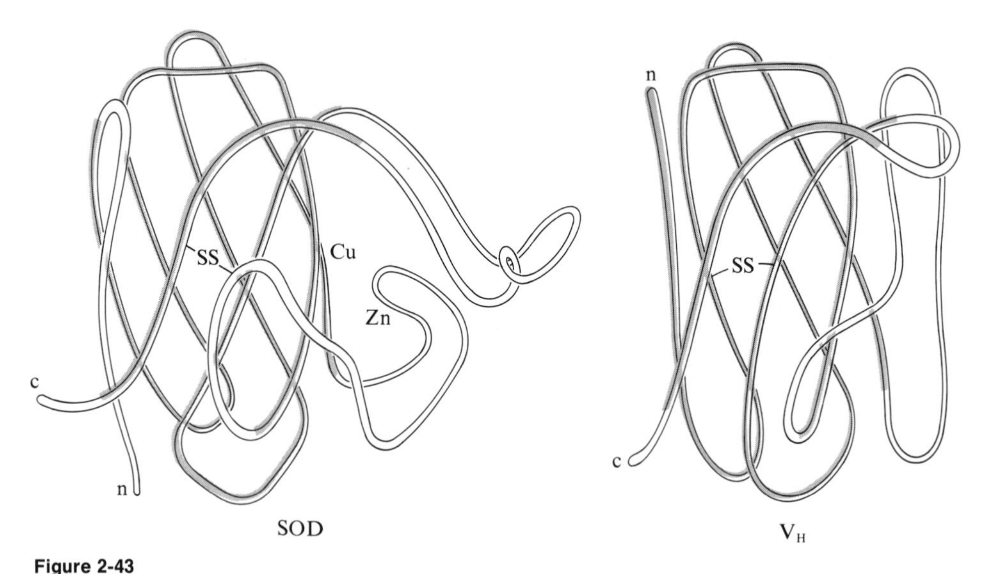

Figure 2-43

A comparison of the schematic tertiary structures of bovine superoxide dismutase (SOD) *and the variable domain of the heavy chain of immunoglobulin* G(V_H). The principal structure feature, a β barrel (shown in color) composed of seven strands of β sheet, is essentially identical in these two proteins. [Drawings provided by David Richardson.]

They raise questions of divergent versus convergent protein evolution. It may really be that there are very few catalytic mechanisms open to proteins that do not employ cofactors or prosthetic groups. This could be the result of the fact that the twenty normal amino acids represent only a minute fraction of the known chemical potentialities of small organic molecules.

Perhaps the most astounding case of structural similarity known to date is the tertiary structure of superoxide dismutase and one domain of the F_{ab} fragment of immunoglobulin G. As you can see from Figure 2-43, the entire tertiary structure pattern of chain folding (except for one region) is virtually identical in the two molecules. This could not have been predicted by any rationale we currently have available. These two proteins have totally different functions. In fact, the structural difference corresponds to the part of the chain that is involved in catalytic activity in the superoxide dismutase and the antigen binding site in the IgG. Has nature taken a basic stable structural framework, kept it constant even as primary structures differed drastically, and made just a subtle modification to direct the structure toward entirely different functional aims? Or does this similarity show that there is only a relatively small repertoire of ways in which segments of peptide chains can fold in space to yield a compact globular structure? One last possibility, now a challenge to the biochemist, is that one or both of these proteins may be polyfunctional and can perform some other activity (in addition to its known properties) that is similar enough in the two proteins to provide the rationale for the similarity in structure.

2-6 QUATERNARY STRUCTURE

Many proteins, as isolated from the cell, consist of noncovalent assemblies of subunits. The range of complexities of subunit composition and arrangement is quite broad. The simplest possible case is a protein with only two identical subunits. Each has the same primary and tertiary structure, and the two are related within the quaternary structure by a rotational symmetry axis. An example is liver alcohol dehydrogenase, whose tertiary structure has already been discussed. At a slightly more complex level are proteins with one or two each of several different subunits. Examples are hemoglobin ($\alpha_2\beta_2$) and *E. coli* RNA polymerase ($\alpha_2\beta\beta'\sigma$). Many more complex cases are known. Some proteins have large numbers of each of several subunit types, still with a fixed total size and stoichiometry—e.g., pyruvate dehydrogenase $[t_{24}(p_2)_{12}(f_2)_{12}]$. Others are polymeric structures where the relative composition may be fixed but the overall size is not—e.g., microtubules $[(\alpha\beta)_n]$. Important questions for such proteins are the following: What is the subunit composition? What is the spatial arrangement of subunits? What stabilizes the subunit arrangement? How are the subunits assembled in vivo? What functional roles necessitate (or are expedited by) having subunits? We sketch some of the answers here; others are discussed later in the book.

Symmetry considerations

For quaternary structures with a small number of subunits, it is possible to gain from symmetry arguments some insight about what sort of structures to expect. Table 2-8 summarizes some typical protein subunit compositions. You can see immediately that numbers of copies of identical subunits occur as multiples of 2 or 3 rather than of any other higher prime numbers. This is merely the result of the most likely symmetric ways of packing solid objects in three dimensions. (For a review of symmetry operations, see Box 2-5.)

Let's examine the simplest case of proteins, with only one type of subunit arranged in one of the perfect point-group symmetries. The examples that follow can always be generalized to take care of more complex cases. Each subunit, in general, is an asymmetric object. Furthermore, all of its constituent α-carbons are asymmetric. Therefore, no protein quaternary structure can contain mirror reflections or inversion symmetry axes. The only possible point symmetry operations for proteins are rotations. All possible arrangements of subunits containing only rotational symmetry axes fall into three classes of point groups (Fig. 2-44).

1. *Cyclic symmetry*. These structures possess only a single rotation axis. If n is the number of subunits, rotations of $360/n$ degrees generate the structure. Hence, all subunit centers can be set at the vertices of a regular polyhedron. The point groups of cyclic symmetry are named either n or C_n, depending on which general nomenclature one prefers. C_2 and C_3 symmetries have actually been seen in oligomeric protein structures.

Table 2-8
Subunit composition of proteins with more than one subunit

Protein	Source	Molecular weight of protein	Number of subunits	Molecular weight of subunits
Nerve growth factor	Mouse	26,518	2	13,259
Luteinizing hormone	Sheep	27,322	1	12,500
			1	14,830
Chymotrypsin inhibitor	Potato	39,000	4	9,800
Superoxide dismutase	*E. coli*	39,500	2	21,600
Hemerythrin	*Phascolosoma*	40,600	3	12,700
Galactokinase	Human	53,000	2	27,000
Hemoglobin	Mammals	64,500	2	16,000
			2	16,000
Tu·Ts complex	*E. coli*	65,000	1	41,500
			1	28,500
Malate dehydrogenase	Rat	66,300	2	37,500
Avidin	Chicken	68,300	4	18,000
Troponin	Rabbit	80,000	1	37,000
			1	24,000
			1	20,000
Alkaline phosphatase	*E. coli*	86,000	2	43,000
Procarboxypeptidase A	Bovidae	88,000	1	40,000
			2	23,000
Seryl tRNA synthetase	*E. coli*	100,000	2	50,000
Nucleoside diphosphokinase	Yeast	102,000	6	17,000
Tubulin	Pig	110,000	1	56,000
			1	53,000
Lactate dehydrogenase	Pig	140,000	4	35,000
Tryptophan synthetase	*E. coli*	148,000	2	45,000
			2	28,700
lac Repressor	*E. coli*	160,000	4	40,000
Methionine tRNA synthetase	*E. coli*	170,000	2	85,000
Qβ replicase	*E. coli*	205,000	1	70,000
			1	65,000
			1	45,000
			1	35,000
Arylamidase	Human	223,500	6	38,100
Leucine aminopeptidase	Swine	255,000	4	63,500
Isocitrate dehydrogenase	Yeast	300,000	8	39,000
Aspartate transcarbamoylase	*E. coli*	310,000	6	33,000
			6	17,000
Nitrogenase	*Clostridium*	330,000	2	59,500
			4	27,500
			2	50,700
Enolase	*T. aquaticus*	355,000	8	44,000
Glutamine synthetase	*Neurospora*	360,000	4	90,000
RNA polymerase core	*E. coli*	400,000	2	39,000
			1	155,000
			1	165,000
Apoferritin	Horse	443,000	24	18,500
Glutamine synthetase	*E. coli*	592,000	12	48,500
Ovomacro globulin	Chicken	650,000	2	325,000
Isocitrate dehydrogenase	Bovidae	670,000	16	41,000
Hemoglobin	*Arenicola*	2,850,000	48	54,000
Pyruvate dehydrogenase complex	*E. coli*	5,000,000	24	91,000
			24	65,000
			24	56,000

SOURCE: Condensed from D. W. Darnall and I. M. Klotz, *Arch. Biochem. Biophys.* 166:651 (1975).

2. *Dihedral symmetry*. This symmetry arises when an array has at least one C_2 axis perpendicular to another n-fold rotational axis. The minimal number of subunits in the oligomer will have to be $2n$, where n can take any value. One accepted nomenclature calls these point symmetry groups D_n, where n is the highest-order rotational symmetry present. An alternative nomenclature explicitly notes each symmetry axis. For example, D_2 is denoted 222 because in fact two perpendicular twofold axes are automatically accompanied by a third perpendicular to each of the two (see Fig. 2-44). D_3 is called 32, and D_4 is 422.

3. *Cubic symmetry*. A perfect cube has three fourfold rotational axes (through the centers of opposite faces), four threefold axes (along the diagonals connecting opposite vertices), as well as six C_2 axes (which pass through the centers of opposite edges). The rotational point groups that are classed as cubic all have at least four C_3 axes; these groups are tetrahedral, octahedral, and icosahedral. Figure 2-44 shows some of the symmetry axes expected for such structures.

A simple tetrahedral array of four identical subunits does not have full tetrahedral (T) symmetry. This is because each individual subunit at a vertex is asymmetric, and the C_3 symmetry axes are nonexistent. Thus a simple four-subunit tetrahedron has only D_2 symmetry. To construct a subunit array with T symmetry requires at least 12 identical monomers, arranged in triangles with local C_3 symmetry at each vertex of the tetrahedron. Similarly, an octahedral arrangement of single subunits has only D_3 symmetry. Squares of four subunits must be placed at each vertex to reach full octahedral (O) symmetry. This symmetry thus requires 24 subunits. Note that the octahedron has four threefold axes, three fourfold axes, and six twofold axes; thus it has the same unit symmetry group as a true cube. To generate a true cubic symmetric array of asymmetric subunits also requires 24 subunits, placed three at a time in a triangle at each vertex of the cube. Similar arguments show that 60 identical subunits would be required for full icosahedral symmetry.

In practice, it is rare to have the large number of identical subunits required for cubic point groups. A few possible examples are shown in Table 2-9. However, most oligomeric proteins that show any symmetry at all fall into the cyclic or dihedral classes. The structural implications of the various point groups become a little clearer when the actual binding or contacts at the subunit interfaces are considered. There are two possible modes of contact between two subunits, symmetric and asymmetric (Fig. 2-45). Symmetric contact implies that two identical contacts are made and demands that the two subunits be related by C_2 symmetry. Asymmetric contact implies that two different contacts are used, and therefore contacts on each subunit remain available to generate a higher oligomer. Thus asymmetric contacts naturally generate lines or helices. Where the pitch of the helix is zero and the geometry of contact is just right, successive asymmetric contacts generate oligomers with cyclic symmetries greater than twofold.

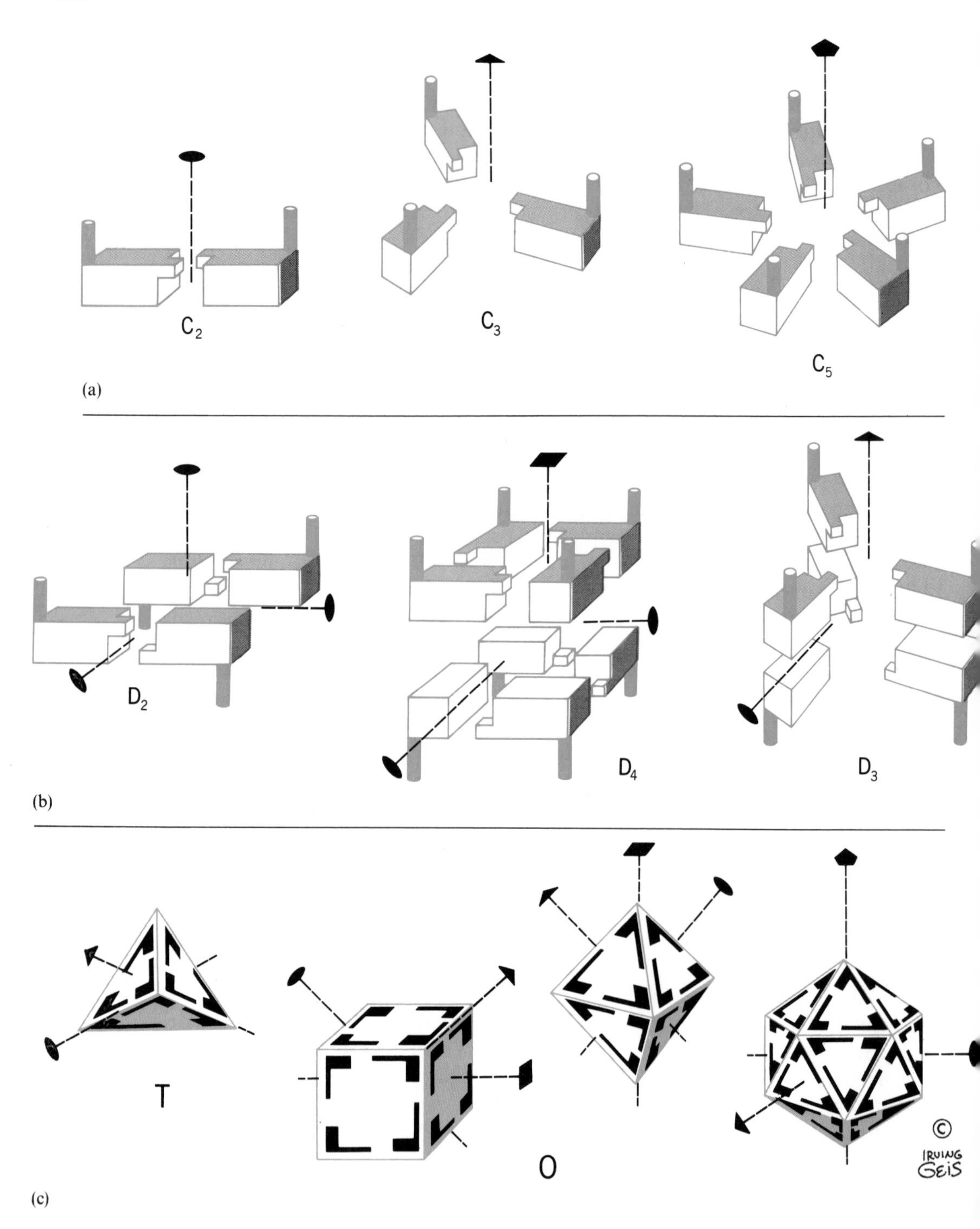

C_2
C_3
C_5
(a)
D_2
D_4
D_3
(b)
T
O
©
IRVING GEIS
(c)

Figure 2-44

Some possible symmetric point groups for arrays of identical subunits. (**a**) Assemblies with cyclic symmetry. Location of a single rotational axis is shown for assemblies with C_2, C_3, and C_5 symmetry. (**b**) Assemblies with dihedral symmetry. Locations of perpendicular rotational symmetry axes are shown for structures with D_2, D_3, and D_4 symmetry. (**c**) Assemblies with cubic symmetry: Some of the symmetry axes present in a cube and in structures with tetrahedral (T), octahedral (O), and icosahedral symmetry. [Drawings by Irving Geis.]

Table 2-9
Nature of subunit interfaces

Protein	Symmetry	Regions in contact	Van der Waals contacts	Hydrogen bonds	Ion pairs
α-Chymotrypsin	C_2	A	443	9	1
		B	57	6	—
Concanavalin A	D_2	‖ to A	—	2	—
		‖ to B	142	14	6
		‖ to C	174	14	—
Hemoglobin					
Oxyhemoglobin	C_2	$\alpha_1\beta_1$	110	5	—
		$\alpha_1\beta_2$	80	1	—
		$\alpha_1\alpha_2$	—	—	—
		$\beta_1\beta_2$	—	—	—
		$\alpha_1\beta_1$	98	5	—
Deoxyhemoglobin	C_2	$\alpha_1\beta_2$	69	1	1
		$\alpha_1\alpha_2$	—	—	2
		$\beta_1\beta_2$	—	—	1
Insulin	D_3	OP	111	8	—
		OQ	99	2	1

SOURCE: Adapted from A. Liljas and M. Rossman, *Ann. Rev. Biochem.* 43:485 (1974).

Box 2-5 SYMMETRY OPERATIONS

Symmetry operations are rotations, translations, or other transformations that leave an object unchanged. Point symmetry operations are a special class of transformations that not only leave an object unchanged, but also leave at least one point in the object stationary. For example, if a 180° rotation exchanges two identical subunits, they are related as shown.

The rotation is called a C_2 symmetry axis. This is a type of point symmetry because the point on the axis itself remains stationary in space. A mirror reflection can also generate a symmetric arrangement of two subunits. This is also a point symmetry because parts of the object at the mirror plane do not move during the reflection.

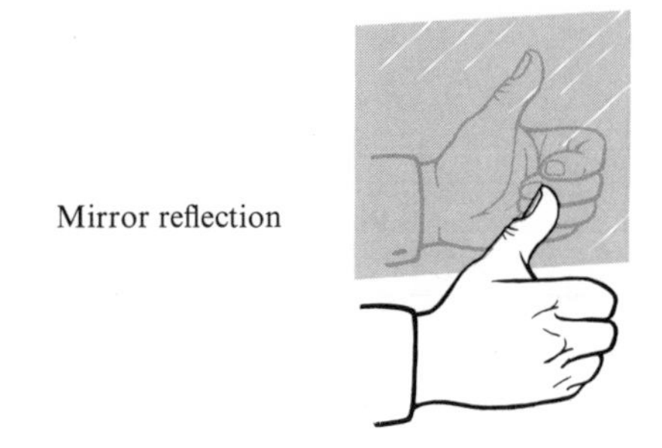

An inversion axis is a simultaneous C_2 rotation and reflection through a perpendicular mirror plane. This too is a point symmetry operation.

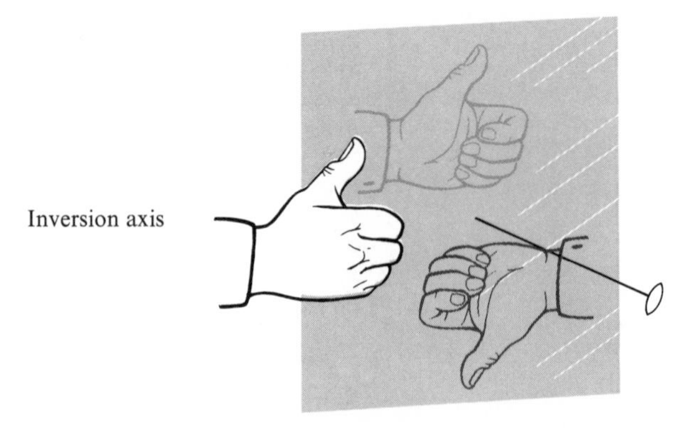

The point symmetry group of an object is a list of all the symmetry operations that leave it unaltered. A completely unsymmetric object is group E. Only the identity operation, which does nothing at all, leaves the object stationary and unchanged. Other examples of point groups are given in the text.

Other symmetry operations involve translations as well as possible rotations or other operations. These are called space symmetry operations. An example is a simple translation along an axis.

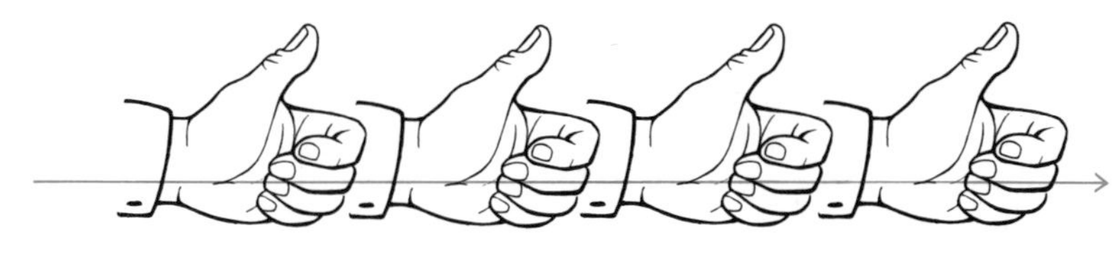

Translation

If a translation is combined with a rotation, the result is called a screw axis. It generates a helix or ribbon. The example shown combines a C_2 rotation with a translation along the C_2 axis. Additional discussion of symmetry is found in Chapter 13. An excellent introduction to symmetry operations is given in P. W. Atkins, *Physical Chemistry* (San Francisco: W. H. Freeman and Company, 1978).

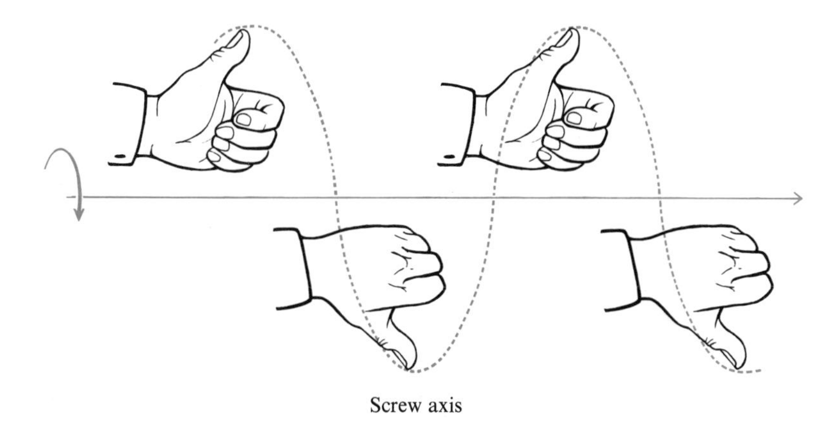

Screw axis

[Drawings by Irving Geis.]

Sometimes the same point group symmetry can be met with different types of subunit contacts. For example, Figure 2-45 shows three arrangements of six subunits, each with D_3 symmetry. However, one structure has two different sets of three symmetric contacts. A second is a tighter structure that has three symmetric contacts and six identical asymmetric contacts forming the two triangular moieties. The final arrangement is the closest packed. It has a total of twelve contacts: six asymmetric and three each of two different symmetric types.

Symmetry arguments are important in describing subunit structures but, when these are examined in fine detail, small but possibly functionally significant deviations from perfect symmetry are sometimes seen for nominally identical subunits. These can appear as small translations of one subunit in the quaternary structure or as small nonequivalent alternations in tertiary structure. Although most proteins consisting of multiple numbers of subunits show symmetrical arrangements, there is nothing a priori that requires this. For example, yeast hexokinase has two identical

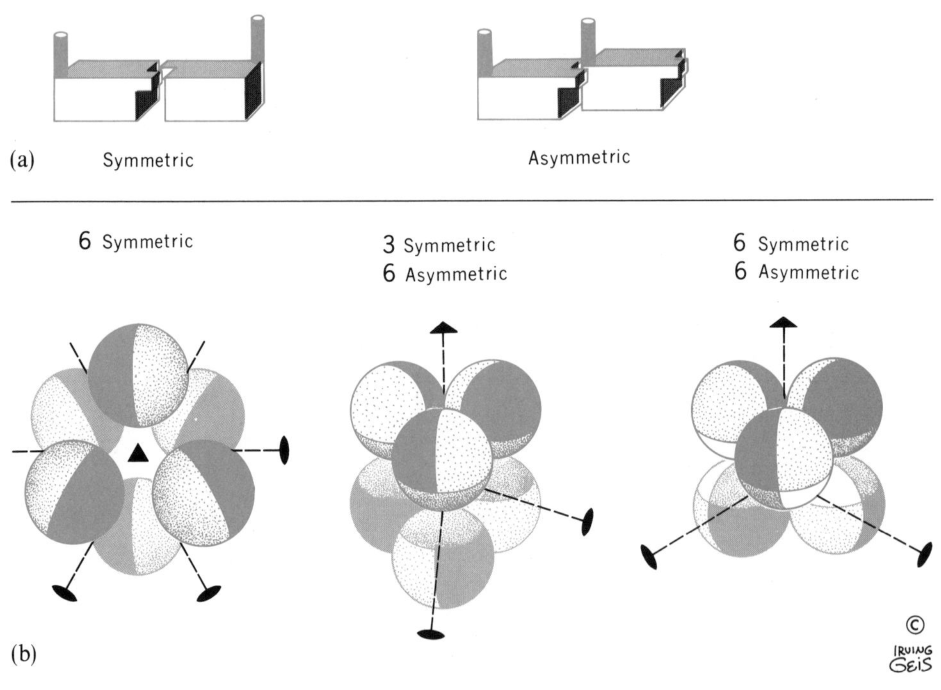

Figure 2-45

Symmetry of contacts between subunits. (**a**) Schematic illustration of a symmetric and an asymmetric contact. (**b**) Three structures with D_3 symmetry, each containing different sets of symmetric and asymmetric contacts. The structures are progressively more compact, and the total number of contacts increases from six to nine to twelve. [Drawings by Irving Geis.]

subunits arranged into a dimer with no point symmetry. The two monomers are related by a screw axis, but their geometry is such that a helical array of many subunits does not form (Fig. 2-46).

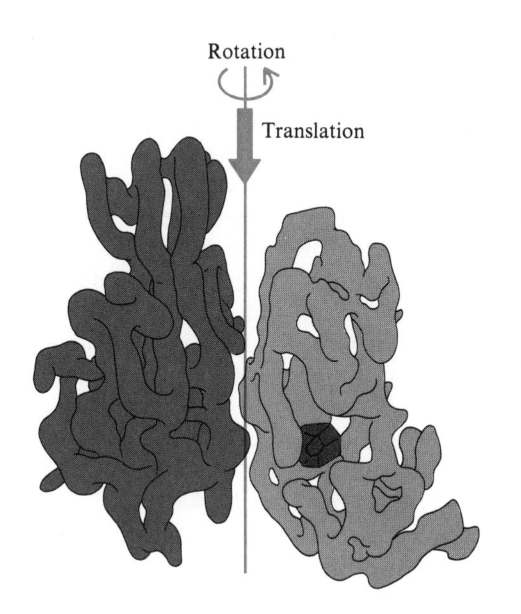

Figure 2-46

Structure of yeast hexokinase at low resolution, showing two subunits related by a screw axis. The polyhedron in the right-hand subunit represents a bound glucose. [After a drawing provided by Thomas Steitz.]

Analysis of the number of subunits and chains

X-ray crystallographic or electron-microscopic evidence, when available, provides the most definitive indication of the arrangement of subunits in the quaternary structure. Low-resolution x-ray data often suffice to give the desired information. Stains specific for individual subunit types, such as protein-specific antibodies, greatly increase the amount of structural information available from electron microscopy.

Much simpler physical and chemical data can aid or replace both types of measurements. For example, a number of the hydrodynamic methods described in Chapters 11 and 12 can provide the molecular weight of the intact quaternary structure (M_q) to accuracies between 1% and 10%, depending on the technique and any complications such as the presence of non–amino acid constituents.[§]

In many cases, one can assume that each subunit will be a single polypeptide chain. Alternative conditions can usually be found that completely separate all the

[§] It is necessary to be sure that no significant dissociation into subunits occurs during the measurements, because this dissociation can increase the errors by an order of magnitude.

individual peptide chains of the quaternary structure. This procedure typically involves rather vigorous denaturants such as sodium dodecyl sulfate and treatment with reducing agents to break any disulfide bonds. Often all the resulting Cys residues are alkylated with iodoacetate or iodoacetamide to prevent the possibility of any disulfides reforming during subsequent measurements. Separation procedures can almost always distinguish the number of different types of chains. For example, electrophoresis frequently can resolve proteins that are absolutely identical save for a single charge difference. The molecular weight of each of the denatured chains (M_c) can be determined by hydrodynamic techniques, although not always with quite the accuracy available for native proteins. An alternative involves chemical methods for determining the number of amino acids per N-terminus. This, together with an amino acid analysis, often can give a fairly accurate chain molecular weight.

To go directly from chain molecular weights to subunit molecular weights is not always straightforward. If there is only a single chain type, then—barring such ambiguities as native subunits consisting of sets of chains cross-linked together by disulfides—there should be a direct correspondence of one chain to one subunit. The number of subunits then is simply M_q/M_c. With proteins composed of more than one chain type, it is necessary not only to know the subunit molecular weight, but also to know the fraction in weight (W_c) of each type in the original quaternary structure. Then the number of chains of each type is $W_c M_q M_c$. This sounds like straightforward quantitative analysis, but sometimes it can be a major stumbling block. For example, one or more subunits could be unstable or difficult to keep in solution, and therefore be preferentially lost in the process of separation or analysis.

The moment there is more than one chain type, it is no longer clear how to assign chains to subunits. Usually it is then necessary to return to the native structure and try to use less violent denaturing conditions to gradually separate it into subunits which retain a native tertiary structure. These must be purified, and then molecular weights and chain compositions must be determined for each subunit type. For example, hemoglobin examined under violent denaturating conditions yields two α and two β chains. However, under mild denaturing conditions, $\alpha\beta$ dimers are a predominant form; free α and β subunits are rarely observed. One can call hemoglobin a four-subunit protein based on its chain composition, but operationally hemoglobin behaves more like a two-subunit protein where each subunit is a two-chain $\alpha\beta$ dimer.

Intersubunit cross-linking is a useful chemical technique for quaternary structure analysis. For simple oligomers in which all the subunits are identical, the protein is treated with an excess of a molecule such as dimethysuberimidate:

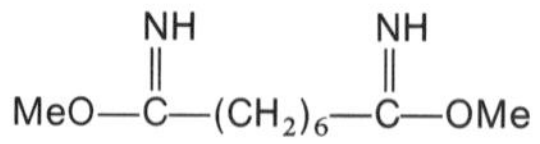

$$\mathrm{MeO{-}\overset{\overset{\displaystyle NH}{\|}}{C}{-}(CH_2)_6{-}\overset{\overset{\displaystyle NH}{\|}}{C}{-}OMe}$$

This molecule will crosslink at lysines with convenient maintenance at neutral pH of positive charges in place of each ε-amino positive charge. The product is

$$R{-}(CH_2)_4NH{-}\overset{\overset{+}{N}H_2}{\overset{\|}{C}}{-}(CH_2)_6{-}\overset{\overset{+}{N}H_2}{\overset{\|}{C}}{-}NH(CH_2)_4{-}R'$$

Many of the cross-links will be intrasubunit, and these will be invisible in most analytical procedures. However, any intersubunit cross-links will lead to species of twice (or higher multiples of) the subunit molecular weight when the protein is treated with vigorous denaturing conditions. Assuming the cross-linking is done on a sufficiently dilute solution to avoid intermolecular reactions, the largest product seen places a lower limit on the number of subunits present.

In some cases, a protein is in dynamic equilibrium with dissociated subunits. This complicates some of the methods we have described, but it facilitates one additional technique that occasionally is quite powerful. Suppose you have two variants of a homo-oligomeric protein. These can be different species, proteins from different tissues—as in the case of heart (H) and muscle (M) chicken lactate dehydrogenase—or a native and a chemically modified species. The important requirement is that the two variants have different charges. In each separate species, there will be equilibria—such as $M_4 \rightleftarrows 4M$ and $H_4 \rightleftarrows 4H$, for example. Mixing the two species will lead to the formation of a whole set of hybrid structures. If there is no preference for self-association of particular variants, the distribution of the hybrids will be binomial. Starting with equal initial amounts of M_4 and H_4, one should see the following ratios of products at equilibrium:

$$H_4\,(1):\;\; H_3M\,(4):\;\; H_2M_2\,(6):\;\; HM\;\;(4):\;\; M_4\,(1)$$

As long as the rate of equilibration is not too rapid, electrophoresis can separate each component of the mixture because of the charge differences on M and H monomers. The number of bands seen is one greater than the number of subunits.

Arrangement of subunits

Solution techniques also can provide information about the arrangement of subunits. For fairly symmetric arrays of only one or two subunit types, a particularly useful approach is to use very mild denaturation to produce fragments that still contain more than one subunit each. Homo-oligomers with cyclic symmetry usually have only one discrete type of intersubunit contact. The intact oligomer will have one contact per subunit, whereas any partial assembly has only $n - 1$ contacts for n subunits. These two considerations imply that dissociation of such an oligomer, caused by any perturbation that weakens subunit contacts, is likely to procede in an all-or-none fashion. That is, one is rather unlikely to observe significant concentrations of subassemblies.

In contrast, oligomeric proteins with more than one subunit type (or with more than one contact type) often show stepwise dissociation. From the composition or organization of the resulting subassemblies, one can sometimes infer features of the subunit arrangement of the original quaternary structure. For example, in hemoglobin, $\alpha_2\beta_2 \rightleftarrows 2\alpha\beta$ occurs readily. In the dimers, the same α is always paired to the same β, implying that one type of α–β contact is more stable than the other. This suggests that the symmetry of the tetramer is C_2. A number of structures consistent with C_2 symmetry and composition $\alpha_2\beta_2$ are shown in Figure 2-47.

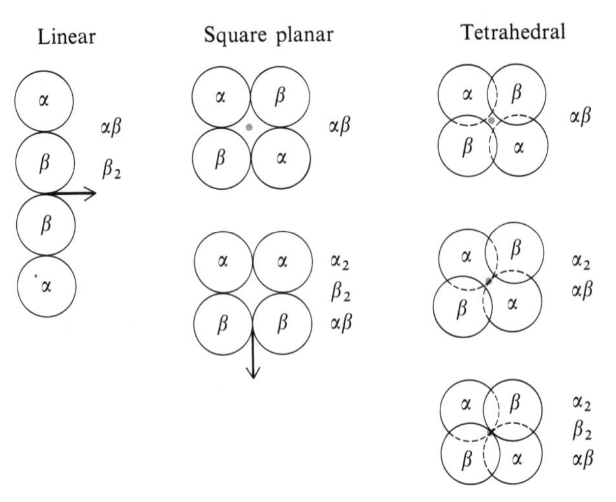

Figure 2-47

Some of the possible subunit arrangements of hemoglobin consistent with the presence of a C_2 symmetry axis (shown as an arrow or dot). Other arrangements can be generated by interchanging α and β chains. To the right of each structure is shown the pattern of dimeric subunits expected from cross-linking with a short reagent.

In more complex cases, such as proteins with six identical subunits, isolation of trimers will immediately exclude many possible structures. With assemblies of many different subunit types, detection of partial assemblies becomes truly important because it can suggest which subunits can be in contact with which others. It is clear though that more direct information usually is desirable. One possibility is to make an accurate determination of the shape. In some cases, this is possible from hydrodynamic measurements (particularly where there is some indication that the isolated subunits have approximately the same tertiary structure as the intact assembly). Then one can analyze only the *change* in hydrodynamic properties upon assembly. We show in Chapter 10 that this can be done with sufficient accuracy to discriminate between linear, square planar, and tetrahedral hemoglobin. It is not sensitive enough to decide among various intermediate cases.

A second approach is cross-linking. With complex assemblies, analysis of cross-linked products can often be difficult, so it is convenient to use a cross-linker that can be broken at will, such as the reducible reagent *bis*-methylthiobutyrimidate:

$$MeO{-}\overset{\displaystyle NH}{\overset{\|}{C}}{-}(CH_2)_3{-}S{-}S{-}(CH_2)_3{-}\overset{\displaystyle NH}{\overset{\|}{C}}{-}OMe$$

If a particular subunit–subunit cross-link can be formed and biological activity maintained, it is fairly certain that the two subunits in question either are in contact or are at least no farther away than the length of the cross-linker. A particularly powerful control is to purify a specific cross-linked subunit pair and reassemble it with unmodified subunits to form a quaternary structure in which every molecule contains the same cross-link. If these molecules are active biologically, it is most unlikely that the cross-link seriously perturbs the structure.

Figure 2-47 shows the pattern of cross-links expected for various possible hemoglobin structures. It is clear that many possible structures can be excluded if a clear set of cross-linked products is produced by a short reagent. Neither subassemblies, shape measurements, nor cross-linking alone can provide a definitive model for the arrangements of subunits, even for a protein as simple as hemoglobin. However, if all three techniques are used in concert, most possible structures can be eliminated.

There are a few techniques that can actually measure specific distances within macromolecules in solution. Two that have a sufficient distance range to be useful for deciphering subunit arrangements are singlet–singlet energy transfer and neutron scattering. These are discussed fully in Chapters 8 and 14, respectively.

The quaternary structure of aspartate transcarbamoylase

It is useful at this point to consider a specific example in detail. *E. coli* aspartate transcarbamoylase (ATCase) is an instructive one because it has the most complex subunit arrangement of all proteins with known tertiary and quaternary structure. The enzyme catalyzes the following reaction.

$$^{2-}OPO_3{-}CO{-}NH_2 + Asp \leftrightarrows NH_2{-}CO{-}NH{-}\underset{}{\overset{\displaystyle COO^-}{\overset{|}{C}H}}{-}CH_2{-}COO^- + HOPO_3^{2-} + H^+ \qquad (2\text{-}4)$$

This is the first step in the biosynthesis of pyrimidines. An interesting feature of the enzyme is that the presence of the pyrimidine CTP dramatically slows the rate of the ATCase-catalyzed reaction for any constant concentration of Asp. This feedback inhibition is an important biological regulatory process. It is sensible that it occurs at the first step in the pathway because this prevents useless (or perhaps even deleterious) intermediates from piling up.

Native ATCase behaves as an 11.3S particle in the ultracentrifuge. A combination of various hydrodynamic measurements produced two completely independent determinations of the molecular weight: 3.0×10^5 and 3.1×10^5 d. This agreement is all one can ask for.[§]

John Gerhart and Howard Schachman used *p*-hydroxymercuriobenzoate to dissociate ATCase into subunits. This reagent covalently reacts with cysteine sulfhydryl groups. This technique generally is useful when, for example, there are sulfhydryls exposed in isolated subunits but sterically inaccessible in the intact quaternary structure. Two products were seen in the ultracentrifuge at sufficiently high mercurial concentrations. These had sedimentation velocities of 5.8S and 2.8S and appeared in an approximate weight ratio of 2:1. When the isolated subunits were examined, a most interesting observation was made. The 5.8S subunit had full catalytic activity in reaction 2.4 but had lost the ability to be inhibited by CTP. The 2.8S subunit had no catalytic activity, but it could bind CTP. This suggested that the 5.8S species is a catalytic subunit (C), whereas the smaller species is a regulatory subunit (R). When C and R were added back together, the original 11.3S species could be regenerated, and the ability of CTP to regulate activity was recovered.

Molecular weights were determined for the separate subunits. C was about 9.6×10^4 and R was approximately 3×10^4. The simplest combination of these weights that was consistent with the native molecular weight and weight ratios of R and C was looked for. It appeared to be:

2C	Total weight	1.92×10^5
4R	Total weight	1.20×10^5
Native C_2R_4	Total weight	3.12×10^5

The agreement with the observed native molecular weight was nearly perfect, and the C:R weight ratio was at least qualitatively reasonable. To maximize symmetry, it was suggested that each catalytic species has two equivalent binding sites for R. This would occur if each 5.8S subunit actually contains two identical 4.8×10^4 molecular weight chains. So, a tentative chain composition of c_4r_4 was proposed.

However, the true situation is more complicated. From primary structure studies, the molecular weight of the r chain was found to be 1.7×10^4, whereas that of the c chain was found to be 3.3×10^4. This means that the C subunit must have a chain composition c_3 and a molecular weight of 9.9×10^4, whereas the R subunit would be r_2 with a molecular weight of 3.4×10^4. Now, recombining these to arrive at the native molecular weight, we find the following.

2C	Total weight	1.98×10^5
3R	Total weight	1.02×10^5
Native C_2R_3	Total weight	3.00×10^5

[§] These values come, respectively, from the Scheraga–Mandelkern equation and from equilibrium ultracentrifugation. (See Chapters 11 and 12.)

The resulting chain composition is c_6r_6. This model is still consistent with the observed hydrodynamic data but, of course, a very different picture of the protein is now emerging.

Two additional experimental findings strengthen this assignment. Several types of ATCase crystals have been obtained. One of these has ATCase in a threefold crystallographic symmetry axis, one has it on a twofold axis, and the most informative one has ATCase placed on a position of 32 (D_3) symmetry in the crystal. This demands that each subunit be present in multiples of six in the quaternary structure. Chemical analysis showed that there are six Zn^{2+} ions per intact 11.3S species. Each Zn^{2+} is associated with one r chain. This information plus the molecular weight data ensure that the chain composition of native protein is c_6r_6 and not some higher multiple.

We can describe the mercurial-induced dissociation of ATCase as $c_6r_6 \rightarrow 2c_3 + 3r_2$. In an elegant series of solution experiments, Schachman and his collaborators asked whether the individual c_3 and r_2 subunits retain their identity in the native quaternary structure. Some of their results are illustrated in Figure 2-48. To prove that the 5.8S subunit has three chains, c_3 was treated with succinic anhydride to form c_3^s, a variant with altered charge. Mixtures of c_3 and c_3^s did not equilibrate in aqueous buffer to give hybrid species, so the subunit structure is stable and the monomer is nonexchangeable. Treatment of c_3^s and c_3 with 2M guanidinium hydrochloride, a strong protein denaturant, produced 1.9S c and c^s monomers. If these were mixed and then the denaturant removed, four products were formed. These could be identified as c_3, c_2c^s, cc_2^s and c_3^s (Fig. 2-48a).

To prove that the two c_3 units retain separate identities in the native ATCase structure, c_3 and c_3^s were mixed with excess regulatory subunits to reconstitute the native quaternary structure. If individual c_3 units became scrambled, one would expect to see seven products, corresponding to compositions c_6r_6, $c_5c^sr_6$, $c_4c_2^sr_6$, etc. These would probably not be resolvable by electrophoresis and, in any event, you can easily show from the binomial coefficients that only 1/64 of the final sample will be pure c_6r_6 or $c_6^sr_6$. What is actually found is three discrete products (in a 1:2:1 ratio) identified as c_6r_6 and $c_3c_3^sr_6$ and $c_6^sr_6$. Therefore, the individual c_3 units retain their identities and are distinct in the quaternary structure (Fig. 2-48b).

Similar experiments were performed on the r subunits. The analysis is somewhat more complex because of two problems. The mercury-treated r_2 is not a stable dimer and exists in a rapid equilibrium with monomers. This is why the original molecular weight determined for r_2 was only 3×10^4. When the original Zn^{2+}-containing form is prepared, conditions can be found where it is more stable, and the actual measured molecular weight by ultracentrifugation is 3.38×10^4. Treatment of isolated r_2 with succinic anhydride led to r_2^s, but this product would not reassemble with c_3 to give native ATCase. Presumably, a lysine residue that is critical for assembly has been lost.

To circumvent this problem, native c_6r_6 was treated with succinic anhydride in the hope that the critical lysine would be protected. This process was successful, and r_2^s subunits isolated from such a preparation will reassemble. However, under reassembly conditions, some dissociation of r_2^s and r_2 still occurs. Therefore, if r_2^s and r_2 are incubated together prior to addition of c_3, a blur of hybrid ATCase electrophoretic variants results. This should represent seven unresolved species. If

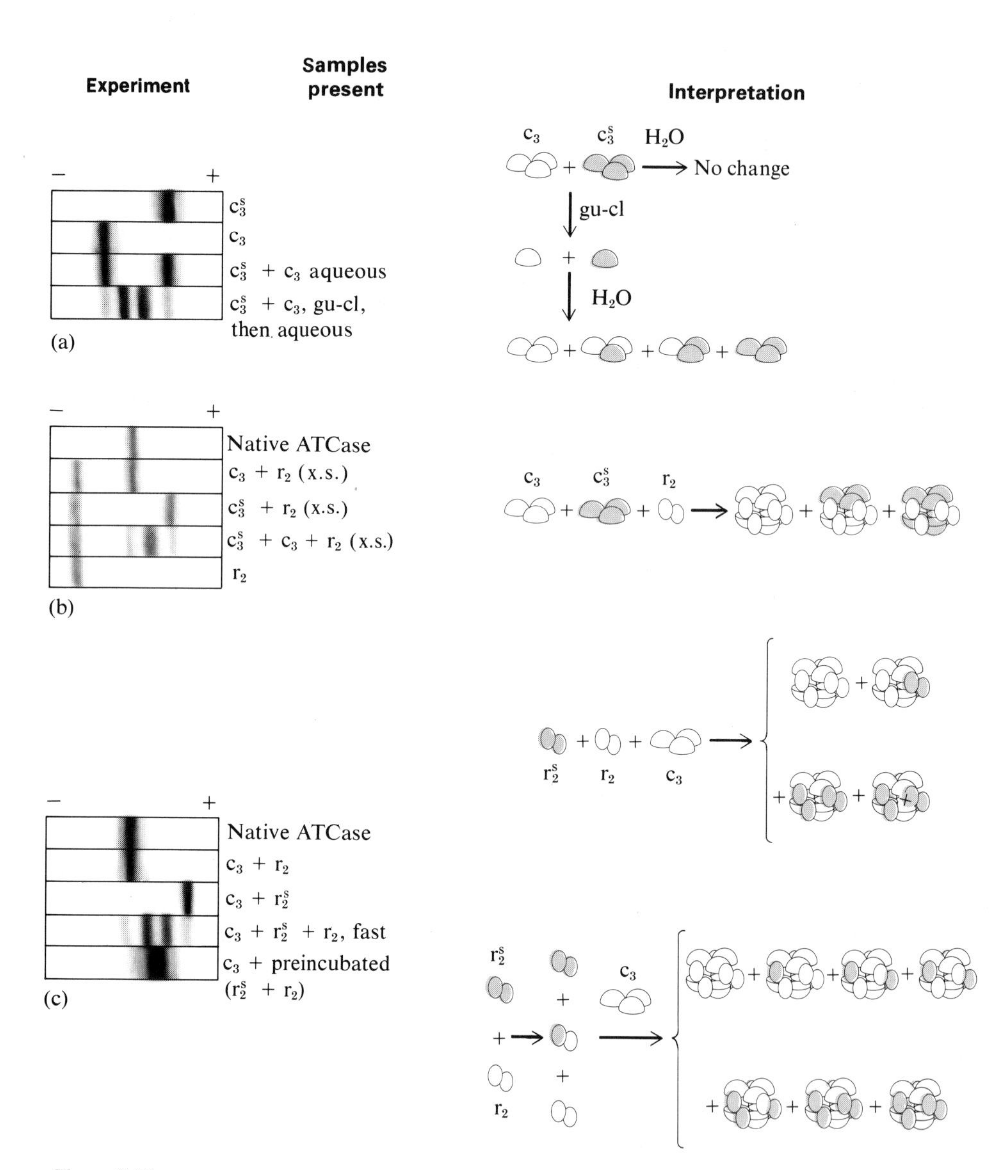

Figure 2-48

Analysis of the arrangement of subunits of aspartate transcarbamoylase (ATCase) by electrophoresis of mixtures of native and succinylated subunits. In each schematic reaction shown, succinylated subunits are shown in color. (**a**) c_3 and c_3^s maintain their identities in aqueous solution, but dissociate in guanidinium chloride. There are three c chains in c_3. (**b**) c_3 and c_3^s maintain their identities during reconstitution with r_2. There are two c_3 units in native ATCase. (**c**) r_2 and r_2^s maintain their identities during reconstitution with c_3, but exchange subunits if incubated together in the absence of c_3. There are three r_2 subunits in native ATCase. [After H. C. Schachman, in *The Harvey Lectures*, vol. 68 (New York: Academic Press, 1974).]

instead r_2^s and r_2 are mixed rapidly with c_3, four discrete electrophoretic bands are seen for the reconstituted enzyme. These correspond to c_6r_6, $c_6r_4r_2^s$, $c_6r_2r_4^s$, and $c_6r_6^s$. This shows that r_2 units retain their identity in the native enzyme and indicates that its subunit composition can be written as $(c_3)_2(r_2)_3$ (Fig. 2-48c). Proof that the two monomers in each r_2 are actually in contact in the native structure came by reassembling active ATCase from pure r_2 dimers that had been covalently cross-linked with a *bis*-imidoester.

The structure of ATCase determined by x-ray crystallography is shown schematically in Figure 2-49. You can see that it fits the solution data extremely well. The two catalytic subunits have C_3 symmetry and are approximately triangular. They are not in contact with each other, explaining why there is no observed tendency for the reaction $2c_3 \rightleftarrows (c_3)_2$ to occur in the absence of regulatory subunits. Similarly, the extent of contact between individual r_2 subunits appears to be rather slight. This is consistent with the failure to detect the reaction $3r_2 \rightleftarrows (r_2)_3$ in the absence of the catalytic subunits. The sandwich structure of ATCase displays a large central hole

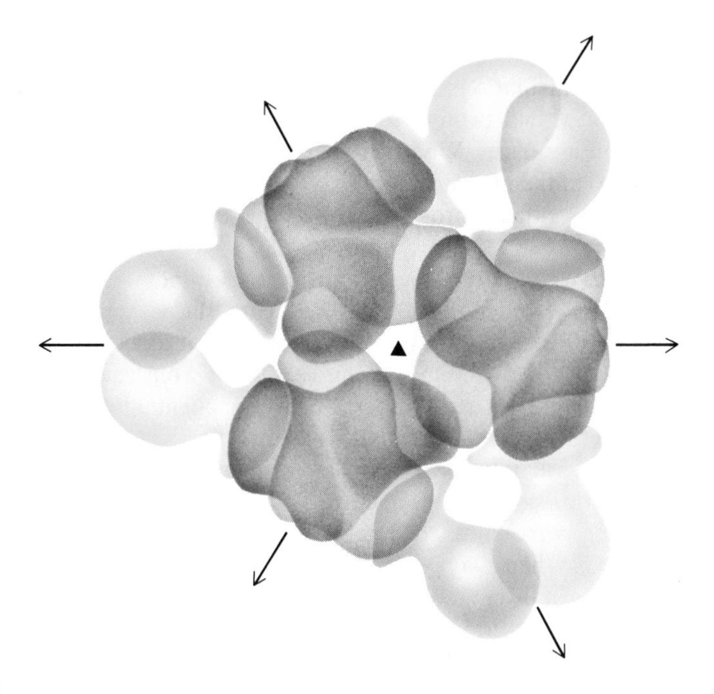

Figure 2-49

Structure of aspartate transcarbamoylase determined by x-ray crystallography. The catalytic subunits extend below and above the regulatory subunits (*colored*). The C_3 axis is shown by a solid triangle; arrows indicate the C_2 axes. For further details see H. L. Monaco, J. L. Crawford, and W. N. Lipscomb, *Proc. Natl. Acad. Sci. USA* 75:5276 (1978).

filled with solvent. This is quite a novel feature. You might infer from the structure that several subassemblies besides c_3 and r_2 ought to be stable or at least metastable. In fact, $c_3(r_2)_3$ can be prepared by adding small amounts of c_3 to a large excess of r_2. The trick is to complex all the c_3 subassemblies quickly to prevent the highly favored reaction $c_3 + c_3(r_2)_3 \rightarrow (c_3)_2(r_2)_3$. When a small amount of r_2 is added to excess c_3, the subassembly $(c_3)_2(r_2)_2$ can be formed. This presumably has a sandwich structure with one-third of the regulatory filling bitten out. Pure $(c_3)_2(r_2)_2$ is readily converted to native ATCase by the addition of one additional equivalent of r_2.

Stability of globular quaternary structures

With ATCase and other examples we have discussed, it is clear that knowledge about subunit contacts is critical to gaining an understanding of what stabilizes quaternary structure. The number of quaternary structures for which high-resolution x-ray data are available is still quite limited. However, it is of interest to examine what is available to get some clues about the forces nature has chosen to use to bind subunits together. Table 2-9 presents a summary of interactions that can be inferred for subunit contacts. You can see immediately that Van der Waals nonbonded contacts predominate overwhelmingly. This observation suggests that there is a good steric fit between the associated subunits. Calculations of packing density were described earlier for internal residues of individual protein subunits. The same approach can be used to examine intersubunit interfaces. When this is done, the results show that the packing density of side chains at the interface is almost the same as packing densities found in amino acid crystals. In other words, the fit between the two protein surfaces is extremely good.

A single Van der Waals contact is worth only a few hundred calories. Because up to a hundred or more contacts form the subunit interface, they may contribute a total of up to 10 to 50 kcal toward the energy of subunit association. However, to bring two subunits together, one must remove solvent from each surface that will form the interface. This causes a net loss of a considerable number of Van der Waals contacts. Thus the net Van der Waals energy of subunit association is too small to be the only significant interaction.

Typical dissociation constants for simple two-subunit systems range from 10^{-8} to $10^{-16}\text{M}\ell^{-1}$. These correspond to free energies of dimerization of -11 to -22 kcal mole^{-1} at 25°C. When two protein subunits are brought together to form a complex, there is a loss of three translational and three rotational degrees of freedom, because the two subunits no longer can move independently. Many side chains free to move on the surfaces of the separate proteins will become immobilized at the subunit interface. These effects will lead to an intrinsic entropy loss that must be overcome by interactions between the two subunits if the complex is to be stable. Estimates of such entropy losses in solution are not especially reliable, but a rough order of magnitude is 70 to 100 cal deg^{-1} mole^{-1}, which corresponds to 20 to 30 kcal of free energy at room temperature. Hence, in order to counteract the entropy loss and still produce the observed negative ΔG^0 of dimerization, the intrinsic interactions between residues of the two subunits must have a strength of -30 to -50 kcal mole^{-1}.

Much of this energy appears to come from the hydrophobic interactions discussed in Chapter 5. It can be estimated that, when two protein subunits come together to form a close fit, 1,000 to 2,000 Å^2 of surface area previously accessible to the solvent becomes buried. If the complementary surfaces consist largely of nonpolar hydrophobic residues, considerable energy can be gained if these are removed from the water. One estimate is an average of 25 cal mole^{-1} for each 1 Å^2 of surface area lost. This leads to hydrophobic contributions of -25 to -50 kcal mole^{-1} for two protein subunits. So, even without a detailed consideration of specific polar contacts between two subunits, one can account qualitatively for the free energy of subunit association mostly from hydrophobic effects.

Sometimes elements of the secondary structure play an important role in the formation of the quaternary structure. In alcohol dehydrogenase (Fig. 2-24), concanavalin A, and insulin, a continuous β sheet passes between pairs of subunits. Note from Table 2-9 the relatively high number of hydrogen bonds found at the interface for the latter two proteins. Sheets also can be used at the interface in a different way. In part of concanavalin A and glyceraldehyde-3′-phosphate dehydrogenase, one surface of a sheet within an isolated subunit is on the surface of the protein. When two subunits come together, their sheets stack on top of each other to form a close-fitting array in which the two sheets are related by twofold symmetry.

Helical quaternary structures

Until now we have been dealing with relatively simple protein quaternary structures. Many of the same considerations apply to more complex assemblies, such as viral heads or helical arrays of subunits found in actin polymers, microtubules, bacteriophage tails, or even in more complex structures such as striated muscle or collagen fibers. However, several new factors must be considered for these polymeric arrays of subunits. To appreciate these factors, it is helpful to begin by considering the thermodynamics of forming a linear array of subunits.

Suppose that the equilibrium constant for adding a single monomer to the end of a chain is k_a, independent of length, as shown schematically in Figure 2-50. Then it is possible to ignore direct interactions between oligomers or polymers.[§] The relative molar concentrations (M) of chains as a function of length can be written

[§] For example, formation of a tetramer from two dimers should have the same equilibrium constant (k_a) as formation of a dimer from two monomers: $(M_4) = k_a(M_2)^2 = k_a^3 m^4$, which is identical with Equation 2-5.

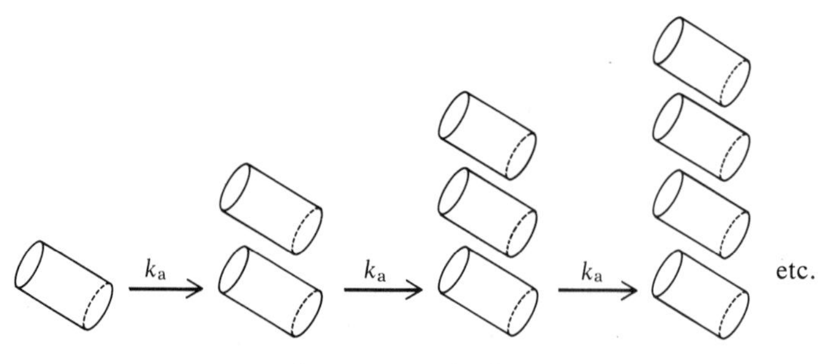

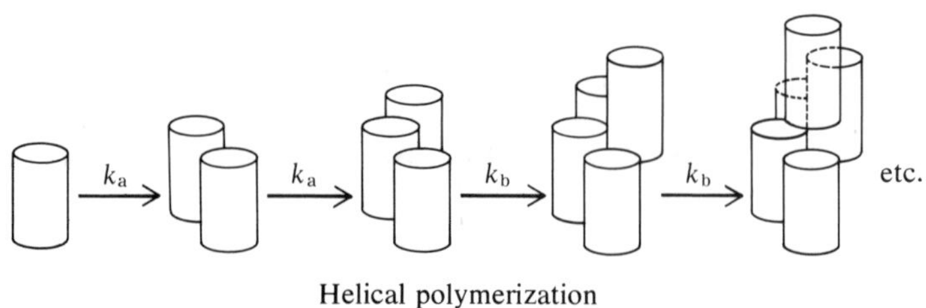

Figure 2-50

Two general mechanisms for polymerization of subunits. The expected behavior for each of these is shown schematically in Figure 2-51 as a function of protein concentration. [After F. Oosawa and M. Kasai, *J. Mol. Biol.* 4:12 (1962).]

simply in terms of the free monomer concentration m.

$$(M_2) = k_a m^2$$

$$(M_3) = k_a m(M_2) = k_a^2 m^3$$

$$(M_i) = k_a m(M_{i-1}) = k_a^{i-1} m^i \tag{2-5}$$

The total concentration of polymer chains (with lengths of 1 or more) is

$$C_{\text{Tot}} = \sum_{i=1}^{\infty} (M_i) = \sum_{i=1}^{\infty} k_a^{i-1} m^i = (1/k_a) \sum_{i=1}^{\infty} (k_a m)^i = m/(1 - k_a m) \tag{2-6}$$

where the sum was evaluated using the fact that the infinite geometric series $\sum x^i$ is just $x/(1 - x)$, when $0 \leqslant x \leqslant 1$ (corresponding to $m < k_a^{-1}$).

The total amount of monomer present in the solution is obtained by summing the monomer content in chains of each length.

$$m_{\text{Tot}} = \sum_{i=1}^{\infty} i(M_i) = \sum_{i=1}^{\infty} i k_a^{i-1} m^i = m \sum_{i=1}^{\infty} i k_a^{i-1} m^{i-1} \tag{2-7}$$

The sum in Equation 2-7 is easily done by recognizing that each term is simply $[d/d(k_a m)](k_a m)^i$. Because the derivative and summation operations commute, Equation 2-7 becomes

$$m_{\text{Tot}} = m[d/d(k_a m)]\left[\sum_{i=1}^{\infty} (k_a m)^i\right] \qquad (2\text{-}8)$$

We have already evaluated this sum in Equation 2-6. Therefore,

$$m_{\text{Tot}} = m[d/d(k_a m)][k_a m/(1 - k_a m)] = m/(1 - k_a m)^2 \qquad (2\text{-}9)$$

Because m_{Tot} usually is known, determination of either m or k_a allows computation of the other from Equation 2-9. Then by Equation 2-5, concentrations of all species of each length can be easily calculated.

The number average degree of polymerization will be the average number of monomers per chain.

$$\langle i \rangle = m_{\text{Tot}}/C_{\text{Tot}} = 1/(1 - k_a m) \qquad (2\text{-}10)$$

We are interested in the behavior of $\langle i \rangle$ as a function of m_{Tot}, the only variable that can be adjusted directly by the experimenter. As m_{Tot} is gradually raised from zero, m increases slowly (as shown by Eqn. 2-9), and therefore $\langle i \rangle$ also increases slowly (as given by Eqn. 2-10). (See the results plotted in Fig. 2-51a.) Any given value of $\langle i \rangle$ will represent a broad distribution of chain lengths.

In contrast, consider the simple model for helical polymerization shown in Figure 2-50. This is drawn for a helix of three monomers per turn, but the same general arguments will apply for any helix. Until the first turn of the helix is completed, each monomer addition can be described by the same association constant as before (k_a). However, once the second turn starts, each new monomer now makes contacts with two preexisting subunits rather than one. Hence, the free energy of association will be altered, and each new addition is now governed by a new association constant k_b. Because subunit contacts usually are energetically favorable, if a structure forms, then $k_b \gg k_a$. Intuitively, you should be able to see that this model will lead to cooperative polymerization. Chains longer than one helix turn should be able to grow at the expense of shorter chains because of the higher binding constants of new monomers to them.

Let j be the length of the shortest helix with tighter monomer binding. Then the equilibria involved in the polymerization reaction are

$$(M_i) = k_a^{i-1} m^i \qquad \text{for } i < j$$

$$(M_i) = k_a^{j-2} k_b^{i-j+1} m^i \qquad \text{for } i \geqslant j \qquad (2\text{-}11)$$

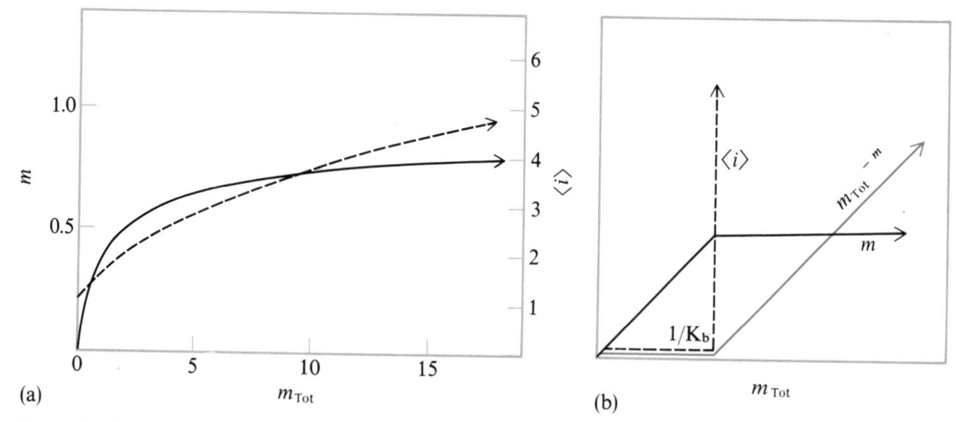

Figure 2-51

Physical features of the two polymerization mechanisms shown in Figure 2-50. **(a)** Linear polymerization. Free monomer concentration m (*solid curve*) and number average degree of polymerization $\langle i \rangle$ (*dashed curve*) rise continuously as the total protein concentration m_{Tot} is increased. **(b)** Helical polymerization. Free monomer concentration (solid curve) rises linearly until a critical concentration and then remains essentially constant. Number average degree of polymerization (*dashed cuve*) is 1 until the critical concentration and then rises sharply to large values. Amount of material in long polymers (*colored curve*) is zero until the critical concentration and then rises linearly with increasing total concentration. [From F. Oosawa and M. Kasai, *J. Mol. Biol.* 4:12 (1962).]

In the example shown in Figure 2-50, $j = 4$. As in the case of linear polymerization, it is useful to compute the total concentration of protein monomers.

$$m_{\text{Tot}} = \sum_{i=1}^{\infty} i(M_i) = \sum_{i=1}^{j-1} ik_{\text{a}}^{i-1}m^i + k_{\text{a}}^{j-2} \sum_{i=j}^{\infty} ik_{\text{b}}^{i-j+1}m^i \tag{2-12}$$

The second term can be put into a convenient form by multiplying and dividing by k_{b}^{j-2}. Then that term becomes

$$(k_{\text{a}}^{j-2}/k_{\text{b}}^{j-2}) \sum_{i=j}^{\infty} ik_{\text{b}}^{i-1}m^i = (k_{\text{a}}/k_{\text{b}})^{j-2}\left(\sum_{i=1}^{\infty} ik_{\text{b}}^{i-1}m^i - \sum_{i=1}^{j-1} ik_{\text{b}}^{i-1}m^i\right) \tag{2-13}$$

The first of these two sums on the right-hand side can be done as shown in Equation 2-8. This procedure lets the total monomer concentration be written as

$$m_{\text{Tot}} = \sum_{i=1}^{j-1} ik_{\text{a}}^{i-1}m^i + \sigma\left([m/(1 - k_{\text{b}}m)^2] - \sum_{i=1}^{j-1} ik_{\text{b}}^{i-1}m^i\right) \tag{2-14}$$

where $\sigma = (k_{\text{a}}/k_{\text{b}})^{j-2}$ is much less than unity.

Now consider what happens as the total monomer concentration of the solution gradually increases from zero. At first, $mk_a \ll mk_b \ll 1$. Thus, the only large term in Equation 2-14 is the first term in the first sum; m_{Tot} is approximately equal to m, which means that all of the added protein remains as monomer. As m_{Tot} is increased up to $1/k_b$, the value of m will increase until mk_b becomes close to one. Note that at this point mk_a is still $\ll 1$, so there is no appreciable concentration of oligomers with a length less than j. The $(1 - k_b m)^{-2}$ term becomes very large as m approaches $1/k_b$. However, m_{Tot} is a finite quantity. Therefore the free monomer concentration can never reach a value as high as $1/k_b$. When σ is very small, m can approach a value so close to $1/k_b$ that experimentally one will measure free monomer concentration as $1/k_b$. At concentrations of m_{Tot} above $1/k_b$, a constant pool of free monomer is maintained at a concentration of $1/k_b$. Any material above this concentration will now incorporate into helical polymers. (These results are illustrated in Fig. 2-51b).

The essential feature of helical polymerization is that a critical protein concentration exists, given approximately by $m_{Tot} = k_b^{-1}$. Below this concentration, all protein exists as monomer. Above it, a constant monomer concentration exists, and all additional protein incorporates into polymers. These polymers will be long. The number average degree of polymerization will be given approximately by $\langle i \rangle = 1/(1 - k_b m)$, by analogy with Equation 2-8. However, because $k_b m \approx 1$, the value of $\langle i \rangle$ is mammoth. The physical reason for this has already been mentioned. Once a critical nucleus of length j is reached, addition of monomers to this nucleus is favored over the creation of new nuclei because $k_b \gg k_a$. In helical polymerization, therefore, one expects to see an equilibrium between long polymers and free monomers. This is precisely what has been observed for systems such as actin and microtubules.

Determination of the length of helical assemblies

The helical polymerization mechanism just described is a simple way to ensure the creation of large aggregates. Other mechanisms with unfavorable initiation steps followed by very favorable propagation steps can produce the same result. None of these mechanisms can produce a very precise distribution of polymer lengths. However, sharp distributions occasionally are observed in vivo in such structures as rodlike viruses and bacteriophage tails. These structures can be reassembled in vitro from their components, and precise length distributions still result. Therefore, to explain the structure formation, it is necessary to consider only macromolecular components actually found in the final assembled structure. A number of different mechanisms can explain such sharp length distributions. A few are shown schematically in Figure 2-52.

In the template mechanism, a single long molecule serves to bind and align proteins. Its length thus is essentially the same as the length of the final structure. This mechanism is clearly operative in the case of many rodlike viruses such as tobacco mosaic virus. Here, each of the three nucleotides of the single-stranded RNA molecule of the virus serves to bind one coat-protein subunit. If the coat protein is

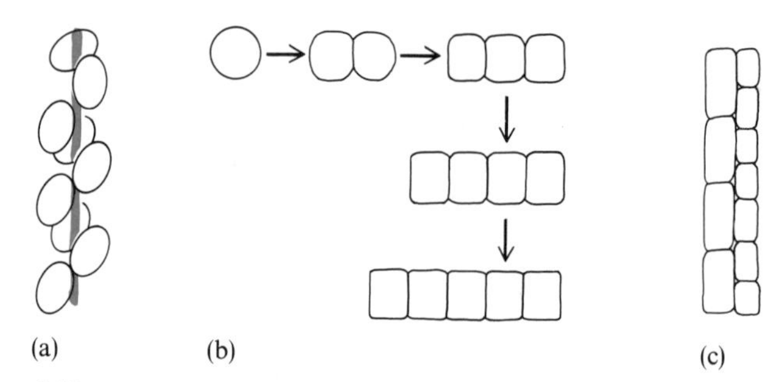

Figure 2-52

Three schematic mechanisms for generating a sharp length distribution in a polymer aggregate. (Each is explained in the text.) (**a**) A core protein or other molecule. (**b**) Accumulated strain. (**c**) A vernier assembly.

allowed to assemble in the absence of RNA, rodlike structures are formed that are very similar to those of the normal viruses. However, the length is now variable. Depending on protein concentration and environmental conditions, average lengths can be longer or shorter than those of the normal virus. In each case though, a broad distribution of lengths is observed if the RNA is absent.

When the self-association of tobacco mosaic virus protein is studied more closely, conditions can be found that stabilize several discrete intermediates in addition to individual monomers and long helical aggregates (Fig. 2-53). Such a figure is called a phase diagram, by analogy with descriptions of macroscopic forms of pure substances or simple mixtures. Because the transitions between various quaternary structures generally occur fairly abruptly with changes in environment, each form can be likened to a macroscopic phase. The results in Figure 2-53 show that the simple helical polymerization model is inadequate to describe tobacco mosaic virus protein association. A more realistic model would have to include additional parameters to account for some of the intermediate states. Nevertheless, any successful model will have to include the principles of nucleation and chain growth to explain the appearance of very long rods.

Two more models capable of producing discrete polymer lengths deserve brief mention. The first model argues that the addition of the first monomer is highly favored, but that each additional subunit binds progressively more weakly. For example, the association constants for the reaction $M_1 + M_i \rightleftharpoons M_{i+1}$ could be of the form $k_i = kx^i$, where $x < 1$. This model is often called the accumulated strain mechanism. The strain free energy (ΔG_s) per subunit contact is just $-RT\ln x$. This energy could be distributed over all the subunits (resulting in a final uniform structure), or it may pile up periodically at certain points. If the increments in strain per subunit added are great enough, a point will be reached where the energy of monomer addition shifts from quite favorable to quite unfavorable. This will produce a rather sharp

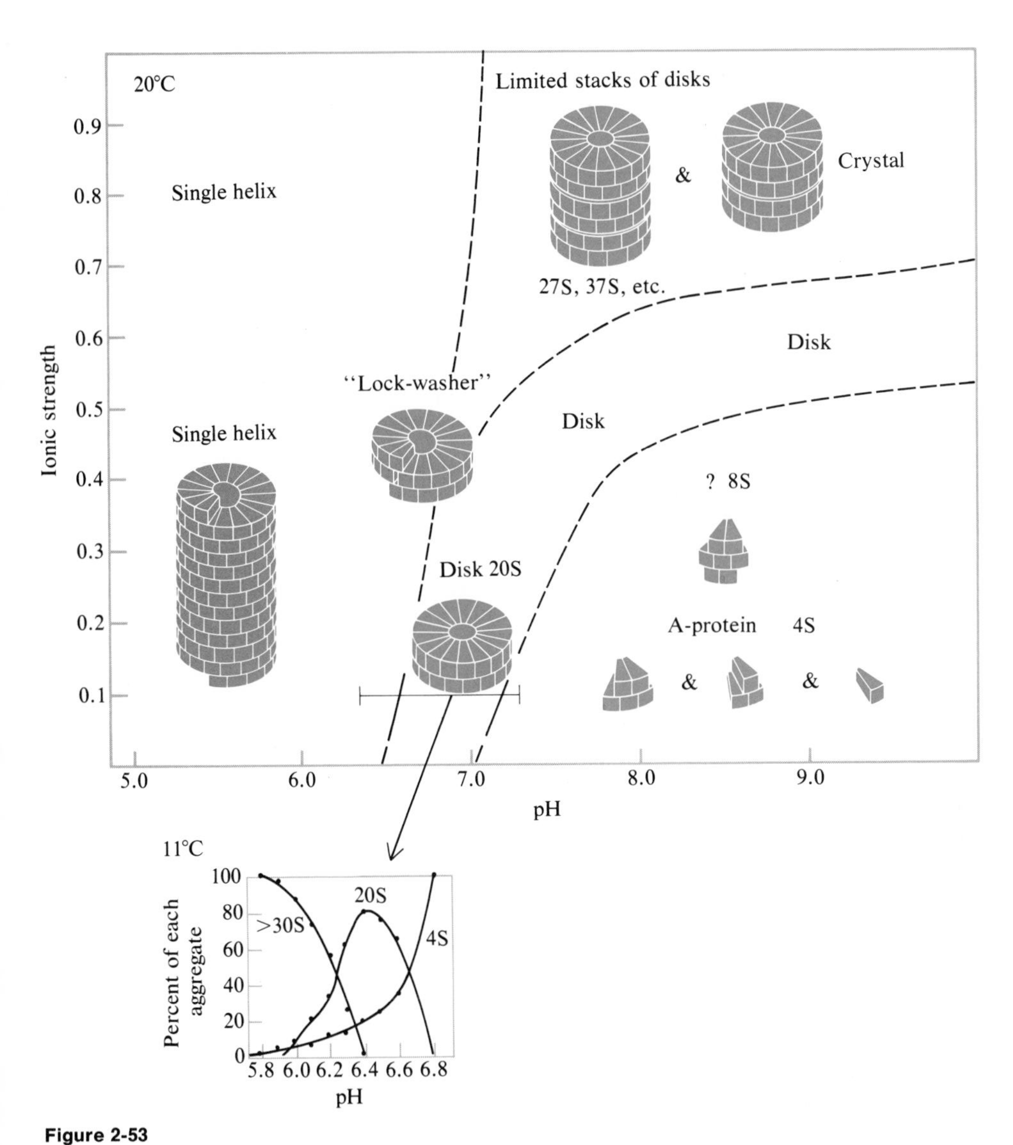

Figure 2-53

A phase diagram for the self-association of tobacco mosaic virus protein at 20°C as a function of pH and ionic strength. The boundaries actually show where a large species becomes detectable, and not necessarily where small species disappear. Thus they are analogous to critical concentrations. The lockwasher form shown on a phase boundary is a metastable form and not a true equilibrium state. Also shown are the results of a typical experiment where the pH is varied at a constant 0.1 ionic strength. The >30S species is a broad distribution of helices with different lengths. [After A. Klug and A. Durham, *Cold Spring Harbor Symp. Quant. Biol.*, vol. 36 (1971).]

length distribution. The strain mechanism may be responsible for the length distribution of some bacteriophage tails where no evidence of a template molecule has been found.

The last model (Fig. 2-52c) is based on the principle of the vernier. If one attempts to align molecules of different lengths, only certain precise numerical combinations will lead to a structure with no dangling ends. Such a model will require that the assembly contain more than one type of subunit—either two or more different proteins, or different tertiary structures of the same protein. Thus far, no system is known with compelling evidence for the vernier model, but it is an interesting hypothesis that must be considered in thinking about some of the more complex biological assemblies.

Many more complex models for length determinations may be operating within a living cell. Fixed points in the cell could serve as initiation or termination regions for polymerization. In addition, the intracellular environment need not be constant; this could favor polymerization in some regions and inhibit it in others. What is clear is that some polymerized structures, such as microtubule and actin filaments, span appreciable distances within cells. They appear to play vital roles in the determination of cell shape, cell motion, and cell division—and possibly in communication between the cell surface and the interior. Factors that regulate the formation or length of such structures clearly are of vital importance in our understanding of cell biology.

Summary

Proteins are built from 20 amino acids, plus occasional extra groups such as metal ions, organic prosthetic groups, and sugars. Amino acids can be roughly divided into two classes: polar and nonpolar.

The amino acid composition differs substantially among various proteins, and some correlation is evident between composition and structure or function. Amino acid sequences are relatively easy to determine and often afford additional insights into structure and function. Furthermore, they are nearly essential if the detailed structure of the protein is to be determined and understood.

Three common secondary structures are found in proteins: α helices, β sheets, and β bends. These collectively make up more than half of the total residues of a typical protein. In proteins where proline is a common amino acid, α helices and β sheets are unlikely, and instead polyproline-type helices may occur.

Tertiary structures of most proteins are organized into a tightly folded globular structure or into several globular structural domains. Within such folded structures, nonpolar groups tend to be internal, and polar residues tend to be external. Extensive networks of hydrogen bonds probably exist between amino acid residues, and between residues and the solvents. Considerable insights into protein structure are available from the comparison of the folding patterns in different proteins. Protein families with similar or nearly identical structures may be fairly common. Particularly striking is the preservation of similar patterns of organization of β sheets and α helices seen among proteins with quite different functions.

Quaternary structures occur in two general types. Globular arrangements of subunits are typical for many proteins. When multiple copies of identical subunits are present, these almost always occur in a symmetric arrangement. Only certain arrangements are allowed because individual subunits are themselves asymmetric. These fall into the cyclic, dihedral, and cubic point group symmetries. Other proteins assemble into helical quaternary structures; here screw symmetries relate the positions of adjacent subunits. Several mechanisms can regulate the length or length distribution of such structures.

Problems

2-1. Proteins known to be equivalent in function and similar in structure are isolated from three sources: aardvark (A), beaver (B), and camel (C). The electrophoretic mobility of each of the three proteins is somewhat different, but each shows a single band. When A and B or A and C are mixed, three bands are seen in electrophoresis; but when B and C are mixed, five bands are seen. Explain.

2-2. Offer a simple justification why the trimer and stacked disk are apt to be relatively stable oligomeric forms of tobacco mosaic virus protein. (See Fig. 2-53.)

2-3. Suppose that a protein composed of a single polypeptide chain and a single subunit is found by x-ray diffraction to possess nearly perfect C_2 symmetry. What can you conclude about the primary structure? about the location of N- and C-termini within the tertiary structure?

2-4. A ribosomal protein with 85 amino acids has been found to have a length of about 150Å on the ribosome. It exists as a single subunit in solution. Comment on the possible secondary structure of this protein when isolated in solution, and when in contact with other proteins in the ribosome.

2-5. A hypothetical protein has the sequence:

N–Ala–Arg–Val–Ser–Met–Lys–Ilu–Glu–Ala–Lys–Gly–Asp–Trp–Thr–Gly–Gly–Gln–Met–Thr–
Gly–Asp–Ala–Asn–Phe–Arg–Ala–Ser–Val–Glu–Leu–C

This protein is found to form a very stable secondary structure and exists in free solution as a dimer. The isolated monomer binds readily to membranes; the dimer does not. Speculate on the secondary, tertiary, and quaternary structure of the protein.

References

GENERAL

Boyer, P. D., ed. 1970. *The Enzymes*, 3d ed., vol. 1. New York: Academic Press. [See especially the following articles: L. J. Reed and D. J. Cox, Multienzyme complexes, p. 213; E. L. Smith, Evolution of enzymes, p. 267; D. E. Koshland, The molecular basis of enzyme regulation, p. 342; E. A. Stadtman, Mechanisms of enzyme regulation in metabolism, p. 395; D. F. Atkinson, Enzymes as control elements in metabolic regulation, p. 461.]

Cold Spring Harbor Laboratory. 1972. Structure and function of proteins at the three-dimensional level. *Cold Spring Harbour Symposia in Quantitative Biology*, vol. 36.

Dickerson, R., and I. Geis. 1969. *The Structure and Action of Proteins*. New York: Harper & Row. [A very clear and well-illustrated introductory treatment.]

Gutfreund, H., ed. 1974. *Chemistry of Macromolecules*. London: Butterworths (Biochemistry Series One, vol. 1). [A very fine set of review articles. See especially the following articles: J. Williams, The primary structure of proteins in relation to evolution, p. 1; H. Muirhead, The three-dimensional structure of proteins, p. 57; M. A. Joynson, The subunit structure of proteins, p. 85; J. P. Knowles, The functions of proteins as devices, p. 149.]

Haschemeyer, R. H., and A. E. V. Haschemeyer. 1973. *Proteins: A Guide to Study by Physical and Chemical Methods*. New York: Wiley. [Covers many techniques and phenomena.]

Neurath, H., and R. C. Hill, eds. 1975– . *The Proteins*, 3d ed. New York: Academic Press. [A multivolume treatise containing many definitive articles of high quality.]

Schulz, G. E., and R. H. Schirmer. 1979. *Principles of Protein Structure*. New York: Springer Verlag. [An advanced text with an exhaustive bibliography.]

SPECIFIC

Chou, P. Y., and G. D. Fasman. 1978. Empirical predictions of protein conformation. *Ann. Rev. Biochem.* 47:251.

Dayhoff, M. O., ed. 1972–1976. *Atlas of Protein Sequence and Structure*. Silver Spring, Md.: National Biomedical Research Foundation.

Fersht, A. 1977. *Enzyme Structure and Mechanism*. Reading, U.K.: W. H. Freeman and Company.

Fessler, J. H., and L. I. Fessler, 1978. Biosynthesis of procollagen. *Ann. Rev. Biochem.* 47:129.

Oosawa, F. 1975. *Thermodynamics of the Polymerization of Protein*. New York: Academic Press.

Perutz, M. F., and H. Lehmann. 1968. Molecular pathology of human hemoglobin. *Nature* 219:902.

Richardson, J. S. 1976. Handedness of crossover connection on β sheets. *Proc. Natl. Acad. Sci. USA* 73:2619.

Rossmann, M. G., and P. Argos. 1977. Taxonomy of protein structure. *J. Mol. Biol.* 109:99.

Schachman, H. K. 1974. Anatomy and physiology of a regulatory enzyme, aspartate transcarbamylase. *The Harvey Lectures* 68:67.

3
Structures of nucleic acids

3-1 PROPERTIES OF NUCLEOSIDES AND NUCLEOTIDES

The monomeric units of nucleic acids are called nucleotides. These consist of phosphate, a sugar, and a purine or pyrimidine base. (Fig. 3-1 shows the structures and names of the common bases and sugars.)

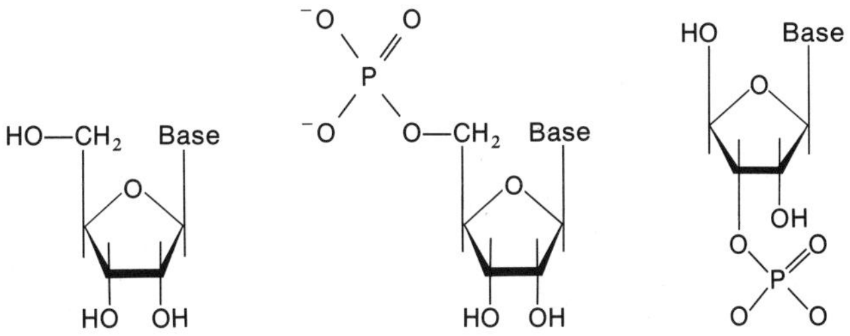

Nucleoside
N

5′-nucleotide
pN

3′-nucleotide
Np

All known sugars in nucleic acids have the D-stereoisomeric configuration shown in Figure 3-1. The common bases and most unusual bases are planar structures with no possibility for stereoisomerism. A few unusual bases (such as the Y base) do have an asymmetric side chain attached. The unit containing just sugar and base is called a nucleoside, and so nucleotides are simply phosphate esters of nucleosides.

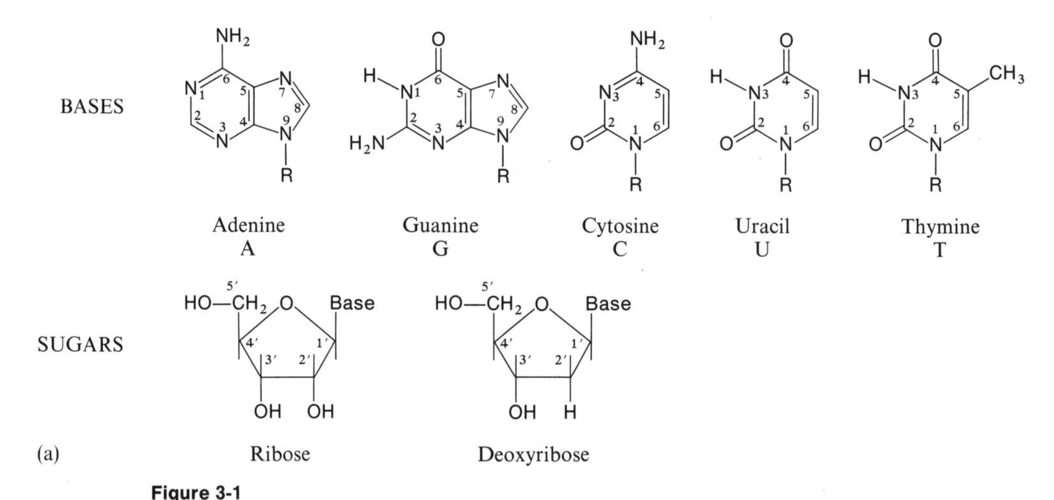

Figure 3-1

Structure of nucleic acid constituents. (**a**) The normal bases and sugars. (**b**) Some of the known unusual bases and nucleosides. Also shown are the standard abbreviations.

Nucleosides are abbreviated by combining an upper-case letter representing the base and a lower-case letter representing the sugar. Thus adenosine (which has a ribose sugar and an adenine base) is called rA, and deoxyguanosine (with a 2′-deoxyribose sugar and a guanine base) is dG. The prefix r usually is omitted where it is clear in context which type of sugar is involved. Nucleotides are abbreviated by adding a p to represent each phosphate, as shown above.

A polynucleotide is abbreviated as

$$\ldots \text{pNpNpNpNpNpNpNp} \ldots \quad \text{or} \quad (\text{pN})_n \quad \text{or} \quad (\text{Np})_n$$

This emphasizes that the backbone of the polymer is made up of alternating sugars and phosphodiesters, as shown in Figure 1-2. In RNA, the sugars all are ribose, and four nucleosides (A, U, C, and G) predominate—although "unusual" nucleosides are quite common, especially in transfer RNA. In DNA, the sugars are all 2′-deoxyribose, and four nucleosides (dA, dT, dC, and dG) make up essentially the total structure of all known DNAs, except for those of a few viruses where a number of unusual nucleosides substitute partially or totally for one normal component. For example, some bacteriophage DNAs contain 5-hydroxymethyl-dC instead of dC, and frequently monosaccharides or disaccharides are attached via the hydroxyl.

Thoughout this book, unless otherwise explicitly noted, we use the convention that a single polynucleotide chain is written as ApBpCpDp . . . pZ so that each phosphodiester bond runs from 3′-linked to 5′-linked sugar hydroxyls, left to right. Therefore, the 5′-end of the molecule is on the left and the 3′-end is on the right. The ends are labeled for antiparallel double strands to remove any ambiguity in strand direction.

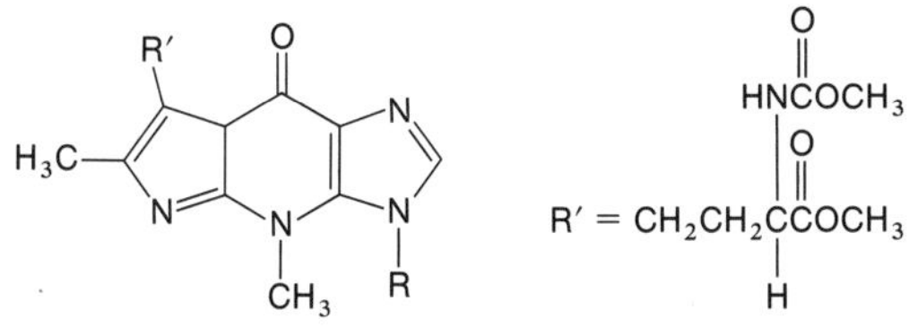

Base Y

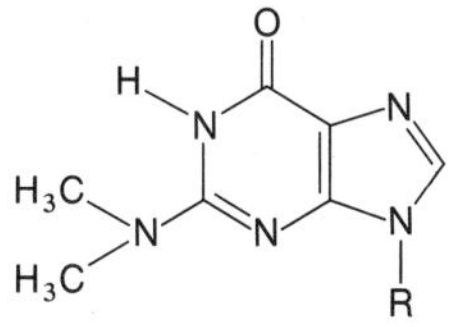

N²-Dimethyl guanosine
2dmG

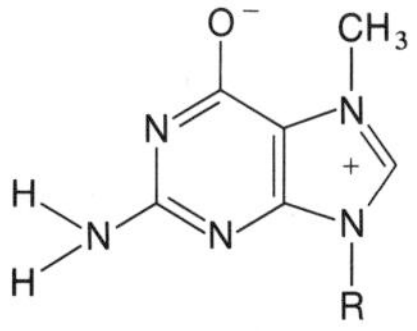

7-Methyl guanosine
7mG

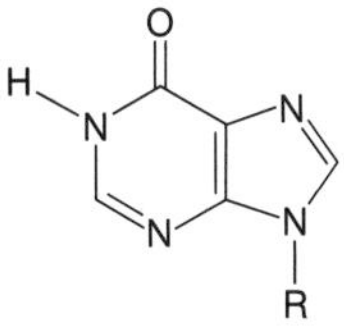

Inosine
I

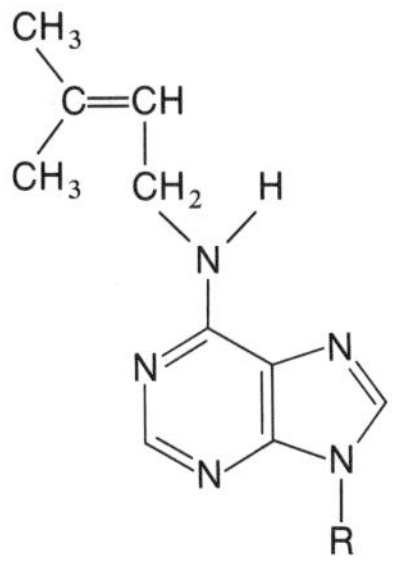

N⁶-Δ² Isopentenyl
adenosine
6iA

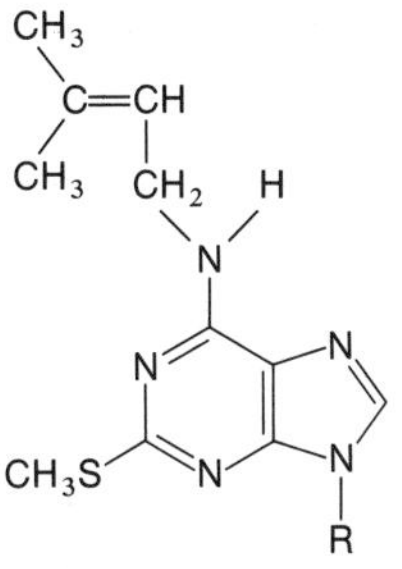

N⁶-Δ² Isopentenyl
2 methylthioadenosine
2ms6iA

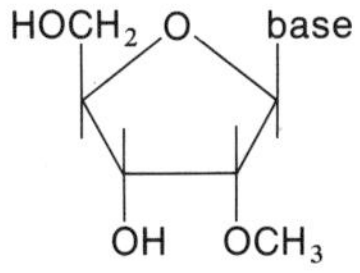

2′O-Methyl nucleoside
2′Om (A, C, G, or U)

(b)

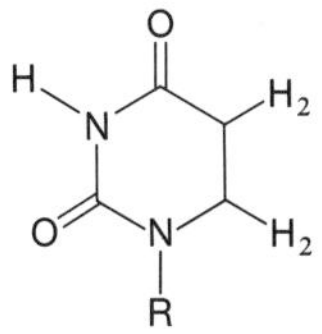

Dihydrouridine
D

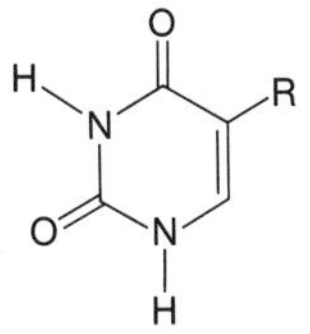

Pseudouridine
ψ

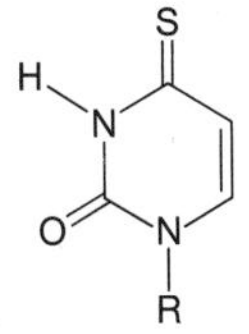

4-Thiouridine
4sU

Ionization equilibria of nucleosides and nucleotides

All of the common nucleosides are uncharged at neutral pH; this is the form shown in Figure 3-1. Most of the unusual nucleosides also are uncharged, although 7-methy-G is positive at pH 7. In contrast, all of the phosphate residues in nucleotides or polynucleotides are negatively charged at any pH at which one normally works with these compounds.

A phosphodiester is a strong acid. The pK_a of the single ionizable proton is around 1:

$$R{-}O{-}\overset{\overset{\displaystyle O}{\|}}{\underset{\underset{\displaystyle OH}{|}}{P}}{-}O{-}R' \rightleftharpoons H^+ + R{-}O{-}\overset{\overset{\displaystyle O}{\|}}{\underset{\underset{\displaystyle O^-}{|}}{P}}{-}O{-}R'$$

Thus in a polynucleotide chain, neglecting end effects, there will be one negative charge for each residue. This high total charge is in striking contrast to typical proteins, in which most residues are uncharged, and in which positive and negative side chains usually are present in comparable (although not equal) numbers.

The phosphomonoester group occurs in isolated mononucleotides or sometimes at the end of polynucleotide chains. It has two ionizable protons with typical pK_a values as shown here:

$$R{-}O{-}\overset{\overset{\displaystyle O}{\|}}{\underset{\underset{\displaystyle OH}{|}}{P}}{-}OH \overset{pk_a \approx 0.9}{\rightleftharpoons} H^+ + R{-}O{-}\overset{\overset{\displaystyle O}{\|}}{\underset{\underset{\displaystyle O^-}{|}}{P}}{-}OH \overset{pk_a \approx 6.2}{\rightleftharpoons} 2H^+ + R{-}O{-}\overset{\overset{\displaystyle O}{\|}}{\underset{\underset{\displaystyle O^-}{|}}{P}}{-}O^-$$

Thus at pH 7, each phosphomonoester has essentially a double negative charge. In the titration of any mononucleotide, pK_a values near 1 and 6 are easily assigned to the phosphate group because these titrations are absent in the corresponding nucleoside. Furthermore, these values are similar to those observed in a simple sugar phosphate. For example, glucose-3-phosphate shows pK_a values of about 0.8 and 5.7.

In nucleosides, a proton with a pK_a of about 12.4 is usually seen instead of the phosphate titrations. This value is due to the ionization of a sugar hydroxyl. In glucose, for comparison, a pK_a of 12.1 has been observed. The negative charge in a nucleotide will shift the sugar hydroxyl pK_a to somewhat higher pH because electrostatic repulsion within a triply negatively charged deprotonated form will be considerable. As a result, the hydroxyl titration probably falls outside the pH range that is accessible for study without risking hydrolysis of the nucleotide. Thus these titrations have not been observed in nucleotides or polynucleotides.

The titration behavior of the base moieties of nucleotides or nucleosides is quite complex because most contain a large number of potential proton donor sites or acceptor sites. Before discussing this, we should note that, even in the un-ionized bases, it is not obvious at first that there should be a unique location of protons on the heteroatoms. Each base in principle will exist as a set of tautomers. A few of the equilibria involved are shown in Figure 3-2. From all available crystallographic, spectroscopic, and electronic data, it seems clear that the keto-amino tautomers shown in Figure 3-1 predominate in aqueous solution, perhaps to the virtual exclusion of most other possible tautomers.

NMR and IR spectroscopy are particularly effective techniques for assigning tautomeric forms because spectral bands from individual protons are often seen.

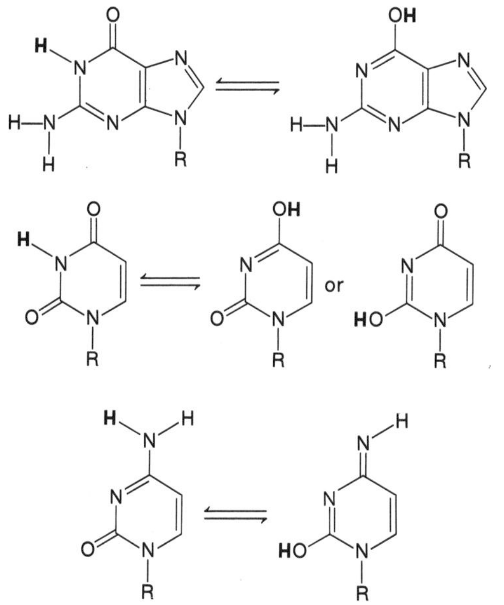

Figure 3-2

Some possible tautomeric equilibria for G, U, and C at pH 7.

These bands can be assigned by comparison with appropriate derivatives where the possibility of tautomerization is eliminated—for example, by replacing some protons with methyl groups. The energy differences between the favored tautomeric forms and some of the alternatives shown in Figure 3-2 are predicted to be small. Therefore, it remains an open question whether unusual tautomers play any role in the structure or function of nucleic acids. At present no direct evidence exists.

When a proton is added to or removed from a nucleic acid base, the product usually still has a range of plausible tautomeric forms. Identifying the most stable of these is equivalent to asking which protons ionize, or which sites accept protons, at pH values correlated with the observed pK_a values. It has taken a substantial amount of experimental work to answer these questions; Figure 3-3 summarizes the results for the four common ribonucleosides. Note that A and C can be found as neutral or positive forms, whereas U is neutral or negative, and G can exist in all three states. The presence of a phosphate shifts all of the pK_a values shown in Figure 3-3 up by a few tenths of a pH unit. This shift is due to the electrostatic attraction between the phosphates and any positive forms, and to repulsion between the phosphates and any negative forms. The observed shifts are larger in 5′-nucleotides than in 2′- or 3′-isomers. This is reasonable because the 5′-position is closer to the ionizable sites on the bases.

Deoxyribonucleosides and deoxyribonucleotides show a close correspondence in all of their titration properties with the corresponding ribo components, except that the pK_a values for the deoxyribo components are generally a few tenths higher.

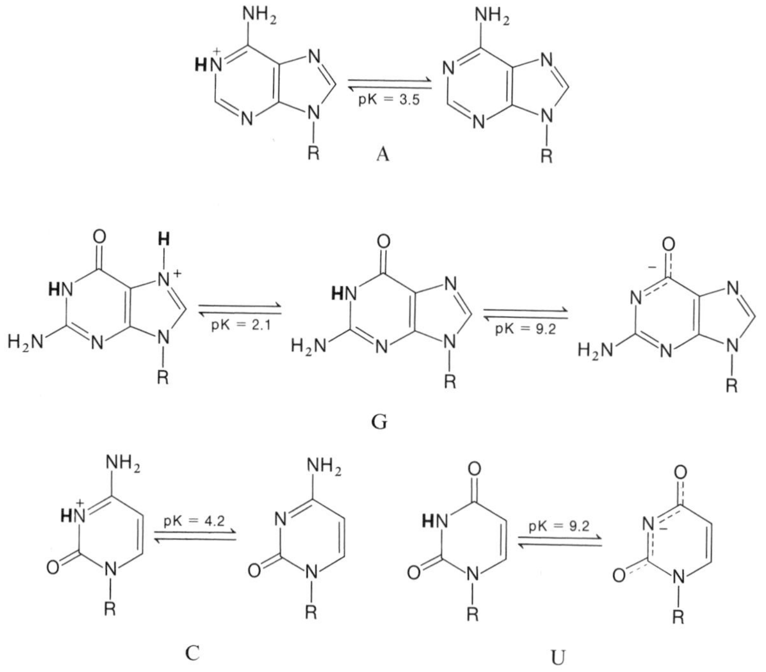

Figure 3-3

Ionization equilibria for the four ribomononucleosides. Equivalent ionizations are found for the deoxymononucleotides.

In oligonucleotides or polynucleotides, all of the observed pK_a values should shift somewhat, due to electrostatic interactions between residues. Larger and more complex effects are observed when the ionizable groups are involved in secondary structure formation through base pairing. Then, as we discussed for proteins, proton titrations become coupled to conformational transitions, and very large shifts in apparent pK_a values are possible. Some examples of this are described in Chapter 22.

The common nucleic acid bases are all similar in a number of basic chemical and physical properties. For example, all are observed or predicted to have rather large dipole moments ranging from 2.5 to almost 7 debyes (D). Thus, from electronic considerations alone, each must be called quite polar. In spite of this, all the common bases except uracil are rather insoluble in water. Purines are less soluble than pyrimidines by a factor of 5 to 10, but this is a smaller range of properties than that seen for amino acid side chains. Thus, loosely speaking, it may be fair to describe all the bases (slightly paradoxically) as polar, hydrophobic moieties. In any case, there is no point in trying to make classifications of nucleotide residues as we did for amino acids in Chapter 2.

3-2 COMPOSITION OF NUCLEIC ACIDS

Chargaff's rules and DNAs

Most DNAs are double-stranded. This imposes the constraint on the base composition that $\chi_A = \chi_T$ and $\chi_G = \chi_C$, where χ represents mole fraction. This fact, first noted and studied extensively by Erwin Chargaff, was one of the important pieces of data that led to the conception of the double-helical structure of DNA. It means that in a fully double-stranded DNA only a single variable, χ_{G+C} $(= 1 - \chi_{A+T})$, is needed to describe the composition. Natural DNAs have quite a range of G + C contents. Such extremes as 75% G + C exist in some bacteria. Presumably, the degeneracy of the genetic code (plus the fact that some of the DNA may not code for protein sequences) allows such a wide range to occur without seriously hampering an organism's ability to store its necessary genetic information. In mammals, χ_{G+C} is around 40% to 45%, and it generally is found that closely related organisms have similar DNA base compositions. Quite a few minor components of eukaryotic DNA can be purified with very extreme base compositions. These are called satellite DNAs; the crab satellite, for example, has a mole fraction G + C of only 5%. The biological role of such satellites still is not well understood but, where they have been examined in detail, the sequences generally are found to consist mostly of regularly repeating short units.

Some DNAs have abnormal bases replacing one (or more) pyrimidine or purine. For example, in T-even phage DNAs we have already mentioned the prevalence of glycosylated 5-hydroxymethyl-C. This still can base-pair as C, and Chargaff's base-composition rules still are obeyed if all C analogs are included in χ_C. Phage-SP2 DNA has dU instead of dT, and a few other DNAs with unusual pyrimidines are known. Occasional bases on some DNAs are found to be methylated at very specific sequences. This occurs as a protection mechanism against intracellular restriction nucleases that, in the absence of methylation, would cleave the DNAs at these positions.

A number of DNAs deviate from Chargaff's rules. For example, the DNA of bacteriophage ϕX174 has the base composition

$$\chi_A = 0.25, \quad \chi_T = 0.33, \quad \chi_G = 0.24, \quad \chi_C = 0.18$$

This means that a maximum of (0.18 + 0.25)2 = 0.86 of the total bases can potentially be involved in double-stranded structures formed from normal A·T and G·C base pairs. In actuality, ϕX174 DNA contains only a single covalent strand, unlike most DNAs. There is no evidence that the sequence is organized to allow highly efficient intrastrand base pairing to occur. In practice, considerably less than the 0.86 maximum base pairing seems to occur. Note that, just because a DNA obeys Chargaff's rules, it need not necessarily be double-stranded. That ability will depend on the actual sequence, and it can be tested by various physical techniques (Chapter 22).

RNA base compositions

The analog of Chargaff's rules for RNA would be $\chi_A = \chi_U$ and $\chi_G = \chi_C$. None of the major types of cellular RNA—such as transfer RNA (tRNA), messenger RNA (mRNA), ribosomal RNA (rRNA), or heterogeneous nuclear RNA (hnRNA)—obey this rule. This is to be expected because all have single-stranded covalent structures. A few RNAs of viral origin do obey Chargaff's rules, and some of these turn out to be perfectly double-stranded molecules. The base compositions of most RNAs imply that intrastrand folding cannot result in the base pairing of all nucleotide residues. Furthermore, certain of the unusual bases are so modified that normal base pairing is chemically impossible. It is found in practice that, although most RNAs contain a considerable amount of base pairing, they rarely reach the maximum possible extent of such pairing.

RNA biosynthesis begins at the 5′-end by condensation of two pppN structures (where N is G or A). It proceeds stepwise toward the 3′-end until termination. Thus an initial RNA transcript should contain a 5′-triphosphate and a free nonphosphorylated 3′-end. In practice, the 5′-triphosphate is virtually never observed because most RNA species present in large quantities in vivo are synthesized by the degradative processing of larger RNA precursors. It is more difficult to draw conclusions about the 3′-end. Were a phosphate to be observed here, one could be fairly sure that processing had occurred at this end also. But a 3′-phosphate is almost never seen in biologically active samples. However, many known nucleases can cleave a phosphodiester bond to yield a free 3′-hydroxyl, so one cannot be sure whether finding such an end implies original synthesis or degradative processing.

Viral RNAs seem to consist totally of the four normal bases, as do prokaryotic mRNAs. The mRNAs from eukaryotes and their viruses have occasional methylated bases, and in addition many of them have an unusual capped structure at the 5′-end:

$$G^{5'}pppNpMp\ldots(pA)_n pA^{3'}$$

This structure contains a 5′–5′ triphosphate bridge. It is synthesized after the mRNA has been transcribed, and it appears to be important for translation. Many of these mRNAs also have an unusual feature at the 3′-end: a stretch of 100 or more continuous pA residues. These also are added after transcription in most (if not all) cases.

Ribosomal RNAs have normal unmodified ends but have a larger fraction of unusual bases. In transfer RNA, the fraction of strange bases and nucleosides can approach 20% of the total. It is found that more evolutionarily advanced organisms tend to have a larger fraction of unusual nucleosides than do more primitive ones. All available evidence suggests that the whole spectrum of modifications of the normal bases occurs after the precursors of tRNA or rRNA are transcribed.

There is no evidence that the overall base composition of RNA or DNA correlates in any significant way with biological function. However, some of the physical properties of double strands, such as thermal stability and density, do vary roughly linearly with mole fraction G + C. This relationship frequently is exploited for a variety of analytical and preparative techniques (Chapter 22). In addition, it seems reasonable to expect that nature has learned to exploit the relative ease of strand separation of high χ_{A+T} regions of double strands and uses such regions for specific functions. If this idea has any general validity, it remains to be proven.

Other constituents of nucleic acids

Nucleic acids appear to have few constituents other than sugars, phosphates, and bases. Some RNAs show a number of tight binding sites for divalent cations. In vivo these are presumably occupied by Mg^{2+} and polyamines. As polyanions, nucleic acids are expected to associate tightly with polycations of the appropriate dimensions Polyamines such as spermine and spermidine may play important roles in the action of DNA or RNA in vivo, and some evidence exists for highly specific localization of these substances on tRNA in vitro. Besides these cations, nucleic acids seem totally devoid of specially synthesized and attached prosthetic groups. However, it must be kept in mind that, within the cell, nucleic acids rarely exist as free molecules. They are almost always tightly associated with specific proteins. We discuss later the basic characteristics of these nucleoprotein complexes.

There is one known specialized covalent interaction between amino acids and nucleic acids. This is the aminoacylation, or charging, of tRNA. Each tRNA species can be recognized by a particular aminoacyl-tRNA synthetase that carries out reactions such as

$$\text{ATP} + \text{tRNA—CpCpA}\begin{matrix}\diagup\text{OH}\\ \diagdown\text{OH}\end{matrix} + \text{NH}_3^+\text{—CHR—COO}^- \rightleftharpoons \text{AMP} + \text{PP}_i + \text{tRNA—CpCp}\overset{\text{OH}}{\overset{|}{\text{A}}}\text{—O—CO—CHR—NH}_2$$

Acylation is known in some cases to occur specifically at either the 2′-or the 3′-hydroxyl of the terminal adenosine (the two OHs shown explicitly above). However, acyl migration between 2′- and 3′-positions of the sugar is fairly rapid and may be exploited as part of tRNA's involvement in the mechanism of protein synthesis. It is interesting that several plant viral RNAs have covalent structures essentially identical to tRNA at their 3′-ends, and some synthetases can aminoacylate these viral RNAs. The significance of this reaction in vivo is unclear.

3-3 PRIMARY STRUCTURE

Covalent chain structure

The number of polynucleotide strands in a nucleic acid is determined by subjecting the native molecule to conditions that disrupt all base pairing—such as high-temperature organic solvent (or, in the case of DNA, alkaline pH[§])—and then measuring the molecular weights and amounts of each of the resulting single-strand pieces. Some of the types of covalent chain structure that have been seen in natural nucleic acids or have been prepared synthetically are shown in Figure 3-4.

Almost all RNAs consist of a single covalent strand. One possible exception is the 28S ribosomal RNA in eukaryotes, which appears to contain a short 5.8S strand bound noncovalently. Another exception is the RNA of some tumor viruses, which seems to exist as a dimer of two identical strands. Evidence for this is shown later.

The most common DNAs contain two single strands of the same length with complementary sequences allowing for full base pairing along the entire length of the molecule. Sometimes, the difference between the base compositions of these strands allows them to be separated physically and studied individually. A few DNA duplexes can be shown to contain single-strand breaks when the strands are separated in vitro. Best studied is T5 DNA, where one strand is whole but the complementary strand is really four different pieces ordered perfectly along the intact strand. Some natural DNA strands are circular and some are linear. Even interlocked DNA circles have been observed occasionally.

No natural DNA has yet been found to exist as a hairpin structure folded back on itself to make a perfect double strand except for a few bases in the loop. However, synthetic DNA hairpins can be prepared easily. Base-composition measurements are not nearly accurate enough to distinguish such structures from duplex double strands. However, measurements of molecular weight of the denatured form would easily reveal the difference.

The size of nucleic acid strands varies enormously. The shortest natural RNAs are transfer RNAs (tRNAs) with 75 to 84 residues. The longest RNAs are in the heterogeneous nuclear RNA (hnRNA) class and may contain up to 2×10^5 residues.

The smallest DNA strands thus far found have only a few thousand residues, whereas the longest may run the length of a whole eukaryotic chromosome and contain up to 10^8 residues. Double-stranded molecules are known in which one strand is RNA and the other is DNA. These are formed as intermediates when reverse transcriptase synthesizes a DNA complement to an RNA strand. They are also presumably a transient intermediate over at least a short region in the transcription of RNA from DNA. RNA–DNA hybrids can readily be made in vitro by annealing two strands with complementary sequences. A few cases are now known in which RNA and DNA sections on the same strand are covalently attached. This occurs at the start of the synthesis of DNA strands by reverse transcriptase because this enzyme uses an RNA primer to initiate the synthesis of new DNA.

Do you wonder why nature has not chosen to use molecules in which single ribonucleotides and deoxyribonucleotides are interspersed in a single strand? As discussed later, ribo and deoxyribo compounds prefer to form double helices with quite different geometries.

[§] Because of the 2′-OH *cis* to the 3′-phosphodiester linkage, alkali hydrolyzes RNA chains. Thus alkali can be safely used only for DNA.

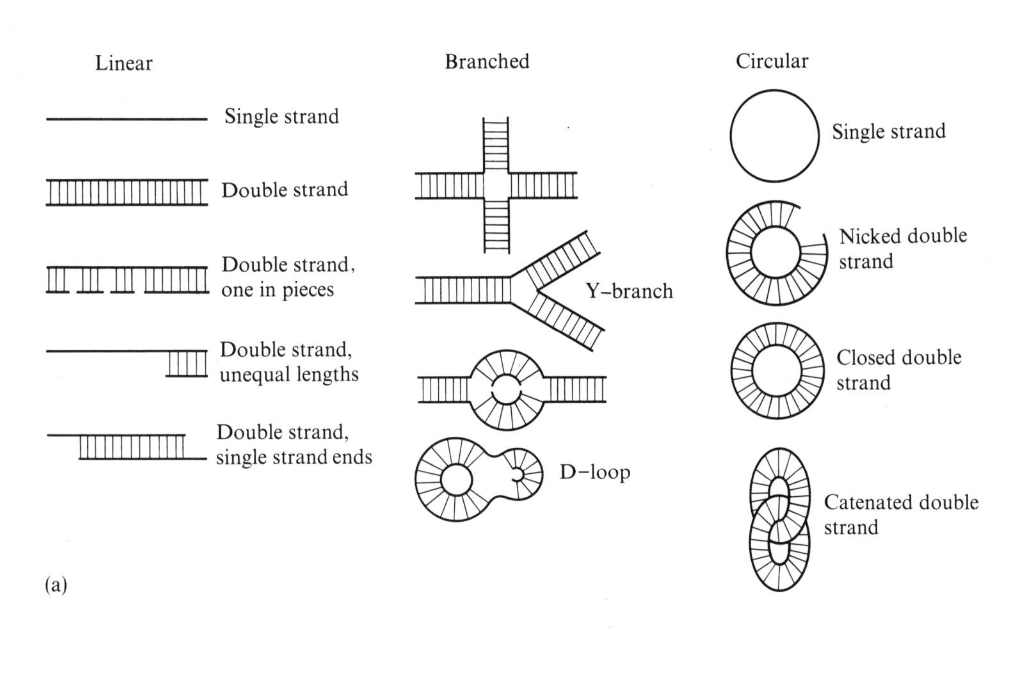

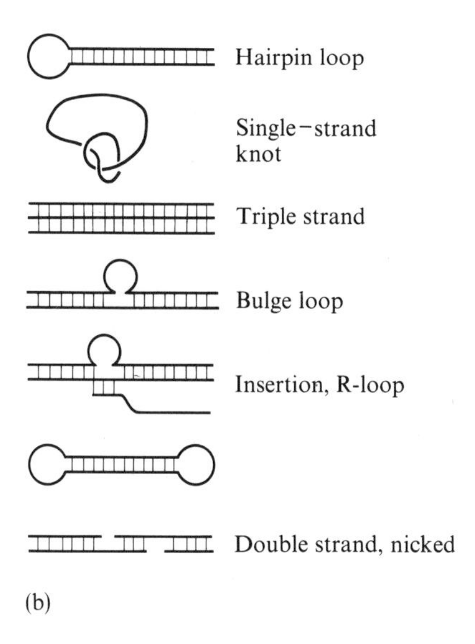

Figure 3-4

Some covalent nucleic acid chain structures. Regions with perfect base pairing are shown hatched. (**a**) Structures that have been observed in naturally occurring nucleic acids. (**b**) Structures that have been prepared synthetically.

Nucleic acid sequences

The actual sequence of bases is known for quite a few of the shorter RNAs, and the sequences of several larger RNAs have also been determined. All of this can be accomplished by the tedious classical methods of digestion with nucleases of overlapping specificity, exactly analogous to the way most protein sequences have been learned. However, new and elegant techniques expedite matters considerably.

Until very recently, DNA sequencing was an almost impossible task, not just because of the size of the molecules but also because suitable nucleases were unknown. Often one had to resort to transcribing sections of the DNA into RNA, or somehow incorporating occasional ribonucleotides into DNA in order to use available ribonucleases to obtain the necessary partial digests. However, in the past few years, a host of highly specific restriction endonucleases have been discovered. They cut double-stranded DNA at unique four- to six-base-long sequences and allow long molecules to be dissected into manageable pieces. New and very powerful techniques have been developed to sequence these restriction-nuclease fragments. The key feature of these methods is the ability to separate (by electrophoresis) DNAs that differ in chain length by only a single nucleotide in a chain of several hundred. This allows the location of chemical or enzymatic cleavages to be pinpointed simply by measuring the length of the resulting DNA. So promising are these new sequencing methods that, in a short time, available DNA sequence information easily could surpass the body of RNA and protein sequences that took decades to accumulate.

Nucleic acid sequence information is of direct help in understanding both the biological and the physical properties of RNA and DNA. Because the genetic code is known, it is almost always possible to decide what protein sequences can be synthesized under the direction of a particular nucleotide sequence. The only ambiguity is selecting the correct reading frame or phase of the triplet code. However, there are only three possibilities, and usually only the correct one has relatively long stretches between start and stop codons. Because there are several of the former and three of the latter out of a total of 64 codons, a random sequence or one chosen with wrong phase is likely to code for only short peptide stretches between each start and stop.

In addition to direct structural genes, DNAs have regions involved in regulation of gene expression. A full understanding of the properties of these regions is still lacking, but the ones known have special symmetries, such as palindromes on opposite strands. This leads to points of local C_2 or pseudo-C_2 symmetry, as shown by the example in Figure 3-5. Other regions, which may simply serve as spacers between genes, sometimes appear as slight modifications of a simple short repeating sequence. Such features give us a good chance of recognizing noncoding regions within sequences, but we are not yet in a position to predict the functional properties of control regions from sequence information alone.

The nucleotide sequence of a single strand immediately provides plausible models for the approximate secondary structure. Base-pairing patterns are quite easy to recognize visually, and several simple rules can guide selection of folding patterns. For example, long continuous stretches of complementary base pairs will almost

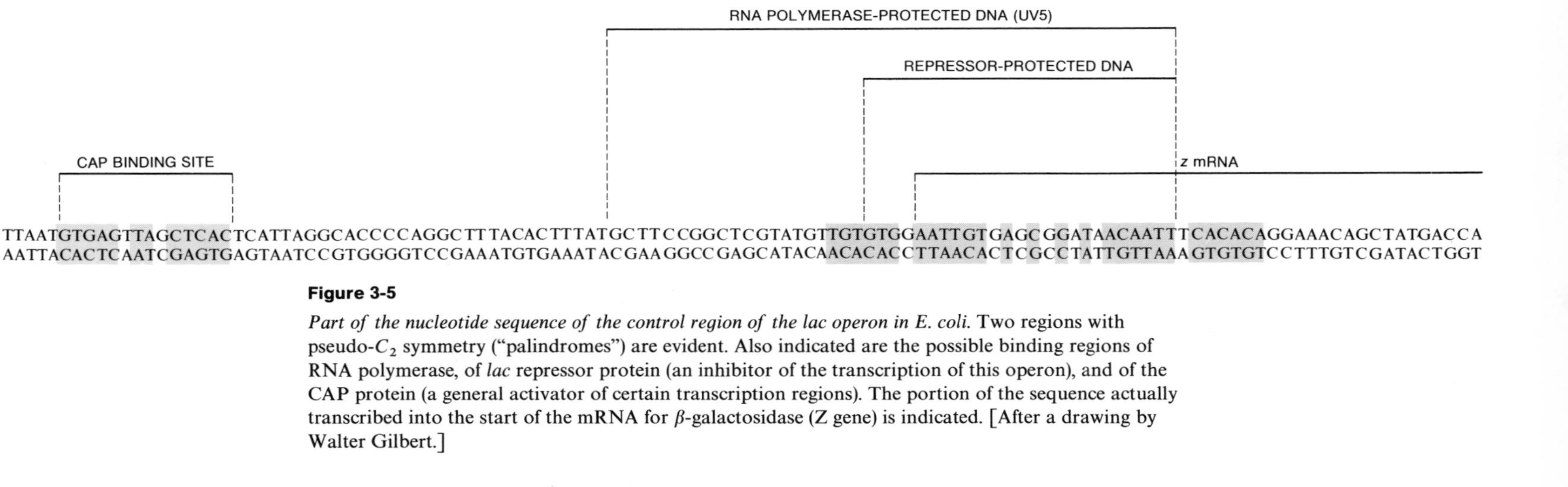

Figure 3-5

Part of the nucleotide sequence of the control region of the lac operon in E. coli. Two regions with pseudo-C_2 symmetry ("palindromes") are evident. Also indicated are the possible binding regions of RNA polymerase, of *lac* repressor protein (an inhibitor of the transcription of this operon), and of the CAP protein (a general activator of certain transcription regions). The portion of the sequence actually transcribed into the start of the mRNA for β-galactosidase (Z gene) is indicated. [After a drawing by Walter Gilbert.]

always form double-stranded hairpins. One long duplex region is preferable to two or more shorter separate regions with a total duplex length equal to the long region. Given a choice between two equivalent and mutually exclusive pairings, the more stable is the one with the highest G-C content. If both base pairings have equal G-C content, the more stable is the one between residues closer in the primary structure (as long as a certain minimum loop or hairpin site is exceeded).

These rules may seem arbitrary now, but we develop the physical basis for them in Chapter 23 in considerable detail. Furthermore, enough is known about nucleotide interactions that such rules can be used to predict the quantitative stability of particular patterns of base pairing. The effectiveness of such simple rules highlights a major difference between proteins and nucleic acids. In proteins we know of no simple patterns of complementarity that allow secondary structures to be assigned merely by considering pairwise interactions between residues.

Secondary structure inferences from tRNA sequence comparisons

When several sequences are available for different species of an RNA believed to have similar structures or functions, the task of assigning a plausible pattern of base pairing is greatly simplified. For example, the first known tRNA sequence (determined by Robert Holley) could be folded into three different patterns that are fairly equivalent according to the rules stated above. The second known tRNA sequence (established by Hans Zachau) was consistent with only one of three patterns. This was the cloverleaf, which is now known to be a good approximation to the pattern of secondary structure within tRNA. There were several reasons for assuming that all tRNAs have similar structures. Various pure or impure species of tRNA have strikingly similar physical properties. All tRNAs have to share many of the same functions and binding sites, including sites on the ribosome, on protein synthesis factors, and on enzymes that can modify tRNA. Therefore, on this basis the cloverleaf was selected long before there was any direct physical or chemical evidence for its existence.

As more tRNA sequences were determined, it became clear that there are actually several families of cloverleafs that resemble one another overall but differ in a few details. Figure 3-6 illustrates three families of secondary structures that essentially account for all of the known tRNAs having a common function on the ribosome. This figure also defines the nomenclature we use throughout this book to refer to various tRNA regions. Some features are common to all of the families, such as the length of the acceptor and anticodon stems, the size of the anticodon loop, and a portion of the detailed sequence of the pseudouridine loop. Other features—such as the size of the dihydrouridine stem, the extra loop, and the details of the dihydrouridine loop—are conserved only within each family.

A few tRNAs with specialized functions do not fall into any of the basic families. Three such sequences are illustrated in Figure 3-7. The tRNA responsible for initiation of protein synthesis in *E. coli* has an altered acceptor stem length. This tRNA is incapable of many of the normal functions of other tRNAs. LaDonne Schulman has

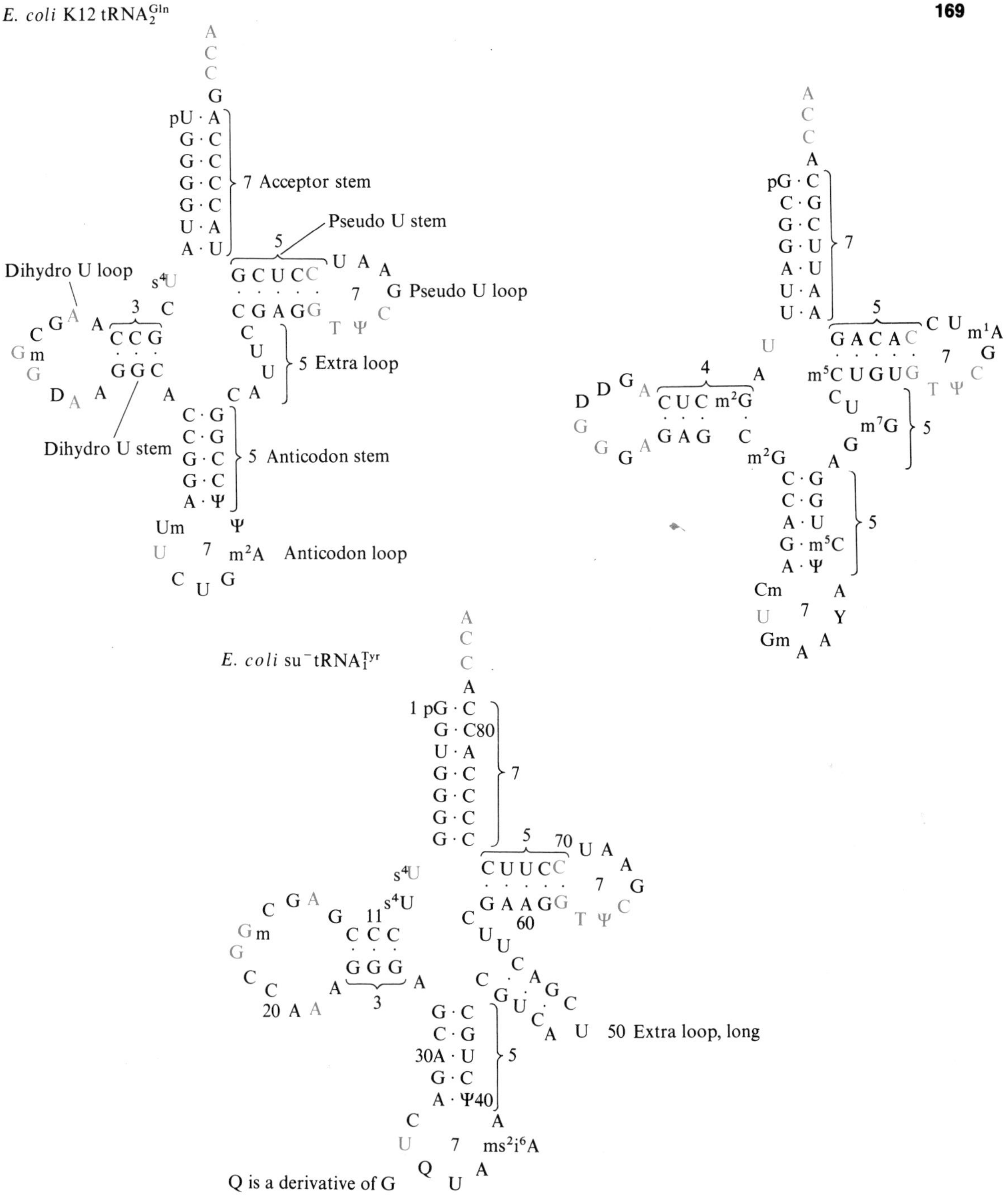

E. coli K12 $tRNA_2^{Gln}$

7 Acceptor stem

Pseudo U stem

Dihydro U loop

Pseudo U loop

5 Extra loop

Dihydro U stem

5 Anticodon stem

Anticodon loop

E. coli su⁻$tRNA_1^{Tyr}$

50 Extra loop, long

Q is a derivative of G

Figure 3-6

Nucleotide sequences and cloverleaf secondary structure patterns of three tRNAs, representative of each of the major classes of known tRNA sequences. Indicated on *E. coli* K12 $tRNA_2^{Gln}$ is the name of each tRNA region. Residues and structural features constant in all normal tRNAs are shown in color. Constant structural features for each tRNA class also are shown.

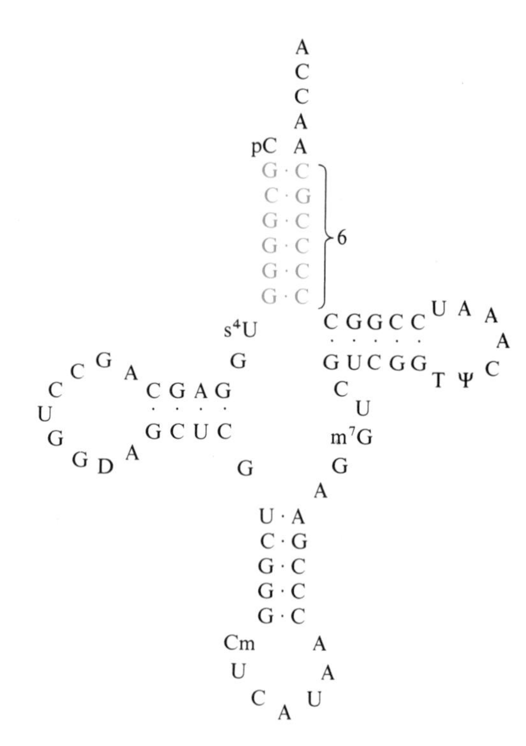

Figure 3-7
Nucleotide sequences and cloverleaf structures of three unusual tRNAs. Shown in color are bases or structural features that violate the rules set down in Figure 3-6. The two tRNA$_f^{Met}$ species are initiator tRNAs, and presumably they interact with ribosomes differently than do normal tRNAs. The *S. aureus* tRNA is involved in cell-wall peptidoglycan biosynthesis and need not interact with ribosomes at all.

shown by elegant chemical modification experiments that all these functions are regained if the acceptor stem is restored to a normal duplex length. Eukaryotic initiator tRNAs have normal acceptor stems but do not have the GTΨC sequence found in all normal tRNAs. Figure 3-7 also shows a tRNA that is involved in the synthesis of cell-wall peptidoglycan structures rather than in the synthesis of proteins on the ribosomes. Here, a number of features common to other tRNAs have been lost.

Sequence information and the analysis of structure and function

Nucleotide sequences are a prerequisite to many other experimental techniques. They must be known before chemical modification can be used to best advantage to explore classes of accessible or inaccessible residues. (See the discussion of tRNA tertiary structure later in this chapter.) Sequences also are needed to interpret various physical or chemical probes of the spatial proximity of particular residues such as intramolecular cross-linking. The sequence can be used in conjunction with spectroscopic or other physical data to attempt to prove which predicted duplex regions actually occur in the native molecule. These techniques are discussed in detail in Chapter 24.

Myeloma, rabbit liver, sheep mammary gland $tRNA_f^{Met}$ *S. aureus* H $tRNA_1^{Gly}$

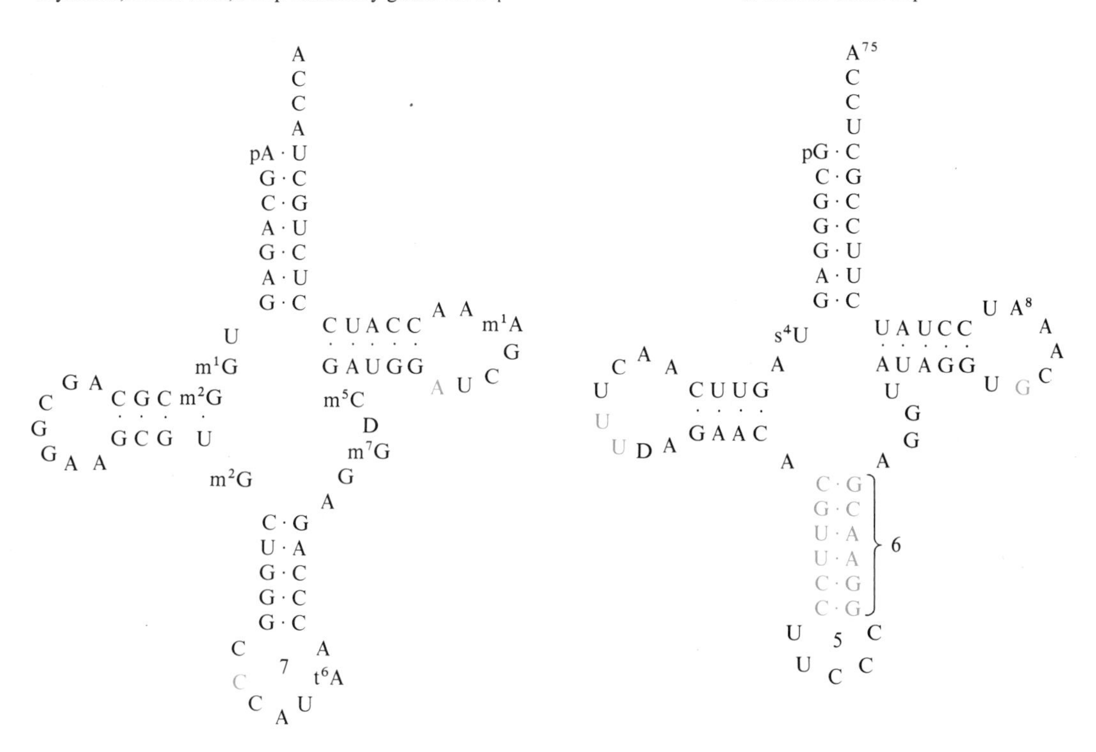

Nucleotide sequences also provide substantial clues to the way in which different nucleic acids interact with one another. For example, from the sequence of a tRNA anticodon loop, one can predict which of the mRNA triplet codons stimulate tRNA binding to the ribosome by virtue of their ability to base-pair with the tRNA anticodon triplet. Here, it is known that the strict rules of base pairing are relaxed somewhat. The 5′ and central positions of the coding triplet (corresponding to the 3′ and central positions of the anticodon) have normal base pairing, except for initiator tRNAs, but the third position allows a broader range of interactions:

tRNA anticodon, 3′-base	G	A	C	U	I
mRNA codon, 5′-base	U, C	U	G	A, G	A, U, C

The ability of I (inosine) to base-pair with three different complements is well supported by physical studies on simple model systems. The significance of the G–U interaction is still somewhat uncertain. It is clear that a G–U pair within an otherwise perfect duplex is not nearly so destabilizing as other mispairings (such as A·C or G·A) would be. However, it is not clear whether any net favorable thermodynamic interaction can be gained by juxtaposing a G with a U, at least in model complexes in solution.

A second example of nucleic acid interactions was predicted first from sequence examination, and was confirmed later by experimental studies. The 3′-end of the 16S rRNA of *E. coli* shows some complementarity with sequences near the 5′-end of protein-synthesis initiation regions in numerous viral RNAs that act as mRNAs (Fig. 3-8). Apparently, this interaction is used to help recognize an initiation region and to position the mRNA for a proper start. Different initiation regions on the

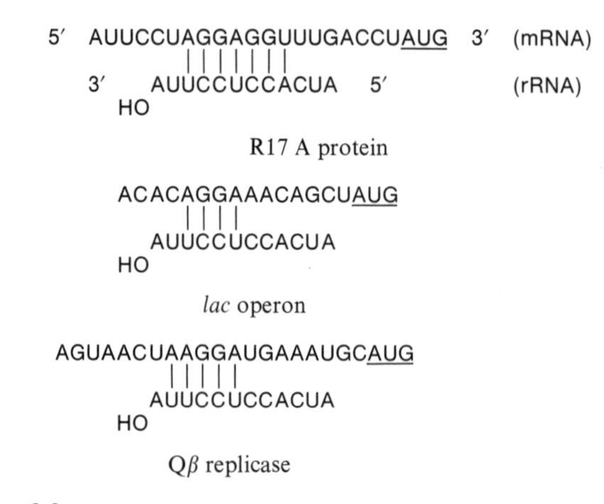

Figure 3-8
Complementary sequences near the 3′-end of the *E. coli* 16S rRNA and just to the 5′-side of AUG protein-synthesis initiation regions in a number of coliphage and *E. coli* mRNAs. [Adapted from the work of Joan Steitz and Karen Jakes.]

mRNAs show varying extents of complementarity with the 16S rRNA. This implies that they may have different probabilities of recognition. Therefore this interaction may play a role in recognizing the frequency of use of particular mRNAs. However, this interesting suggestion remains to be proven.

Techniques for direct sequence comparison

Sequence comparison is a powerful technique. It often is very useful to be able to make such comparisons, even when the detailed sequences are not known. Hybridization measurements, which gauge the ability of separate single strands to combine into duplexes, can be used to examine approximate sequence homology. Various techniques can indicate whether isolated single strands of DNA or RNA from different samples will recombine to form stable duplex regions. The kinetics of this recombination are particularly informative (Chapter 23).

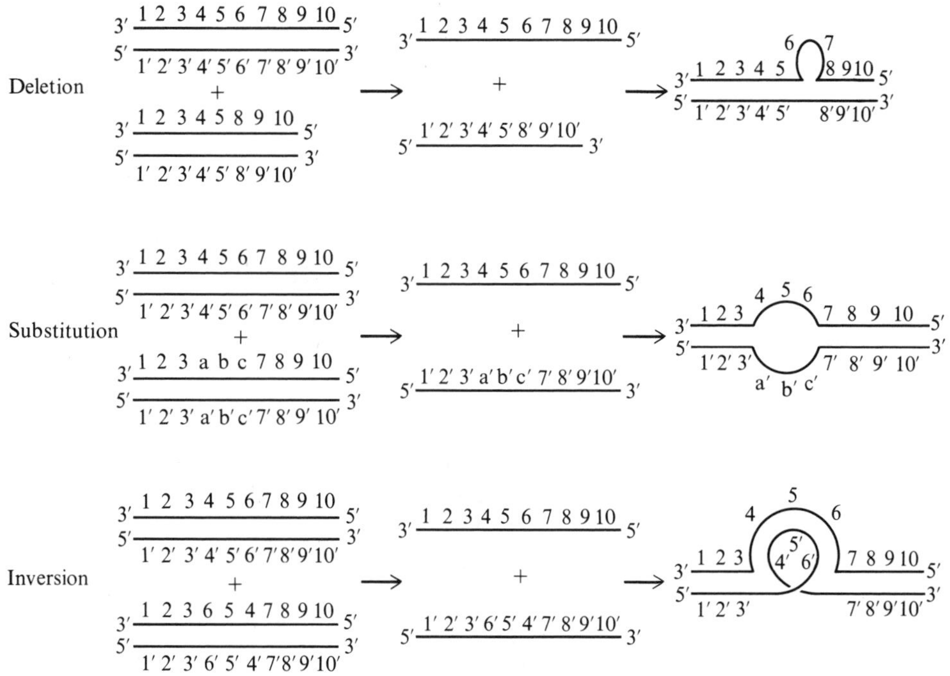

Figure 3-9
Schematic illustration of DNA melting and heteroduplex formation during renaturation. Shown is only one of two equivalent pairs of heteroduplexes that can be generated by each mixture of DNAs.

Low-resolution studies of sequence similarities in hybrid DNA or of sequence homologies between various natural DNAs can be made by electron microscopy. Although current techniques are limited to a resolution of 50 to 100 base pairs, many interesting questions can be explored at this level. This technique is far easier than chemical sequencing of thousands of nucleotides.

Suppose that two DNAs are identical except for one continuous sequence region 100 or more base pairs in length. If the two DNAs are mixed, denatured, and renatured (as shown in Fig. 3-9), some hybrid structures called heteroduplexes will result. These are less stable than normal homoduplexes because they are imperfectly base paired. In principle, heteroduplexes should eventually dissociate and the separated strands recombine to form homoduplexes. However, if the sample is maintained at temperatures where duplex regions are generally stable, this process would entail such a huge activation energy that the rate is significant only on the cosmological time scale. The regions of nonpaired bases thus remain as they are and can be distinguished from normal duplex in the electron microscope as loops when the sample is spread out.

Figure 3-9 shows three different types of loops that may be seen, depending whether the regions of sequence difference are due to sequence deletion, substitution, or inversion. Hybrids can be formed between two DNA molecules even if they have only one short region in common. Thus, electron microscopy can be used, for example, to see if particular DNA pieces carry the same gene, and to locate this common gene along the DNA.

Figure 3-10 shows an electron micrograph of a DNA–RNA hybrid. In this case, the molecule has been visualized by treating it with gene-32 protein of bacteriophage T4. This protein binds strongly only to single-stranded DNA. Therefore, the single-stranded regions appear thick due to their protein coat. This lets single and double strands be distinguished in the microscope. When RNAs are annealed with DNA, electron microscopy of the resulting duplex regions allows the positions of the genes coding for this RNA to be mapped. Very short RNAs, such as tRNA, do not produce a long enough duplex region to be clearly visible in the microscope. In these cases, an electron microscope marker (such as the electron-dense protein ferritin or a large polystyrene latex bead) can be attached to the tRNA to allow the position of the DNA-bound molecule to be visualized.

In most cases, effective use of the heteroduplex mapping techniques we have just described requires that accurate length measurements of single-stranded and double stranded DNA regions be made. The lengths between various loop or hybrid features allow a physical map of the DNA to be constructed. If the genetics underlying deletions or the location of specific genes are known, the physical map can be used to build a genetic map. The technique has special power because it allows genes to be located if the RNA transcript is isolated, even if no mutant phenotypes can be identified. In principle, it even allows mapping of DNA regions that are not transcribed at all. Any DNA piece that can be isolated or synthesized can be formed into a duplex by hybridization and then identified in the microscope. Length measurements can be made in an absolute manner in the microscope as long as a size standard is available to calibrate the magnification. Under some conditions, DNA contour lengths determined by microscopy are in excellent accord with the dimensions one would calculate from the known geometry of the double helix. However, contour lengths tend to be a function of the experimental procedures used to place the DNA on the grid that holds the sample in the microscope. Therefore, in practice, it is more accurate to add DNAs of known length to each sample and use these for internal standardization.

Many aspects of DNA electron microscopy probably are still as much art as science. Care must be taken to try to spread out a molecule as evenly as possible. Where a DNA strand crosses itself, tracing of the chain can become ambiguous. Tangled condensed molecules therefore usually cannot provide much information. Many techniques do not provide as clear a distinction of single strands and double strands as the example in Figure 3-10.

Keep in mind that in the electron microscope one examines a single molecule at a time. A particular structural feature could be just a statistical fluctuation represented by the configuration that molecule assumes on the grid. A chain crossing could

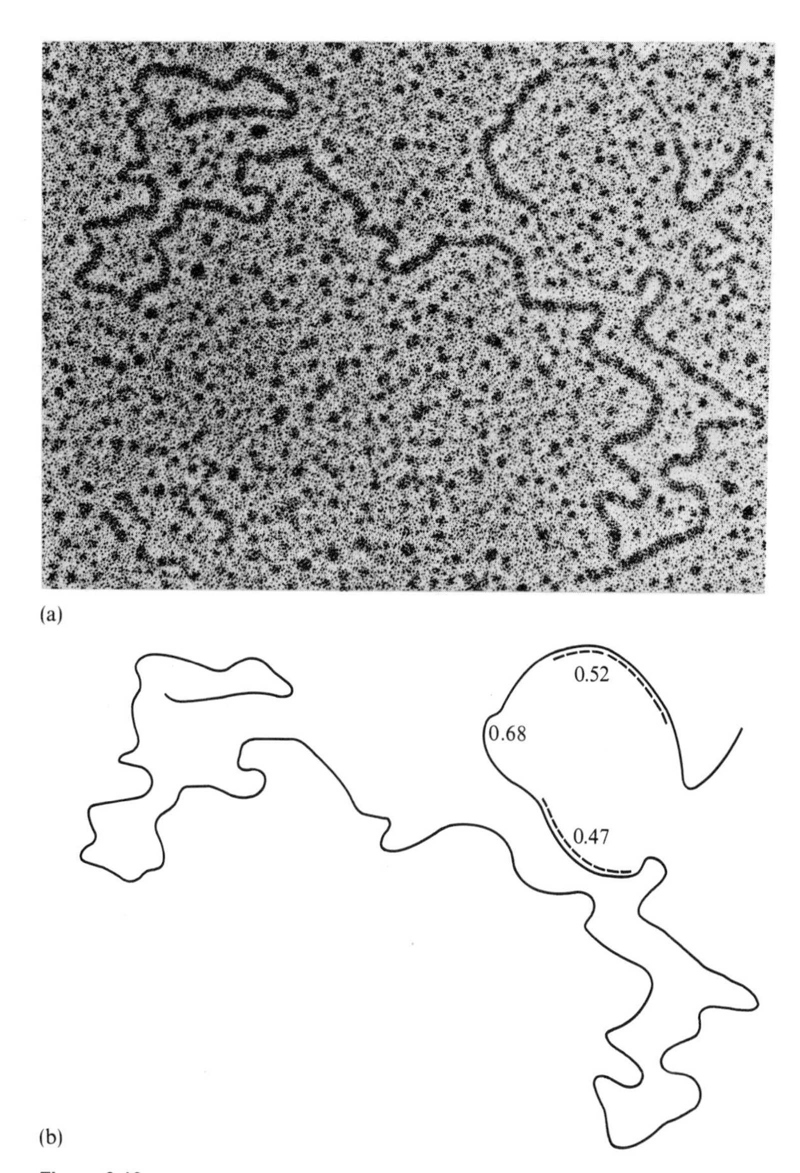

Figure 3-10
An RNA–DNA hybrid. **(a)** Electron micrograph. **(b)** Schematic drawing of the same molecule. The DNA is the single coding strand of a plasmid pSp17 containing two adjacent sea-urchin histone genes. The RNAs are two histone mRNAs that were hybridized to the DNA. The duplex regions appear thin because they do not bind gene-32 protein, which saturates the rest of the long DNA. Numerals indicate the lengths of the two genes and the space between them in units of kilobases. [Provided by Madeline Wu and Norman Davidson. From *Cell* 9:163 (1976). Copyright © MIT. Published by the MIT Press.]

simulate a true loop. Tangles just at borderline resolution can easily be missed, so that the molecule will appear shorter than it actually is. Therefore, it is necessary to examine many molecules and to use statistical criteria to evaluate any structural features seen.

One common ambiguity involves the chemical identities of the two ends of a linear molecule. This is equivalent to lack of knowledge of the physical direction of the sequence in a circular molecule. For example, suppose that one has mapped a deletion loop half the distance from one end. There is no way of knowing *a priori* whether that end is the 3′-end or the 5′-end of the DNA strand that codes for the missing gene. To resolve this ambiguity, either genetic information can be used or, more directly, specific enzymes can be added that partially degrade a DNA strand specifically from one end. When gene locations are mapped using RNA–DNA hybrids, sometimes RNA precursors are available with extra bases on either the 3′- or the 5′-end. These then allow the physical direction of the DNA strand to be determined.

3-4 SECONDARY STRUCTURE

In talking about nucleic acid primary structures, we could not avoid using base-pairing patterns as a way of ordering or comparing sequences. It is quite common, when a new single-stranded RNA sequence is determined, to fold it into some esthetically pleasing base-pairing scheme roughly consistent with the rules outlined earlier. Regardless of the accuracy of the scheme as a representation of the molecule, it is useful as a mnemonic device for the general features of the sequence. If the rules of base pairing are strictly obeyed, the amount of information needed to specify the single-stranded RNA sequence is reduced by roughly a factor of two.

Here, we are interested in some of the more quantitative aspects of nucleic acid secondary structure. Most of what is known derives from studies of the diffraction of x rays by ordered fibers of double-stranded nucleic acids (see Chapter 14). Single-stranded compounds generally do not make good fibers, and so knowledge of their secondary structure is less direct (see Chapter 22).

Base pairing as a central feature of nucleic acid double strands

Several aspects of polynucleotide double strands immediately indicate that their basic organization is very different from protein secondary structures. In a double strand, certain positions on nucleic acid base residues are rendered almost uniformly inaccessible to various chemical reagents. This suggests that the base side chains are located in the interior of the structure. All protein side chains, in contrast, were located at the exterior of helices or sheets (see, for example, Fig. 1-1). All nucleic acids, regardless of sequence, appear able to form the same double-stranded structure. Recall that, in proteins, the sequence has a crucial effect on whether a secondary structure can form at all and, if so, what type it will be.

The idea of a nucleic acid structure in which all side chains are internal and yet all side-chain sequences can be accommodated must have seemed somewhat paradoxical at first. Chargaff's base-composition rules provided an important clue, but it had to be recognized that these applied only to total nucleic acid and not to individual strands. To integrate all this information, James Watson and Francis Crick conceived the idea of specific interactions between complementary bases: A with T or U, and G with C. It is important to remember that at the time there was no direct evidence for specific interactions between the bases. Also, a choice of these and not other possible base pairs was contingent on the acceptance of the keto-amino tautomeric forms of the bases, a finding that then was quite new and not universally accepted.

The critical feature of A·T and G·C base pairs recognized by Watson and Crick was not simply that good linear hydrogen bonds can be formed. Other base-pairing arrangements can do that equally well and indeed are quite prevalent in complexes between isolated monomers. These alternatives even predominate in some cases (Chapter 6). The special aspect of the Watson–Crick base-pairing scheme is that the geometry of A·T and G·C pairs measured from the points of attachments of bases to sugars is virtually identical. Therefore, the same double-helix geometry can accommodate both base pairs without any distortion or loss in symmetry (Fig. 3-11). A second important feature of Watson–Crick base pairs is that a C_2 pseudosymmetry axis exists in the plane of the base pair. This dyad axis interchanges the N^9–$C^{1'}$ purine glycosidic bond with the N^1–$C^{1'}$ bond of the complementary pyrimidine. The same axis exists in both A·T and G·C pairs (Fig. 3-11). This means that exactly the same geometry accommodates A·T, T·A, G·C, and C·G. All four possible residues can occur at any level in a double helix. Bases on either strand of the double helix have exactly the same environment.

Structure of nucleic acid double helices

Many different helical geometrics can be built around the fixed Watson–Crick base pairs. Table 3-1 summarizes structural parameters of a number of these geometries that have been found experimentally. The molecular structures of DNA A and DNA B are illustrated schematically in Figures 3-12 and 3-13, two structures, DNA B and RNA 11, also are shown as space-filling models in Figure 3-14. These illustrations should be studied carefully in conjunction with this section of the text.

All the structures have certain features in common. All are right-handed helices. All have the dyad symmetry axis, and this turns out to be a true symmetry axis for the phosphodiester backbone, even though it is only a pseudosymmetry for the base pair itself. Because it is a rotation axis, the two phosphodiester backbones must run in opposite directions. The two ends of a double helix thus are chemically and structurally identical because each contains one 3′ and one 5′ strand terminus. Box 3-1 shows a simple chemical proof that base-paired polynucleotide strands are antiparallel.

An antiparallel-strand structure is critically important for base pairing within a single strand (as in tRNA). Parallel-strand pairing does not allow for the formation of hairpin loops. The double helices shown in Figures 3-12, 3-13, and 3-14 have a

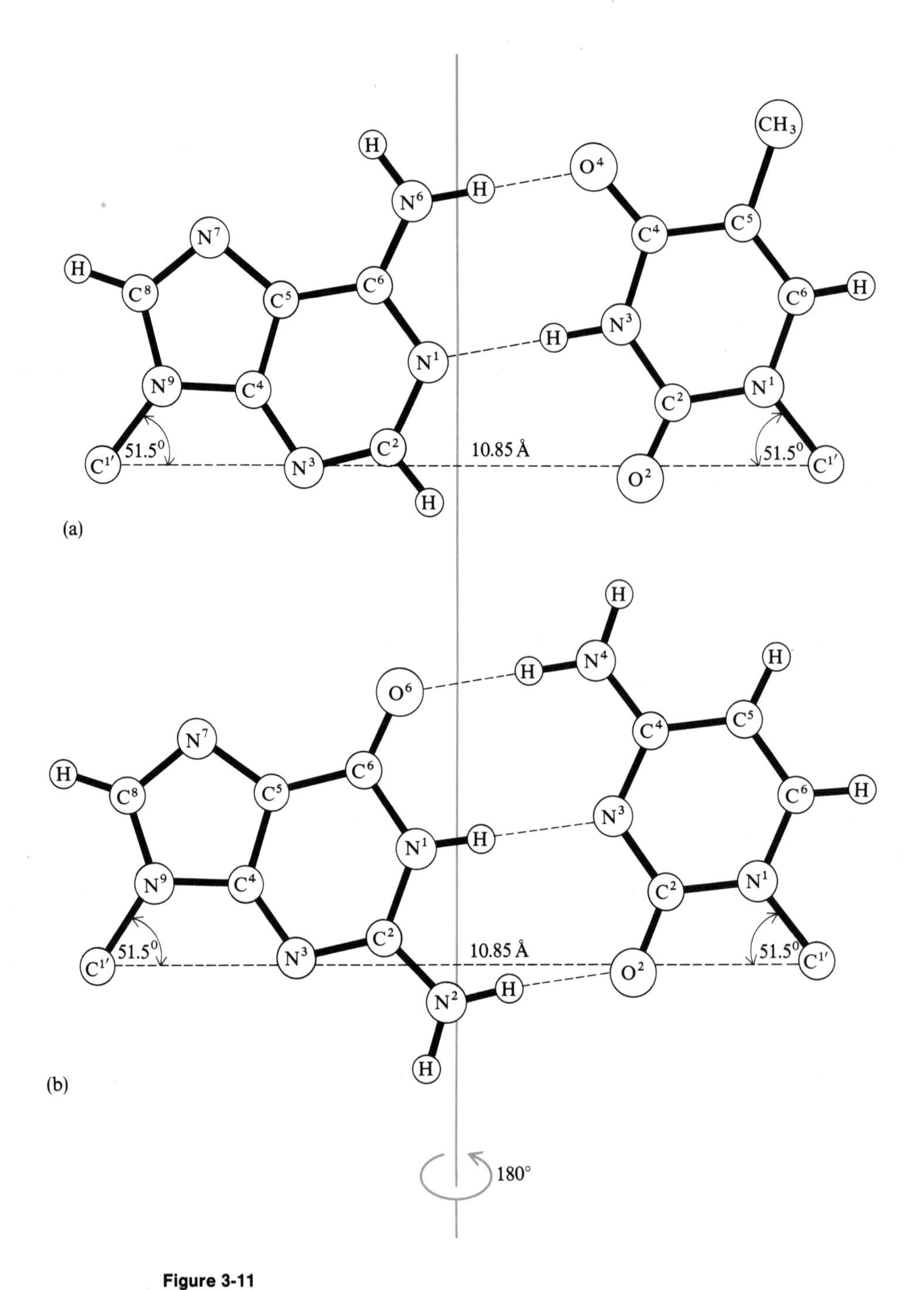

Figure 3-11

Atomic structures of the Watson–Crick A · T and G · C base pairs. Note that the $C^{1'}$–$C^{1'}$ distance and the angles that N^9–$C^{1'}$ bonds and N^1–$C^{1'}$ bonds make with the $C^{1'}$–$C^{1'}$ vector are constant. A pseudo-C_2 symmetry axis is shown as a colored line. [After S. Arnott, S. D. Dover, and A. J. Wonacott, *Acta Cryst.* B25:2192 (1969).]

Table 3-1
Structure of polynucleotide double helices

Sample	Relative humidity (%)	Residues per turn	Translation per residue (Å)	Angle between base planes and helix axis	Dihedral angle between base planes	Sugar conformation
DNA-A, Na^+	75	11	2.55	70°	16°	$C^{3'}$-*endo*
DNA-B, Na^+	92	10	3.46	—	—	—
DNA-B, Li^+	66	10	3.37	88°	5°	$C^{2'}$-*endo*
DNA-C, Li^+	66	9.3	3.32	84°	10°	$C^{2'}$-*endo*
DNA–RNA hybrid, Na^+	75	11	2.62	70°	—	—
Yeast RNA fragments	75	10 or 11	2.9 or 2.64	—	—	—
Reovirus α or β forms	75	10 or 11	3.0 or 2.73	—	—	—
poly rA·rU		10 or 11	3.1	—	—	—
poly rI·rC		12	3.0	—	—	—
poly rCH^+·rC*		12	3.11	—	—	—
poly rAH^+·rAH^+*		8	3.8	—	—	—

* These have parallel strands. All others have antiparallel strands.

Box 3-1 HOW TO DEMONSTRATE ANTIPARALLEL DOUBLE STRANDS

One important feature of all the double-stranded helices we have described thus far is antiparallel strands. This structure can be shown to occur in DNA or RNA in solution by a simple yet elegant sequence technique developed by Arthur Kornberg. DNAs or RNAs are synthesized enzymatically from 5′-triphosphates. The following scheme shows how synthesis from α-^{32}P-labeled precursors can be used to identify the average nearest neighbors of each of the four types of nucleotide residues. For example,

$$\text{ppp*A} + \text{pppT} + \text{pppG} + \text{pppC} \xrightarrow{\text{DNA polymerase I}} \text{pGp*ApCpTp*Ap*ApCpCp*A}\ldots$$

$$\xrightarrow{\text{DNase}} \text{Gp*} + \text{Ap} + \text{Cp} + \text{Tp*} + \text{Ap*} + \text{Ap} + \text{Cp} + \text{Cp*} + \cdots$$

The fraction of A residues preceded by each residue type can be computed from the relative distribution of label into each of the four mononucleotides. Combining this information with the known base composition, and with data from equivalent experiments using each of the four labeled triphosphates in turn, the frequencies of each of the 16 nearest-neighbor sequences can readily be computed. For antiparallel Watson–Crick base pairing, one must satisfy sequence constraints such as

$$\chi_{\text{CpT}} = \chi_{\text{ApG}} \quad \text{and} \quad \chi_{\text{ApC}} = \chi_{\text{GpT}}$$

but for parallel strands these constraints should not hold in general. Instead, different constraints will exist, such as

$$\chi_{\text{CpT}} = \chi_{\text{GpA}} \quad \text{and} \quad \chi_{\text{ApT}} = \chi_{\text{TpA}}$$

Using this principle, it was easy to show that duplex DNAs, RNAs, and hybrids all have antiparallel base pairing.

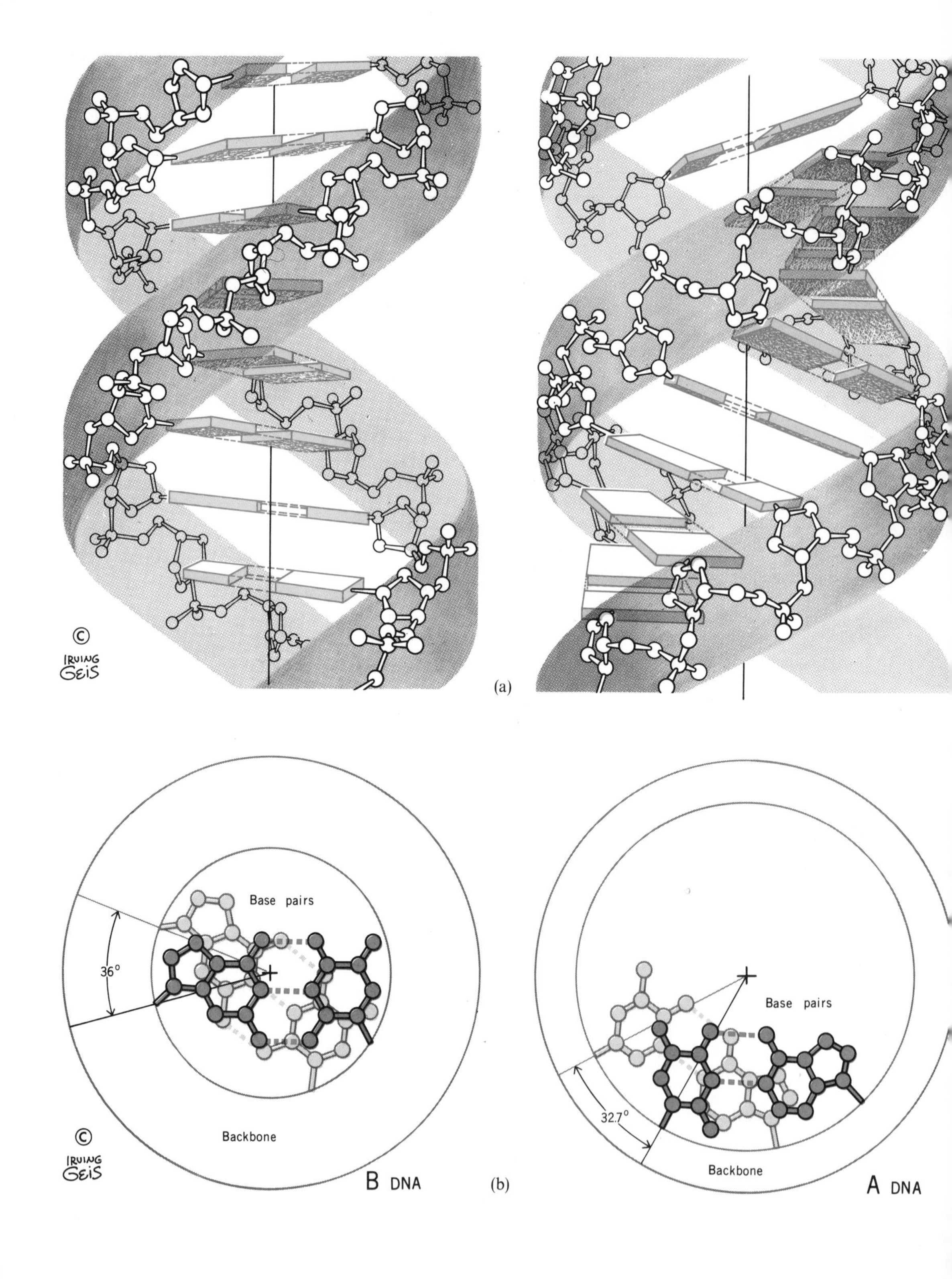
©
IRVING GEIS
(a)
Base pairs
36°
Backbone
©
IRVING GEIS
B DNA
(b)
Base pairs
32.7°
Backbone
A DNA

Figure 3-12

The DNA-B and DNA-A helices. (**a**) A view perpendicular to the helix axis. In these schematic illustrations each base is shown as a board, while all of the backbone atoms except hydrogens are explicitly drawn. The tilting of the base pairs in DNA A is clearly visible. (**b**) Viewed along the helix axis (+). In DNA B the helix passes right between the base pairs, while in DNA A the base pairs rotate around the index axis generating a central hole through the helix. [Drawings by Irving Geis. After computer drawings provided by Wolfram Saenger.]

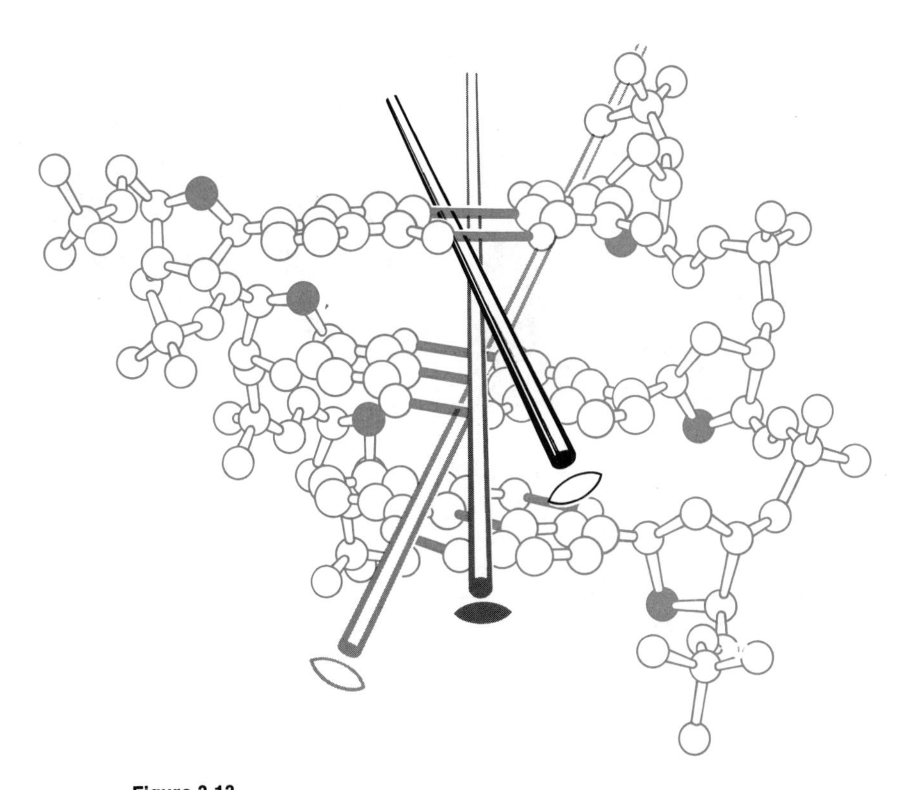

Figure 3-13

Symmetry in the DNA-B helix. The locations of two dyad axes (⬯) and one pseudo-C_2 axis (⬮) are indicated. [Drawing by Irving Geis.]

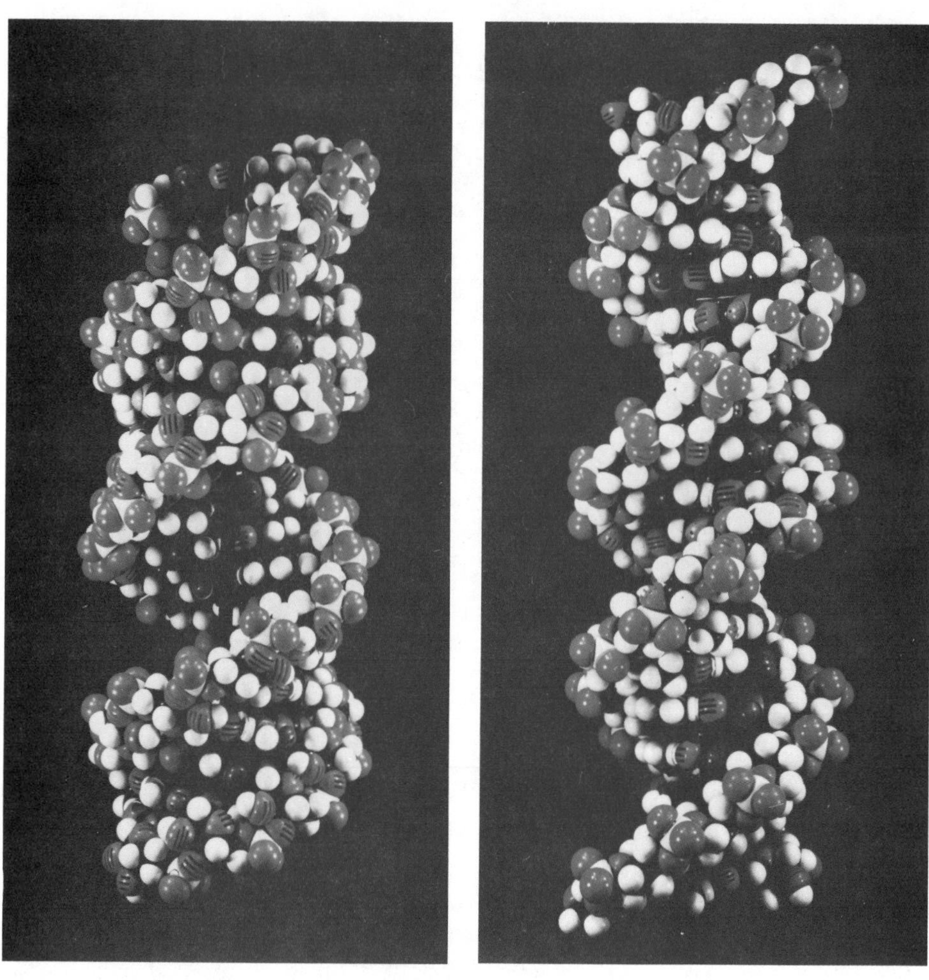

Figure 3-14

Space-filling models of the DNA-B (right) *and RNA*-11 (left) *double helices.* Note the substantial differences in pitch and in the appearance of grooves between the two helices. [Photograph provided by Sung-Hou Kim.]

second pseudo-C_2 axis that passes between each adjacent pair of hydrogen-bonded bases. For self-complementary sequences such as GpC or ApT (or "palindromic sequences" such as that shown in Fig. 3-5), the C_2 axis that runs between the central two base pairs is actually a true symmetry axis because the 180° rotation moves not just the backbone, but also all the bases, into an indistinguishable position.

A third common feature of all of the double helices is base stacking. Although the geometry of adjacent base pairs varies, in each case the distance between neighboring base-pair planes is about 3.5 Å. This is equal to the Van der Waals radius of planar aromatic compounds. It is the closest distance of approach one would expect before repulsive nonbonded interactions begin to occur.

In spite of all the above similarities, the known helical structures of double-stranded nucleic acids differ in a number of important features. These include the number of base pairs per turn, the tilt of the base pairs (the angle between the base pair and the helix axis), the twist of the base pairs (the angle between planes containing two paired bases), the diameter of the helix, and the nature of the grooves of the helix. Some of these differences are visible in stick models, but the grooves are best illustrated by the space-filling models of Figure 3-14.

Let us consider the common structures one at a time. The B form of DNA is found in Na^+-containing fibers at high relative humidity. This suggests that it should correspond to the form present in normal aqueous solutions. This is indeed the case (some supporting evidence is given in Chapter 22). In the B form, there are 10 residues per turn. The base pairs are virtually planar, and they are perpendicular to the helix axis. The double helix has two distinct grooves: a minor groove that lies between the 1′-carbons of the sugars on two strands, and a major groove on the opposite side of the helix. The diameter of the helix is roughly 22 Å.

The B form normally is regarded as the Watson–Crick structure. However their original model, although based on data from a B-form DNA, differed in that only two hydrogen bonds were formed between G and C; also a number of detailed structural features were actually closer to the A form than the B form. Better data and more refined structural analysis were required before the currently accepted geometry of the B form emerged.

The A form of DNA occurs in fibers of the sodium salt at 75% humidity. It has 11 residues per turn, and the diameter is a few angstroms larger than that of the B form. The base pairs are again almost planar, but in this form they are tilted 20° with respect to the helix axis. Thus, if the B form is like a spiral staircase, the A form is like a spiral staircase with each step slanting toward the center. The A and B forms can be interconverted in a fiber simply by changing the humidity. Because the process occurs rapidly and without destroying the ordered packing of the fiber, the A to B conversion must be able to occur without strand separation. This is strong evidence that both A and B duplexes have the same handedness. In solution, the normal B form can be converted to the A form by reducing the water activity—for example, by adding appreciable mole fractions of ethanol to an aqueous solvent.

Whether the A form occurs in nature in duplex DNA is unknown, but RNA–DNA hybrids definitely assume an A helix rather than a B helix, even in normal aqueous buffer. X-ray fiber diffraction of a number of RNA duplex helices has been studied. The structure most consistent with observed results is the RNA 11 helix. This has 11 base pairs per turn, tilted 13° to 14° away from a direction perpendicular to the helix axis. The RNA 11 structure thus is very similar to the DNA A helix. However, diffraction data on most RNA samples are not complete or accurate enough to exclude an alternative: the RNA 10 helix. This has 10 residues per turn, tilted 10° from the perpendicular to the helix axis. The RNA 10 structure is essentially halfway between the A and B forms of DNA.

One other form of polynucleotide double helix to consider briefly is the C form of DNA. This duplex has 9.3 base pairs per turn, a slight 6° tilt away from the perpendicular to the helix axis, and a slight twist of the bases. The C form, found in Li^+

salts at low humidity, is really a variant of the B form. Other variants exist of both A and B forms, so these actually represent structural families rather than unique accessible conformations of the double helix.

Structural differences between RNA and DNA

RNA has never been observed to take on a B double helix. This is understandable. The presence of the 2′-hydroxyl alters the conformation of the ribose ring in such a way as to eliminate the possibility of a B helix. From x-ray crystallographic studies on a variety of mononucleotides, it is known that four of the five atoms in the pyranose ring tend to be in a plane, whereas the fifth (which can be either $C^{2'}$ or $C^{3'}$) is puckered out of the plane. This leads to the four possible sugar-ring conformations shown in Figure 3-15. In the $C^{2'}$-*endo* structure, the 2′-carbon lies above the pyranose ring on the same side as the base and the 5′-carbon. The $C^{3'}$-*exo* structure is quite similar. In contrast, in the $C^{3'}$-*endo* structure it is the 3′-carbon that lies above the pyranose ring. The $C^{2'}$-*exo* structure is quite similar to $C^{3'}$-*endo*.

The preferred ring puckers differ in x-ray structures of polynucleotide helices and in structures observed for mononucleotides. The ribose in RNA helices is almost exclusively $C^{3'}$-*endo*. Monomeric ribonucleotides are found to be $C^{3'}$-*endo* or $C^{2'}$-*endo* in about equal numbers of structures studied in crystals by x-ray diffraction and in solution by NMR. Deoxyribose has been found in both $C^{2'}$-*endo* and $C^{3'}$-*endo* conformations in DNA helices but is almost always found as $C^{2'}$-*endo* in the structure of monomeric deoxyribonucleotides. The B-form double-helix structure requires a

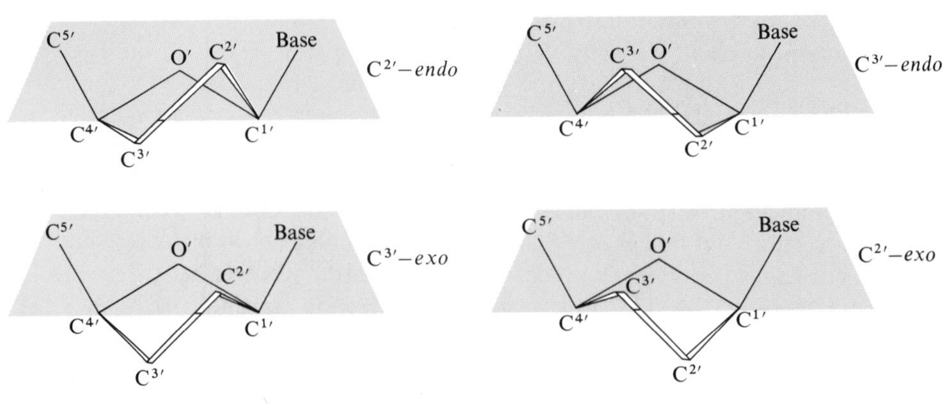

Figure 3-15

The four puckered conformations of the pyranose ring of ribose and deoxyribose. Four atoms ($C^{1'}$, $C^{4'}$, $0^{1'}$, and either $C^{2'}$ or $C^{3'}$) are located almost in a plane. The DNA-B helix has $C^{2'}$-*endo* sugars, whereas DNA-A and RNA-11 helices have $C^{3'}$-*endo* sugars. [After V. A. Bloomfield, D. M. Crothers, and I. Tinoco, Jr., *Physical Chemistry of Nucleic Acids* (New York: Harper & Row, 1974).]

$C^{2'}$-*endo* pyranose conformation. The A-form and RNA 10 or 11 structures are constructed with $C^{3'}$-*endo* sugars. Although the exact barriers to rotation from $C^{3'}$-*endo* to $C^{2'}$-*endo* are not known for ribonucleotides, the overall cost of doing this for each residue of a double helix in an attempt to make a B-form RNA could be mammoth because the unfavorable energies must be summed over all residues in the chain. Apparently RNA cannot form a B helix because of the combined energy cost of converting all sugars to $C^{2'}$-*endo* and because of the steric interference that a $C^{2'}$-*endo* would cause in the structure.

Some of the physical properties of DNA change gradually in solution with small environmental perturbations. It is clear that these changes represent neither breaking of base pairs nor other helix-denaturation phenomena. Instead, they probably reflect the ability of changes in salt concentration or temperature to shift the equilibrium of double-stranded conformations among various members of a given structural family. Certain particular structures appear to be favored by some regular base sequences, such as synthetic DNAs where one strand contains all dA and the other all dT. It is possible that, when such sequences occur within a natural DNA, the local structure becomes altered. This could help to explain how proteins that bind to DNA may recognize particular regions without separating the two strands to interact extensively with the bases.

No example is yet known in which A and B forms are proven to coexist within the same double-helical molecule. Because of the change in base-pair tilting, Struther Arnott has pointed out that, when an A–B helix interface occurs, the helix must bend at that point if base-stacking continuity is to be maintained. It has been suggested that DNA synthesis might employ RNA primers. If this were true, the following structure would exist at early stages in the synthesis of DNA:

```
                 3′  dNpdNpdNpd Npd Npd NpdNpdNp
                 5′  r MprMprMpdMpdMpdM
preferred helix          A    i    B
```

where M is complementary to N; d and r represent deoxyribose and ribose, respectively; and i indicates the interface between A and B as preferred helical forms. Here, a potential A–B interface exists as one passes from the RNA–DNA hybrid section (resulting from the RNA primer) to the pure DNA duplex of the newly synthesized DNA. One would like to be able to say whether a bend occurs at the RNA–DNA interface or, if not, how far the disturbance of the structure due to the A form propagates through the helix. Studies on model sequences suggest that the interactions in one region of a duplex might be felt quite a few base pairs away.

The fact that DNA can exist in either A or B structural families but RNA only in the A family could have important biological consequences. It at least opens the way for nature or the researcher to design substances that either recognize both molecules are double strands, or recognize only the duplex B-form DNA. Such interactions could be general and not sequence-dependent, because of the large difference in the overall helix structure of the A and B forms. To recognize an RNA in such a general way, one could look for single-stranded regions or hairpin loops because these are common in all known natural RNAs, yet probably absent from DNA at all stages, except possibly during replication.

Other polynucleotide helices

A few polynucleotide double helices or multiple helices are known in which parallel strands occur. These structures, however, do not contain Watson–Crick base pairs. The best-studied examples are double-stranded polymers in which both strands are homopolymers of the same base. Table 3-1 lists some of the structural features of the double helix formed by protonated poly A and the duplex formed by half-protonated poly C. One can see immediately that the overall geometry of these helices is quite different from that of the A or B structural families.

The structure of the base pairs formed by protonated polynucleotides is shown in Chapter 22. These base pairs and the helices that contain them have a C_2 axis parallel to the helix axis. Poly I forms a triple helix with a C_3 axis parallel to the helix axis. Such symmetries are possible only with parallel polynucleotide chains. Many other triple-stranded structures are known in which an extra strand base-pairs with one of the two normal strands of an antiparallel Watson–Crick duplex. These structures are seen mostly in homopolymer complexes. Base-pairing schemes in two such triple strands are shown in Figure 22-14.

The structure of one triple-strand helix, poly dT–poly dA–poly dT, is shown in Figure 3-16. The dA strand and the dT strand form a normal base-paired double helix except that the geometry is of a DNA–A structure rather than the more common DNA–B. The second dT strand fits nicely into the major groove of the A-type structure, forming an additional base pair with the NH_2 and N^7 of the adenine.

In general, triple-strand formation between most sequences does not seem likely. There is no known way, with an arbitrary sequence, to arrange three strands to form a continuous symmetrical triple helix. Short triple-stranded regions may well be common for particular sequences within RNA molecules. For example, the triple strand $A \cdot U_2$ (Fig. 22-14) is very stable under the high-salt or high-Mg^{2+} conditions known to favor tightly folded structures within tRNAs, rRNAs, and viral RNAs. However, fortuitous sequences and elaborate folding of a single strand are required to generate triple helices. As discussed below, triple base pairs have been found in tRNA.

3-5 TERTIARY STRUCTURE

Tertiary hydrogen-bonded interactions in tRNA

The three-dimensional structure of only a single RNA (the yeast $tRNA^{Phe}$) is known in detail at this time. It will be discussed extensively here and in other chapters. One

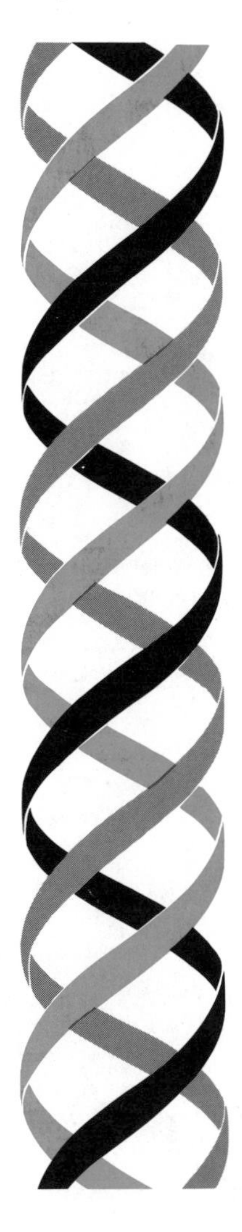

Figure 3-16

Structure of the triple helix poly dT·poly dA·poly dT, as determined by x-ray fiber diffraction. The dA and one dT strand form a normal DNA-A helix. The other dT (*colored*) fits into the major groove of the DNA-A helix. [See S. Arnott et al., *Nucleic Acids Res.* 3:2459 (1976).]

must be careful to avoid making too many sweeping generalizations from this single example of an RNA tertiary structure. For example, the first protein of known tertiary structure, myoglobin, is seen in retrospect to be atypical in many respects. Yet, it is impossible to resist dwelling on a number of fascinating features uncovered in the structure of $tRNA^{Phe}$.

Several groups of researchers independently have determined this tRNA tertiary structure by x-ray crystallography. Two different crystal forms containing different intermolecular packing arrangements of $tRNA^{Phe}$ molecules have been studied. The resulting tertiary structures appear to be nearly identical except for the position of the CpCpA end (Fig. 3-17). This is strong evidence that the structures seen are relatively

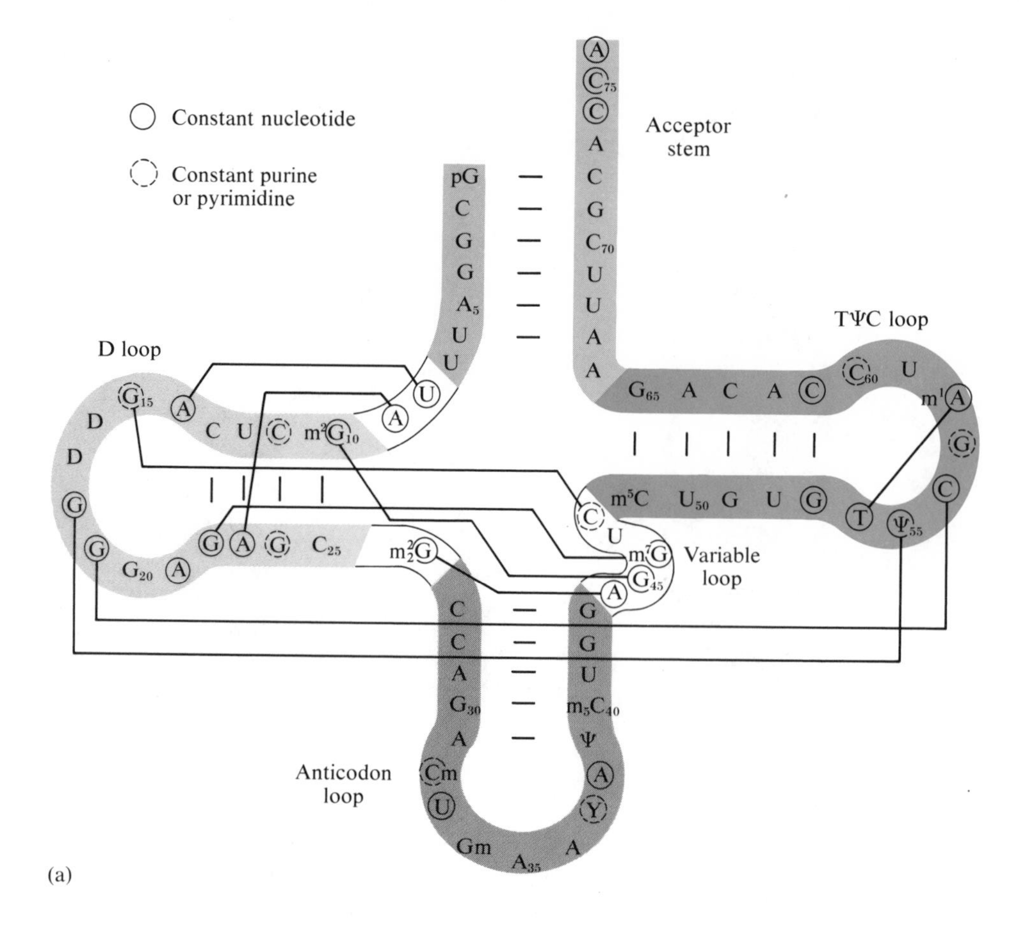
Constant nucleotide
Constant purine or pyrimidine
Acceptor stem
D loop
TΨC loop
Variable loop
Anticodon loop

(a)

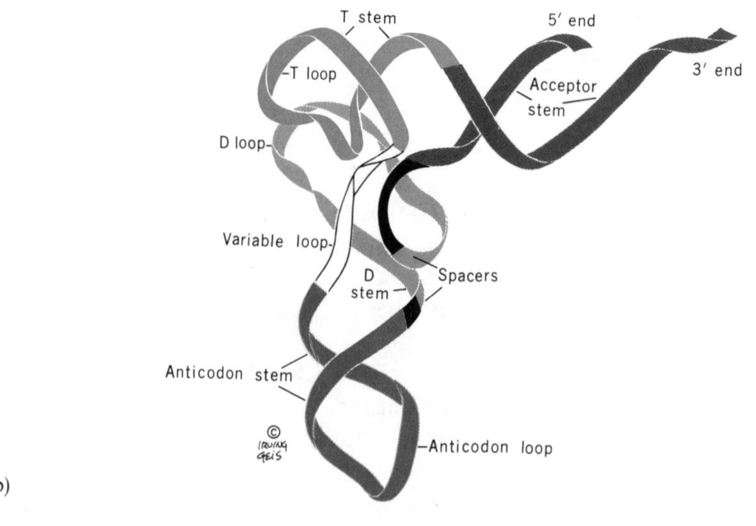
T stem
5′ end
T loop
3′ end
Acceptor stem
D loop
Variable loop
D stem
Spacers
Anticodon stem
Anticodon loop

(b)

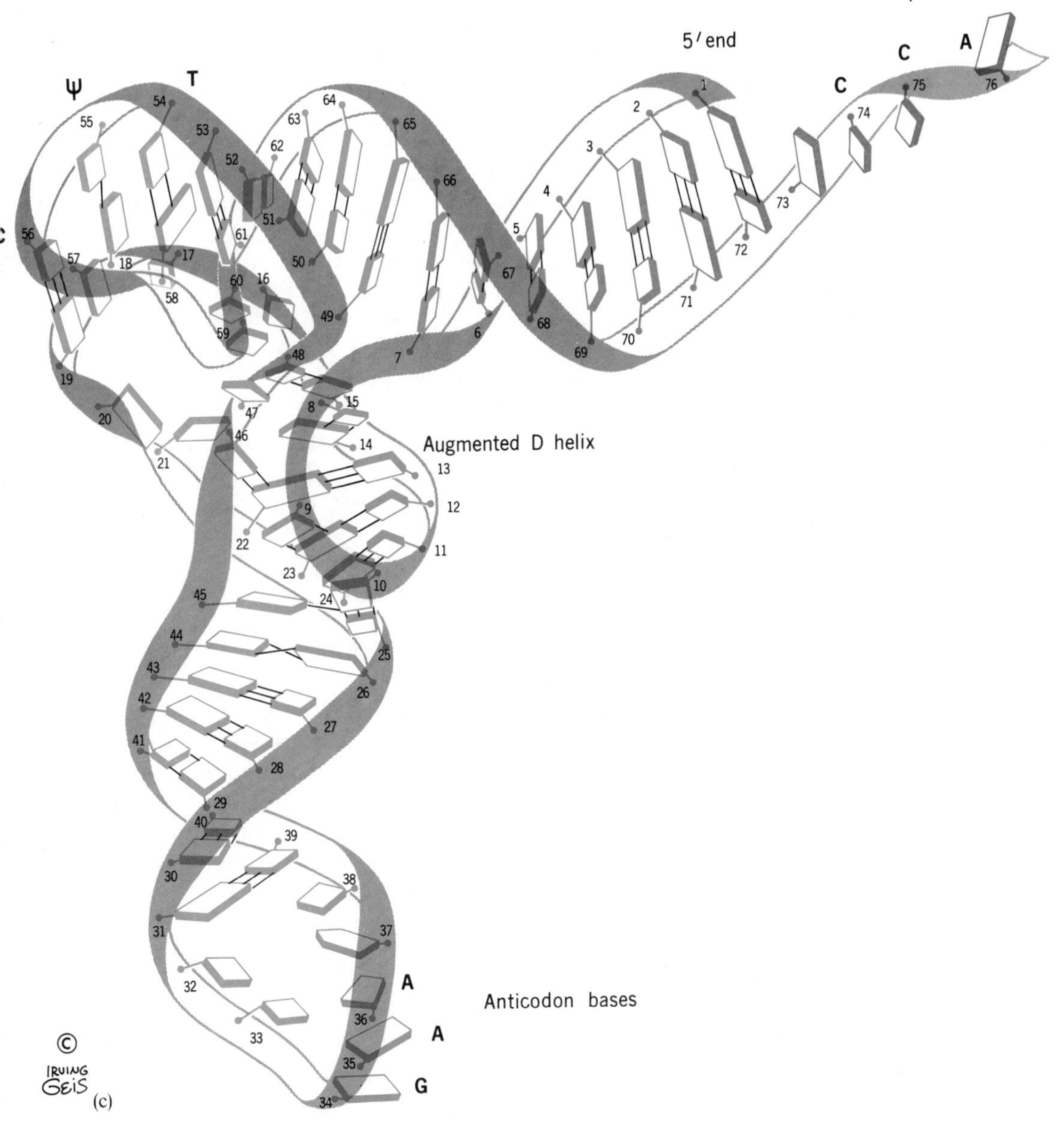

Figure 3-17

The tertiary structure of yeast phenylalanine tRNA. (**a**) Nucleotide sequence. Constant residues and residues that appear to be constantly a purine or a pyrimidine are indicated. Residues involved in tertiary base pairing are shown connected by solid lines. [After G. J. Quigley and A. Rich, *Science* 194:791 (1976).] (**b**) Backbone structure of the orthorhombic form as determined by S. H. Kim, A. Rich, and collaborators. The structure of the monoclinic form as determined by J. Robertus, A. Klug, and collaborators is nearly identical. Various loops and stems are indicated in corresponding colors in parts a and b to facilitate identification. (**c**) The full tertiary structure. Purines are shown as rectangular slabs, pyrimidines as square slabs, and hydrogen bonds as lines between slabs. [Drawing by Irving Geis.]

unperturbed by crystal packing forces and could be an accurate representation of the structure assumed by isolated molecules in solution.

Let us first summarize some of the major features of the structure of $tRNA^{Phe}$. The overall molecular shape is an extended L. Some regions are no thicker than an isolated RNA duplex. It is a much more extended structure than has been seen for typical globular proteins with the same molecular weight as tRNA. The cloverleaf secondary structure is contained within the three-dimensional structure of $tRNA^{Phe}$ (Fig. 3-17a). However, it is supplemented by a number of additional base-pairing interactions. Because these are between distant regions of the cloverleaf, they have been called tertiary base pairs.

The cloverleaf stems are folded in the tertiary structure to make two longer helical regions. The acceptor-stem helix continues directly into the pseudourindine stem helix to form one arm of the L-shaped molecule. The other arm of the L is generated by the anticodon stem and dihydrouridine stem in an almost parallel orientation. Continuity between these two helices is provided by the unusual m^2G^{26}–A^{44} base pair shown in Figure 3-18a. Tertiary interactions augment both arms of the L. For example, one additional base pair occurs within the pseudouridine loop. Two more base pairs are formed between the pseudouridine loop and the dihydrouridine loop. The overall result is to extend the pseudouridine stem helix by three base pairs.

The most novel feature of the $tRNA^{Phe}$ tertiary structure is labeled the augmented D helix in Figure 3-17. It is here, and here only, that tRNA shows the kind of complex multiresidue interactions characteristic of tertiary structure features in proteins. The augmented D helix extends and continues the dihydrouridine-stem helix. However, it is a four-stranded structure involving (1) bases from the dihydrouridine loop, (2) the extra loop, (3) the two bases between the acceptor and dihydrouridine stems, positions 8 and 9 (sometimes called the miniloop), and (4) the two strands of the dihydrouridine stem. The augmented D helix contains several base triples. For example, the guanine at position 22 is hydrogen bonded directly to two other bases C^{13} and m^7G^{46} (Fig. 3-18b). Note, however, that the base triples do not generate a triple helix because the strands involved in the pairing are not constant.

There are six possible ways that four strands can link in pairs. The base triples, plus other tertiary base pairs in the augmented D helix, cover five of the six possibilities. This indicates that the region is very tightly stitched together, and it is likely that this part of the structure will be quite rigid.

Another novel feature of the $tRNA^{Phe}$ tertiary structure is the presence of numerous hydrogen-bonded interactions between bases and the phosphodiester backbone. Notice the interaction between the phosphate of A^9 and the amino groups of C^{13} and m^7G^{46} shown in Figure 3-18b. Another example is the hydrogen bond between the 2′-OH in the ribose of U^8 and nitrogen-1 of A^{21} (Fig. 3-18c). There appear to be at least five such interactions between bases and 2′-groups of ribose in the tertiary structure. These bonds appear to play a significant role in stablizing the tertiary structure. Furthermore, the occurrence of such interactions immediately suggests a major reason why nature uses both DNA and RNA. The backbone of

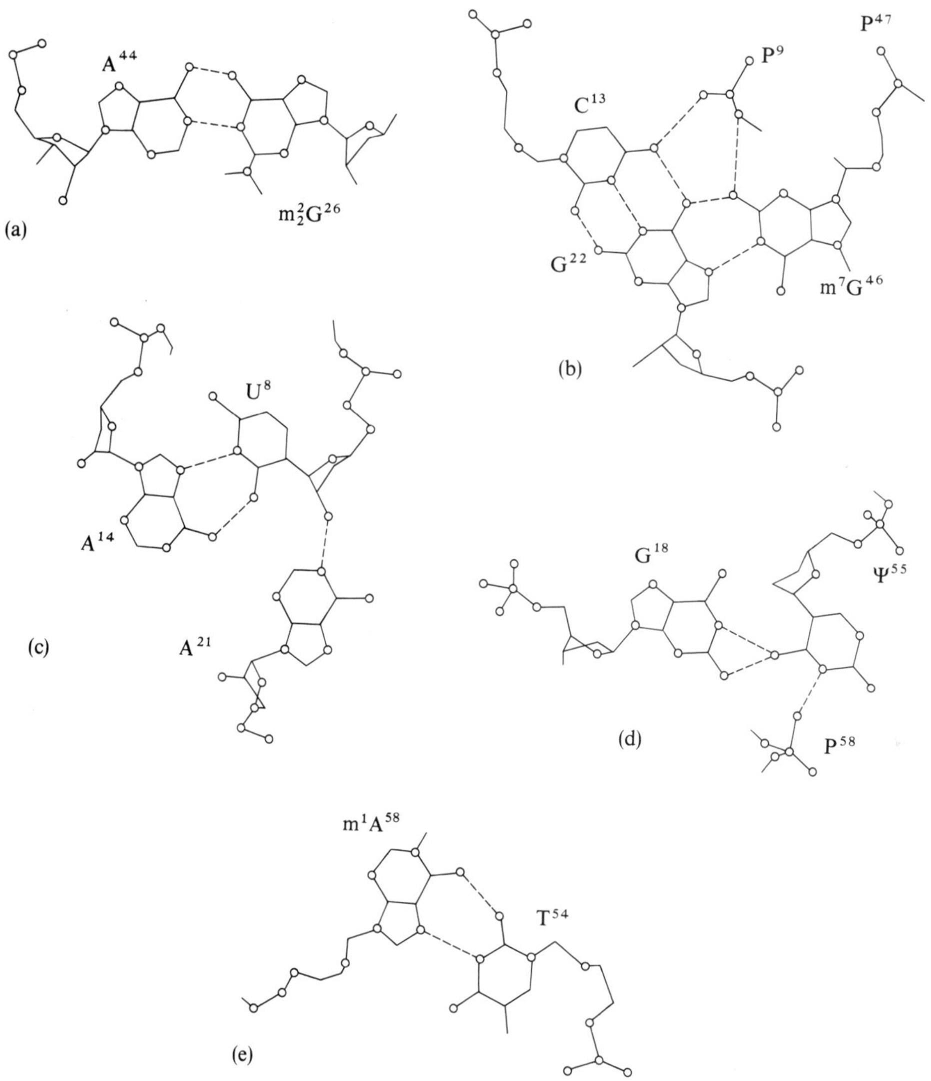

(a) A^{44} $m_2^2G^{26}$

(b) C^{13} P^9 P^{47} G^{22} m^7G^{46}

(c) U^8 A^{14} A^{21}

(d) G^{18} Ψ^{55} P^{58}

(e) m^1A^{58} T^{54}

Figure 3-18

Some of the tertiary hydrogen-bonded interactions found in yeast tRNA^{Phe}. Oxygen, nitrogen, and phosphorus atoms are shown as hollow circles; hydrogen bonds are shown as dashed lines. (See text for discussion.) [After G. J. Quigley and A. Rich, *Science* 194:791 (1976).]

DNA is chemically much more stable than that of RNA. But because it lacks 2′-OH groups, it probably is less able to form tightly folded tertiary structures.

A third interesting feature of the $tRNA^{Phe}$ structure is the occurrence of several sharp bends in the phosphodiester chain. As discussed in Chapter 6, the nucleic acid backbone is fairly rigid because only a restricted range of rotation angles for various backbone atoms are energetically favorable. However, as shown in Figure 3-19, tertiary hydrogen bonds appear to stabilize the formation of a bend in the backbone in the anticodon and pseudouridine loops. This has been called the uridine turn because a U or Ψ residue plays a critical role in the tertiary hydrogen bonding. It may play a general role in nucleic acid tertiary structures similar to the role played by β bends in proteins.

The $tRNA^{phe}$ cloverleaf consists of 20 base pairs containing 52 hydrogen bonds. This is supplemented in the tertiary structure by at least 40 additional hydrogen-bonded interactions. Most of these are not normal Watson–Crick base-paired arrangements. One strong argument for the significance of tertiary interactions is that they always occur among residues that either have a constant identity in all known tRNA sequences, or show coordinated base changes in different species in such a way as to retain equivalent possible tertiary pairing. Thus there is every reason to believe that most features of the tRNA tertiary structure may be common to all species. The structures of a few of these interactions, as they occur in yeast $tRNA^{Phe}$ and as they postulated to occur in other tRNAs, are shown in Figure 3-20.

The G·C pair between positions 15 and 48 in $tRNA^{Phe}$ was predicted by Michael Levitt. He noticed that, whenever the base in position 15 is a G, position 48 is C; and whenever 15 is A, 48 is U. However, the two strands containing these bases run in parallel directions. Thus it is not surprising that the structure of the base pair formed there is not a Watson–Crick structure. Instead, one base is rotated roughly 180° with respect to the normal Watson–Crick position. The resulting structure can be formed in essentially the same way by G·C or A·U (Fig. 3-20a). However, there no longer is a dyad axis, and so C·G or U·A pairs here would require different positions of the phosphodiester backbone. Because these combinations are never observed, the backbone structure in this region probably remains constant.

As a second example, consider the C^{13}–G^{22}–m^7G^{46} base triple (Fig. 3-20c). Here, the first two bases are Watson–Crick paired, whereas the third is not. These three bases are constant in most other species of tRNAs of the same general class. In one case, yeast $tRNA^{Arg}$, the bases are U^{13}–A^{22}–A^{46}. However, this sequence can assume a virtually identical base-triple structure. In still another case, the bases are U^{13}–G^{22}–G^{46}, and here the same base-triple structure is possible if one allows a U·G pair between the first two (Fig. 3-20d, e).

The bases U^8, A^{14}, and A^{21} are common to all tRNAs, and so the interaction formed by these bases also could be a constant structural feature (Fig. 3-18c). Similarly, the sequence G^{18}–G^{19} is a constant feature in all tRNAs, as is the sequence Ψ^{55}–C^{56}; thus the tertiary interaction between these two sequences appears to be conserved (Fig. 3-18d).

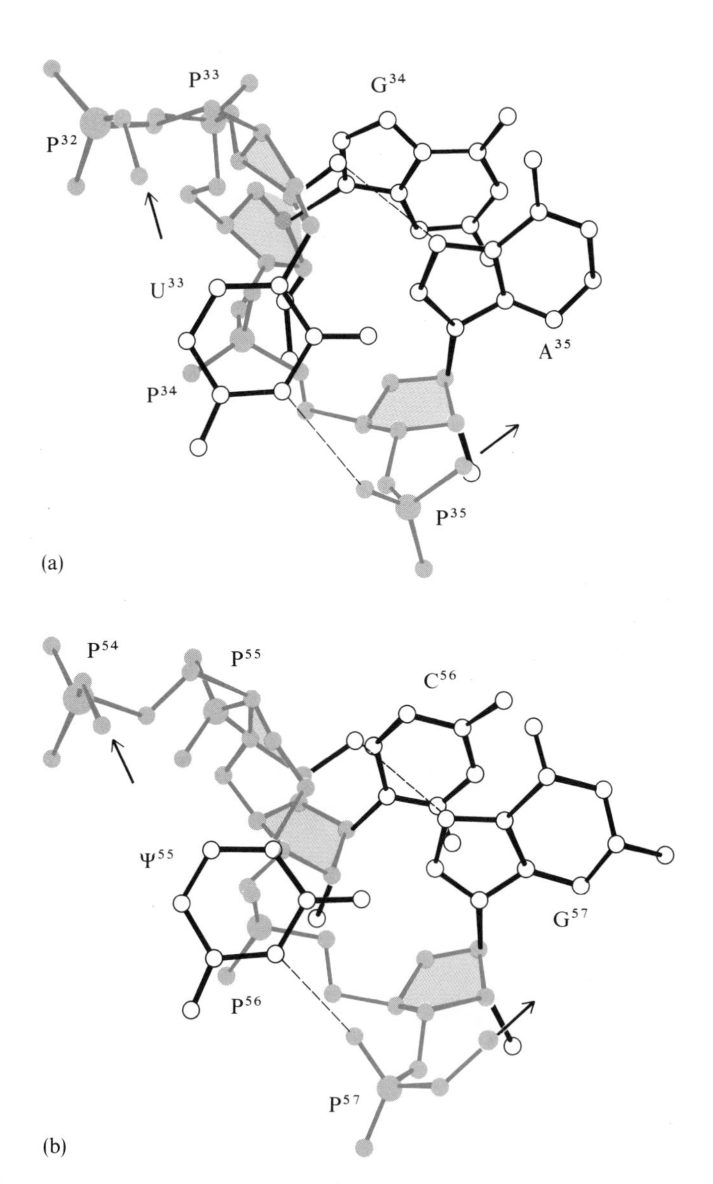

Figure 3-19

Two examples of uridine turns found in yeast $tRNA^{Phe}$. For clarity, the phosphodiester backbone is shown in color and pentose rings are shaded. Phosphates are numbered as the 3′-esters of nucleotides. Note the hydrogen bonds between the 2′-OH of U^{33} or Ψ^{55} and the N^7 of A^{35} or G^{57}. Note also the hydrogen bonds between N^3–H of U^{33} or Ψ^{55} and the phosphate of A^{35} or G^{57}. These tertiary interactions presumably stabilize the turns. [After G. J. Quigley and A. Rich, *Science* 194:791. Copyright © 1976 by the American Association of Science.]

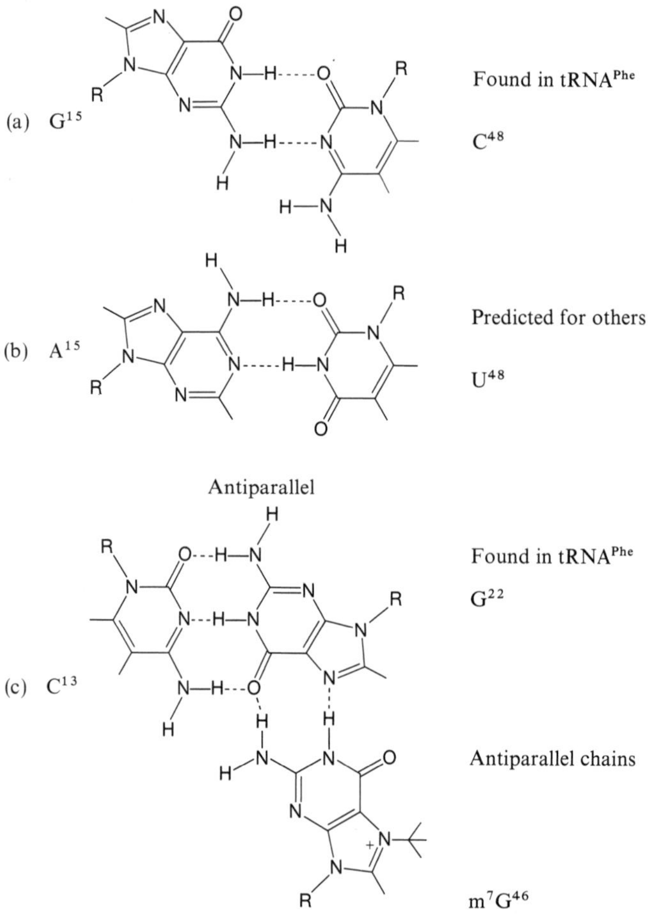

(a) G^{15} Found in tRNA^{Phe} C^{48}

(b) A^{15} Predicted for others U^{48}

Antiparallel

(c) C^{13} Found in tRNA^{Phe} G^{22}

Antiparallel chains

m^7G^{46}

Figure 3-20

Two tertiary base-pair interactions found in yeast tRNA, and similar interactions predicted to occur in other tRNAs. (See text for discussion.) [After A. Klug et al., *J. Mol. Biol.* 89:511 (1974).]

The base pair between T^{54} and m^1A^{58} (Fig. 3-18e) should also be a common structural feature of all tRNAs. Note that a Watson–Crick base pair is impossible for m^1A because the normal N^1 acceptor position is replaced by a methyl. In the tRNA^{Phe} structure, the T^{54}–m^1A^{58} base pair uses the N^7 acceptor nitrogen of the adenine ring rather than N^1. These kinds of base pairs are called Hoogsteen structures after the crystallographer who first observed them in complexes of nucleic acid constituents.

There is reason to expect that all tRNAs, and not just those of the same sequence class as yeast tRNA^{Phe}, can adopt a similar tertiary structure. These other classes have large extra loops, different sized dihydrouridine loops, and so forth. In each case,

Antiparallel

(d) U^{13} — A^{22} — Predicted for others

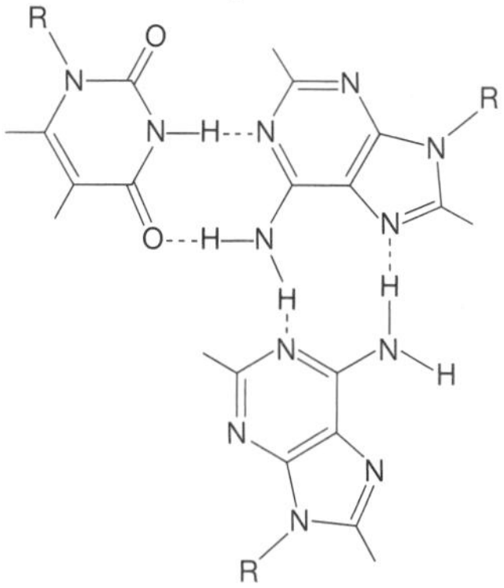

Antiparallel chains

A^{46}

(e) U^{13} — G^{22} — Predicted for others

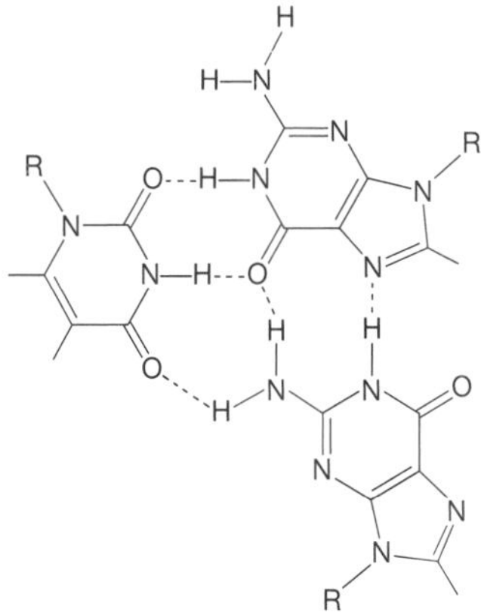

Antiparallel chains

G^{46}

though, the variations start at a region of the structure where residues are near the surface in such a way that extra ones can be added without requiring a major perturbation of the basic tertiary structure.

Base stacking in tRNA

We have concentrated thus far on the base pairing in $tRNA^{Phe}$, but there is another very important aspect to the structure. Planar aromatic compounds, such as purines and pyrimidines, interact strongly to form parallel stacked structures. This stacking can be observed in crystals and in solution. The thermodynamic origins of stacking are explored in detail in Chapter 6. Here it is sufficient to note that stacking interactions, although lacking the specificity of base pairing, are energetically at least as

Figure 3-21

Chemical modification of yeast $tRNA^{Phe}$ by reagents believed to be essentially specific for single-strand residues. [Data mostly from D. Rhodes, *J. Mol. Biol.* 94:449 (1975); as provided by B. F. C. Clark. Drawing by Irving Geis.]

important in determining nucleic acid conformation. Almost a decade before the first $tRNA^{Phe}$ tertiary structure became known, it was predicted that such structures would attempt to maximize base stacking. This is exactly what is found for $tRNA^{Phe}$ (Fig. 3-17). Only four of the 76 bases in the molecule (U^{47}, D^{16}, D^{17}, and G^{20}) are not involved in stacking. Dihydrouridines are never able to stack because the reduced pyrimidine ring is no longer planar. Therefore, base stacking occurs in the tRNA structure to almost the maximum conceivable extent. Note from Figure 3-17 that all of the stacked bases are arranged in two continuous perpendicular units. Any points of structural flexibility within tRNA are likely to be located at regions where the stacking pattern is broken.

An interesting feature of base stacking in tRNA is that some bases are intercalated between others. Note in Figure 3-17 that residues U^8 and A^9 are spread about 6.5 Å apart in the vertical direction rather than the normal 3.9 Å. In between these two bases, residue C^{13} is inserted in such a way that it can stack with both.

Biochemical and chemical properties and the $tRNA^{Phe}$ tertiary structure

One consequence of the extensive stacking and base pairing in $tRNA^{Phe}$ is that most bases are rendered inaccessible to the solvent. The most conspicuous exceptions to this are the CpCpA terminus and the three bases constituting the anticodon sequence. It is reasonable that these positions should be exposed. The latter must be available for base pairing to mRNA on the ribosome. The former is almost surely involved in the binding of tRNA to the peptidyl transferase center of the 50S ribosomal subunit because tRNA fragments containing just CpCpA-aminoacyl will bind and function quite well under certain conditions.

It has been known for a while that most of the bases in tRNA are protected from reaction with a variety of chemical reagents. In fact, this general unreactivity was one of the first pieces of evidence that tRNA possesses a tertiary structure. Starting from the known tertiary structure, one can now ask how consistent this is with the pattern of reactive and unreactive bases. Figure 3-21 summarizes the reaction pattern of yeast $tRNA^{Phe}$ with reagents specific for cytidine (methoxyamine), guanine plus uracil plus dihydrouridine (a carbodiimide), adenine (perphthalic acid), and guanine (kethoxal). It is evident that only a few positions are readily accessible to this broad spectrum of reagents, and it is easy to rationalize all of the highly reactive sites from the known three-dimensional structure of the tRNA. Most known chemical modifications in tRNAs other than yeast $tRNA^{Phe}$ can be explained by assuming that all tRNAs share a common structure.

Many species of *E. coli* tRNA have a 4-thiouridine residue in position 8, rather than the normal uracil present in yeast $tRNA^{Phe}$. It is known that the sulfur atom can be derivatized, even by rather large reagents, without impairing the ability of tRNA to participate in various steps of protein synthesis. This is reasonable on two counts. The 4-position of U^8 in $tRNA^{Phe}$ lies at the bottom of a broad trough on the surface of the molecule. Although U^8 is involved in a base-triple interaction, its hydrogen bonding involves oxygen-2 and not oxygen-4 (see Fig. 3-18c).

A particularly interesting chemical-modification result in many *E. coli* tRNAs is the cross-link efficiently formed between 4-thio-U^8 and C^{13} by ultraviolet irradiation. The resulting tRNA is still quite active biologically. This indicates that these two positions must be close in the native tertiary structure of the free molecule—as indeed they are, as discussed in the previous sections. They must also be able to remain close throughout all stages of function. The photochemistry of the thio-U–cytidine reaction was studied in model systems and, from the mechanism and the structure of the product, it was possible to make a very precise prediction of the relative spatial position that these two residues must have in a tRNA to allow for efficient photo-crosslinking. The prediction is in excellent accord with the known structure of yeast $tRNA^{Phe}$ (see Chapter 24).

Many of the chemical-modification results were known for tRNAs before x-ray studies at high resolution provided definitive structural information. In addition, a host of physicochemical studies had provided estimates of base pairing, proximity

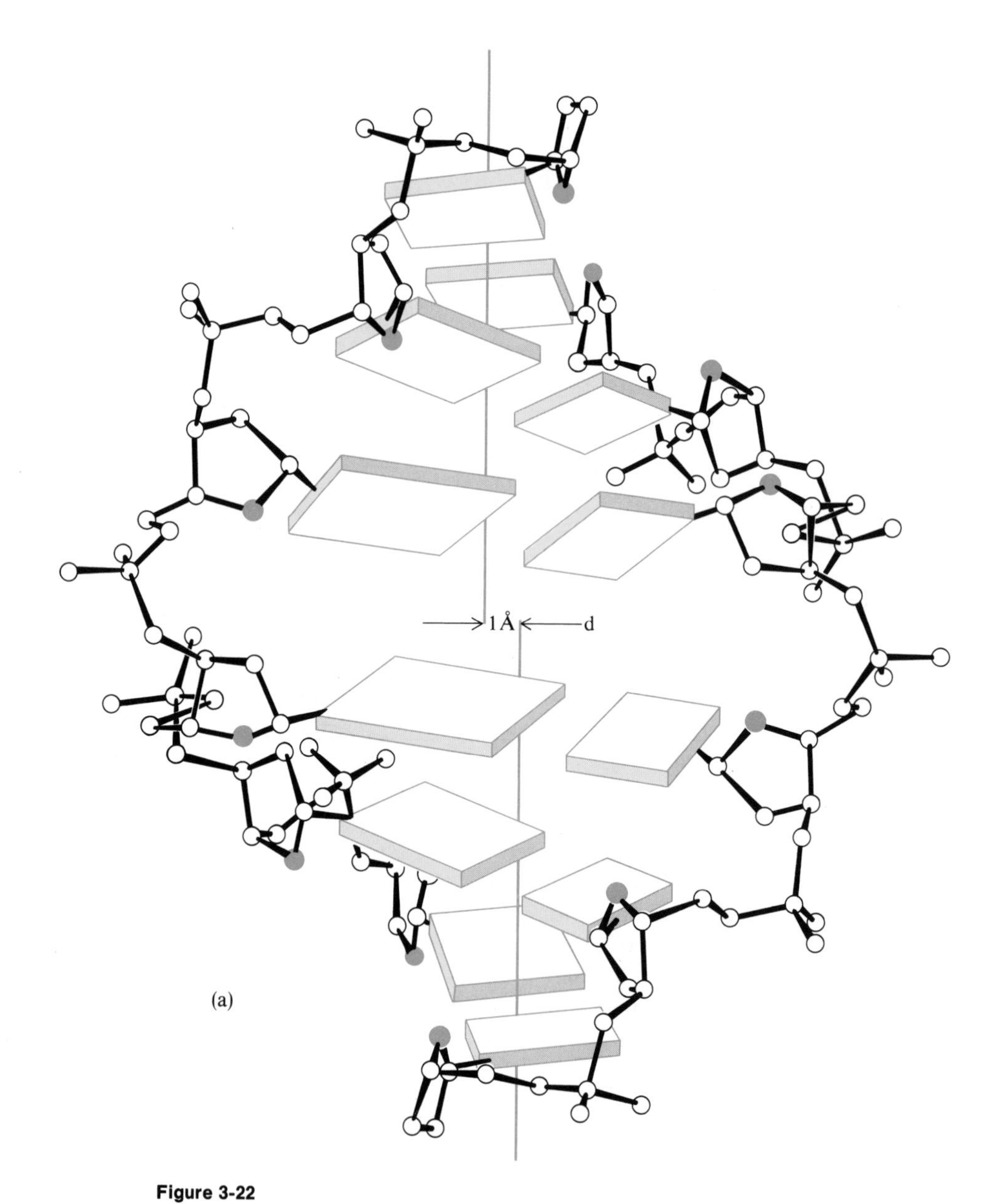

Figure 3-22

Structure of one possible DNA double-strand kink. (**a**) View down the dyad axis. (**b**) View from the side. [From H. Sobell et al., *J. Mol. Biol.* 114:333 (1977).]

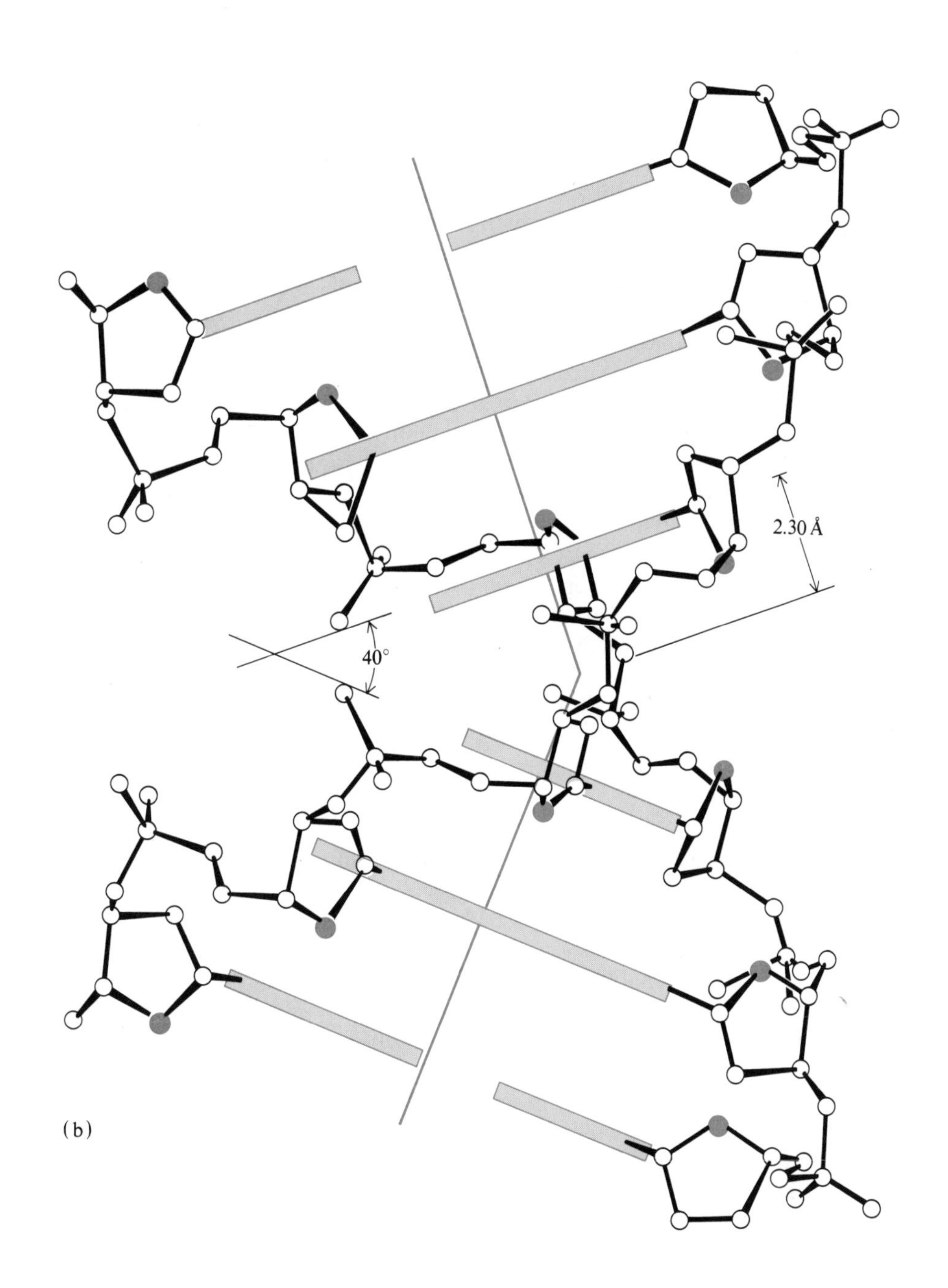

(b)

of various regions of the molecule, and overall size and shape. Sequences were known for more than 30 tRNA species, all of which were presumed to have the same structure; also some mutations were available in which an altered function suggested a structural alteration. It is fair to ask, then, how accurate were attempts to deduce the tertiary structure of tRNA from all of the solution evidence? The answer is that some predictions were fairly accurate. Levitt correctly postulated half of the tertiary interactions, and several models came up with shapes qualitatively similar to the actual structure. However, no prediction correctly provided a really good representation of the path followed by the polynucleotide chain. Part of the problem was that quite a few models could be constructed consistent with the bulk of the available data, and confidence in the rest of the data was not strong enough to allow discrimination among these models. Now that we have seen one RNA tertiary structure, it will be interesting to see if the principles learned from it can be used to enable more accurate prediction of the three-dimensional structures of other small RNAs.

Though very few details are known it seems likely that other small RNAs also have well-defined tertiary structures. For example, 5S rRNA can be changed into several denatured forms by mild treatment. These forms can be separated physically from one another and from the native structure. Larger RNAs surely contain tertiary structure features, but whether there is an unique native tertiary structure is unknown. Cross-linking studies indicate that, in 16S rRNA and 23S rRNA, regions distant in the primary structure come together to form double helices. The pattern of cross-links seen is relatively insensitive to the environment of the RNA, suggesting that these tertiary features are present in all molecules.

Tertiary structure in DNA

No detailed tertiary structure information is available for any DNA. However, there certainly has been considerable interesting speculation on some possible aspects of DNA folding. It is known that, in cells and viruses, DNA is usually packaged into fairly compact structures. A question of great interest is how the rigid DNA duplex can be bent or folded to enable such organized structures to form. To interpret the hydrodynamic properties of large DNA, it is necessary to postulate that the double helix has some flexibility (see Chapter 19). DNA has been modeled as a wormlike coil that is slightly deformable at each residue.

Such a macroscopic treatment is sufficient to explain hydrodynamic behavior, but on a molecular level one would like to know the specific origin of the flexibility. An interesting idea, first proposed by Francis Crick and Aaron Klug, is that kinks can form in the DNA helix by unstacking two adjacent base pairs and untwisting the backbone a bit. Henry Sobell has studied the stereochemistry of possible kinks very extensively through molecular model building. A particularly attractive possible kink structure is shown in Figure 3-22. Here the helix axis bends 40° and is displaced laterally 1 Å. The base pairs are displaced 2.30 Å away from the center of the kink (whereas they are placed 1.70 Å away from the pseudo C_2 axis in DNA-B). In order to make this kink, the pattern of sugar packing is altered from all $C^{2'}$-*endo* in DNA-B to $C^{3'}$-*endo*(3′–5′)$C^{2'}$-*endo* across the kink. This change in sugar pucker may also occur when base pairs separate to allow a planar molecule to intercalate between them (Chapter 23).

A critical feature of the kink shown in Figure 3-22, and of other kinks that have been suggested, is that the kink can occur only in one direction relative to the backbone. Thus the direction of a possible kink will rotate with the helix as one moves away from a fixed point. If kinks are placed into the helix at regular intervals, a superhelical structure will be generated in the DNA. The nomenclature for such regular patterns of kinks is given in Figure 3-23a. For example, κ kinks occur every 10 base pairs. This pattern generates a left-handed superhelix with a diameter of about 100 Å, as shown in Figure 3-23b. Each "residue" of the superhelix consists of 10 base pairs. The superhelix contains approximately 1.5 turns per 140 base pairs.

A second interesting kinked superhelix is shown in Figure 3-23c. This is a β-kinked helix. It is right-handed and has 9.6 base pairs per turn. At this stage, there is no direct evidence that such kinked helices exist in packaged DNA structures. In fact, many researchers favor schemes in which DNA flexibility or bending is generated by small perturbations in the rotation angles of the polypeptide backbone. For example, Wilma Olson has shown that a wide variety of DNA superhelices can be generated by regularly bending a DNA-B reference helix.

More complex examples of DNA tertiary structure occur in circular duplex DNAs. These molecules have some very special tertiary structure features as a result of topological constraints introduced by the presence of linked circles (Chapter 24).

3-6 QUATERNARY STRUCTURE

There are very few examples of quaternary structure in pure nucleic acid systems. One apparent case is tumor virus RNAs that apparently consist of two identical subunits. Electron microscopic evidence for such a structure, the 52S RNA from RD-114 virus, is shown in Figure 3-24. Hydrogen-bonded base pairing between the two RNA subunits presumably stabilizes the quaternary structure. If this is interstrand Watson–Crick base pairing, one must ask why it is favored over intrastrand pairing.

Associations between nonidentical RNAs are much easier to rationalize. For example, tRNA species with complementary anticodons form quite tight intermolecular complexes in vitro. Whether these have any significance in vivo is unknown, but they may be a forerunner of RNA quaternary structure interactions still to be discovered. Some sequence complementarity exists between regions of the 16S and 23S *E. coli* ribosomal RNAs. Quaternary base pairing between these regions may occur when a 70S ribosome is formed from 30S and 50S subunits.

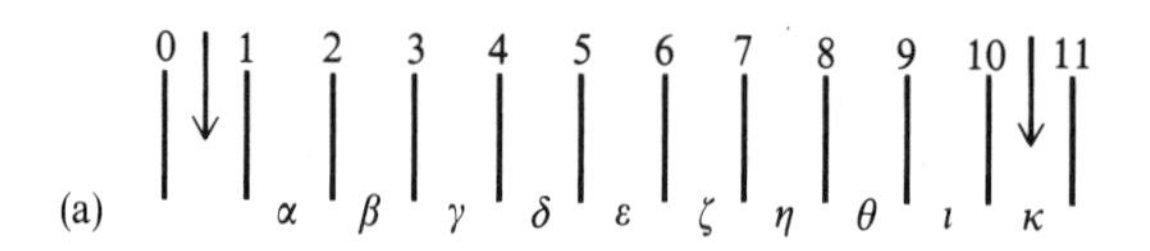

0
1
2
3
4
5
6
7
8
9
10
11
(a)
α
β
γ
δ
ε
ζ
η
θ
ι
κ

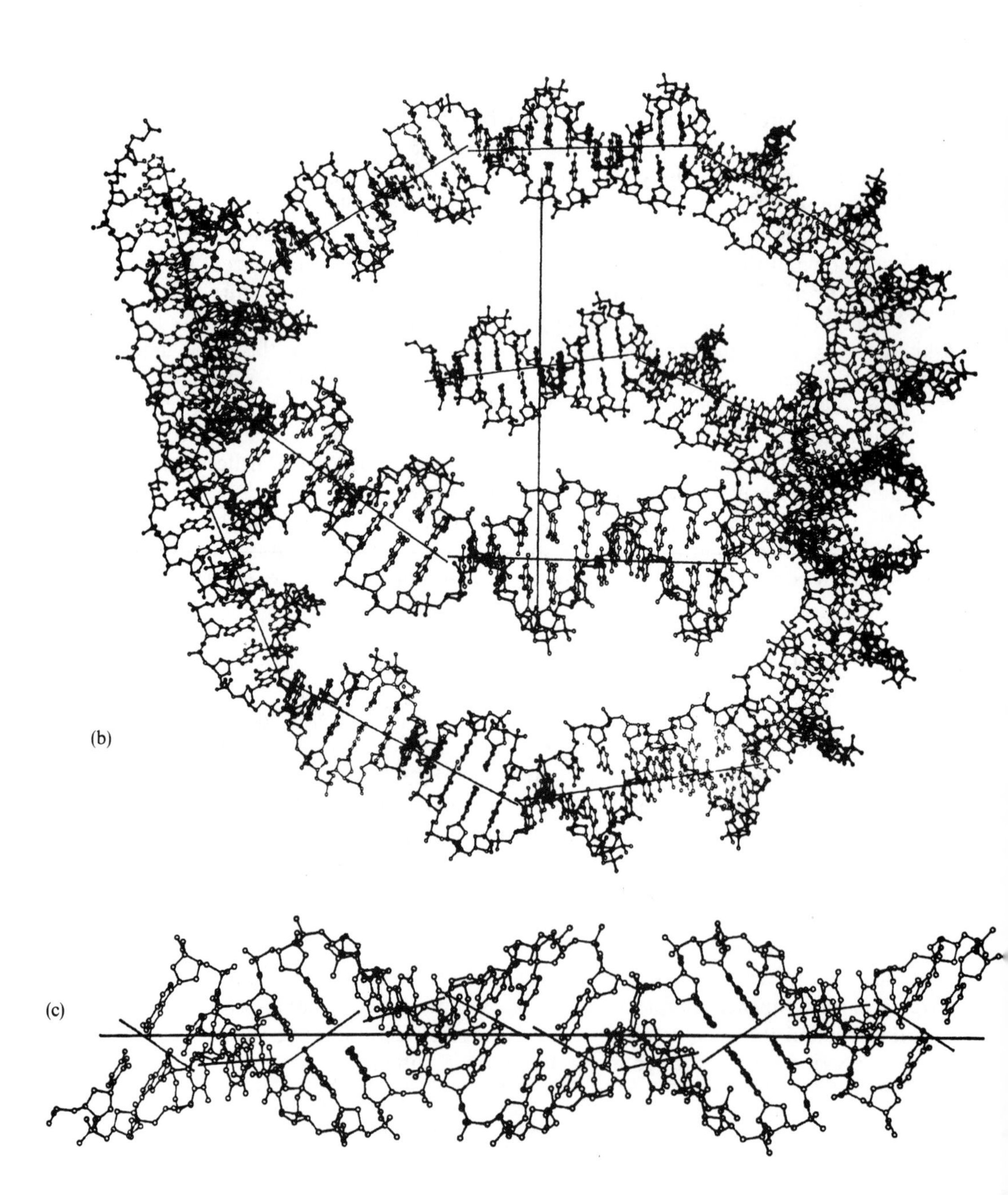

(b)
(c)

Figure 3-23

Superhelices formed by regularly spaced kinks. (**a**) Nomenclature for periodic kinks. (**b**) DNA kinked every ten base pairs (κ-kinked). (**c**) DNA kinked every two base pairs (β-kinked). [From H. Sobell et al., *J. Mol. Biol.* 114:333 (1977).]

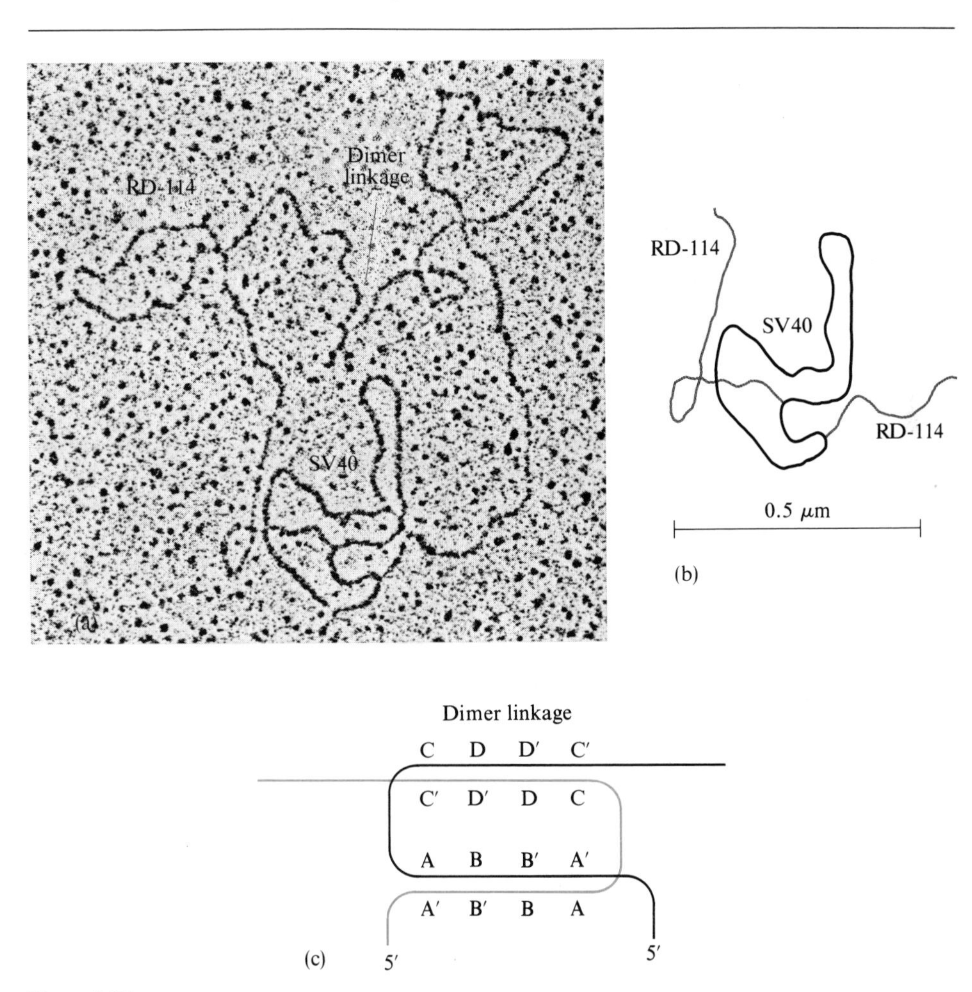

Figure 3-24

A tumor virus RNA, prepared as a complex with SV40 DNA. RD-114 is an endogeneous feline type-C virus. It has two 26S RNA subunits grouped into the intact 52S structure; each subunit ends in a stretch of poly A. To mark the ends of the RNAs, the RD-114 species was hybridized to a circular SV40 DNA chemically modified to add poly dT sequences at randomly placed sites. Thus the structure of the complex is stabilized by two short rA·dT duplexes. (**a**) Electron micrograph. (**b**) Drawing from the micrograph, identifying the RD-114 and SV40 chains in regions where they may not be clear in the micrograph. Note that the two RNA strands have a similar secondary-structure loop about two-thirds of the way toward the 5′-end. (**c**) A plausible possibility for the structure at the 5′-ends. (The actual structure is unknown.) Primes denote sequences complementary to unprimed regions. [Parts a and b from W. Bender and N. Davidson, *Cell* 7:595 (1976). Copyright © MIT. Published by the MIT Press.]

Summary

Nucleic acids consist of the four nucleosides (A, G, C, and T in DNA or U in RNA) linked by phosphodiester bonds.

Base composition is not an especially significant variable in determining structure, but base sequence is a major factor. Strong pairing interactions between A with T or U, and G with C are the predominant feature of all natural nucleic acids. These lead to double-strand helical structures that contain dyad axes in the plane of each base pair. This means that the two strands of the double helix must run in antiparallel directions. DNAs usually contain two perfectly complementary strands, and they thus form a continuous double helix. RNAs are usually composed of a single covalent strand that forms helical regions by folding back on itself into a series of hairpins and loops.

A number of different helix geometries are consistent with the constraints imposed by base pairing. DNA in aqueous solution adopts the B-form helix, in which base pairs are perpendicular to the helix axis, but under some conditions this can be converted to an A-form helix. The A-form helix closely resembles the RNA 11 or RNA 10 helices found for double-stranded RNA. In these structures the base pairs are tilted slightly towards the helix axis.

Yeast $tRNA^{Phe}$ is the only nucleic acid of known tertiary structure. This structure appears to be stabilized by numerous hydrogen-bonded interactions in addition to those involved in normal base pairing. In addition, the structure appears to maximize the possibility of stacking between the planes of adjacent bases. The tRNA structure shows evidence of the multiple-stranded interactions and specific patterns of chain bending that are so typical of protein tertiary structures. The 2′-OH appears to play a critical role in RNA tertiary structure formation. The absence of this group in DNA may explain why the tertiary structure properties of the two basic kinds of nucleic acids are so different. DNA tertiary structure is not yet well understood, but may involve periodic or occasional kinks in the double helix. Only a few examples of nucleic acid quaternary structure have thus far been found.

Problems

3-1. You wish to test whether a DNA strand containing 10,000 bases includes a particular sequence of 20 bases. One way to do this is to see whether the complementary 20-base-pair sequence will hybridize. However, it is difficult to synthesize a sequence 20 bases long. Calculate the minimal-length subsequence that should be used in the hybridization experiment. It must be long enough to insure that there is less than 1% chance of accidental occurrence of the complement somewhere in another region of the DNA just due to chance. You may assume that the DNA contains 25% of each base.

3-2. A DNA has a molecular weight of 2.5×10^8. Calculate the length in cm if it is in the B helix form; in the A helix form. How many helix turns are there in each form?

3-3. What do you think will happen if two complementary circular single-stranded DNAs are mixed in the same solution?

3-4. Suppose a particular polynucleotide prefers to form double helices with parallel strands. It is possible, in principle, for a long single chain to form a parallel-stranded double helix in which all bases are paired. What will this structure look like?

3-5. Consider a single strand of DNA with exactly the same primary structure as yeast $tRNA^{phe}$. In what ways do you expect the tertiary structure formed by this DNA to differ from that of tRNA?

References

GENERAL

Bloomfield, V. A., D. M. Crothers, and I. Tinoco, Jr. 1974. *Physical Chemistry of Nucleic Acids.* New York: Harper & Row. [A very useful advanced survey.]

Cohn, W. E., ed. 1976– *Progress in Nucleic Acid Research and Molecular Biology.* New York: Academic Press. [About two volumes per year of useful review articles.]

Duschensne, J., ed. 1975. *Physical Chemical Properties of Nucleic Acids.* 3 vols. London: Academic Press.

Guschlbauer, W. 1976. *Nucleic Acid Structure.* New York: Springer-Verlag.

Sundaralingam, M., and S. T. Rao, eds. 1975. *Structure and Conformation of Nucleic Acids and Protein–Nucleic Acid Interactions.* Baltimore: University Park Press.

Ts'o, P. O. P., ed. 1975. *Basic Principles of Nucleic Acid Chemistry.* 2 vols. New York: Academic Press. [Many excellent comprehensive chapters.]

SPECIFIC

Barrell, B. G., and B. F. Clark, 1974. *Handbook of Nucleic Acid Sequences.* Oxford: Joynson-Bruvvers.

Jack, A., J. E. Ladner, and A. Klug. 1976. Crystallographic refinement of yeast phenylalanine transfer RNA at 2.5 Å resolution. *J. Mol. Biol.* 108:619.

Kim, S. H. 1978. Crystal structure of yeast $tRNA^{Phe}$. Its correlation to the solution structure and functional implications. In *Transfer RNA*, ed. S. Altman (Cambridge: MIT Press), p. 248.

Ladner, J. E. 1978. Structure of transfer ribonucleic acid. In *Biochemistry of Nucleic Acids*, vol. 2, ed. B. T. C. Clark (Baltimore: University Park Press), p. 1.

Rich, A., and T. Rajbhanday. 1976. Transfer RNA: Molecular structure, sequence, and properties. *Ann. Rev. Biochem.* 49:805.

Sobell, H. M., C. C. Tsai, S. C. Jain, and S. G. Gilbert. 1977. Visualization of drug–nucleic acid interactions at atomic resolution, III: Unifying structural concepts in understanding drug–DNA interactions and their broader implications in understanding protein–DNA interactions. *J. Mol. Biol.* 114:333.

Wu, R. 1978. DNA sequence analysis. *Ann. Rev. Biochem.* 47:607.

Younghusband, H. B., and R. B. Inman. 1974. The electron microscopy of DNA. *Ann. Rev. Biochem.* 43:605.

4
Other biological polymers

Virtually all of the examples to be given in subsequent chapters are drawn from studies on pure nucleic acids or proteins. These are the species best understood at present, and therefore they are the best for illustrating how techniques and principles are applied to understanding biopolymer structure and function. Here the properties of some less-well-studied biological systems will be described briefly. These fall into three general classes. Some are biopolymers composed of monomer units other than amino acids or nucleotides; an example is polysaccharides. Others are biopolymers that contain more than one type of monomer, or else are assemblies of more than one class of biological macromolecules; examples are nucleoproteins, glycoproteins, and glycolipids. The third class includes assemblies of macromolecules associated with lipids, such as the lipid bilayer of cell membranes. Such samples are composed of heterogeneous phases, and so introduce unique complications.

It is not clear at present whether extant techniques and principles, largely developed in studying proteins and nucleic acids, are capable of fully revealing or explaining the behavior of these other systems. Most are now the subject of intense research activity, and we review what is presently known so that the reader can contrast these types of molecules with simple proteins and nucleic acids. (Additional discussion of membranes and protein–lipid interactions will be found in Chapter 25.)

4-1 POLYSACCHARIDES

Primary structure

Polysaccharides are polymers of sugars. The simplest ones possible are linear homopolymers in which the fundamental unit is a single sugar, and all links between sugars are identical. Examples are amylose, a component of starch that serves as a storage reservoir of nutrients in plants; cellulose, a major component of wood or other plant fibers; and chitin, which forms the exoskeletons of insects and other arthropods. Figure 4-1 shows the covalent structures of these polymers.

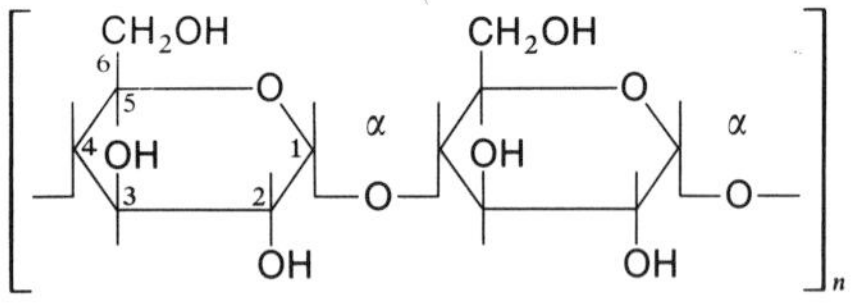

Amylose

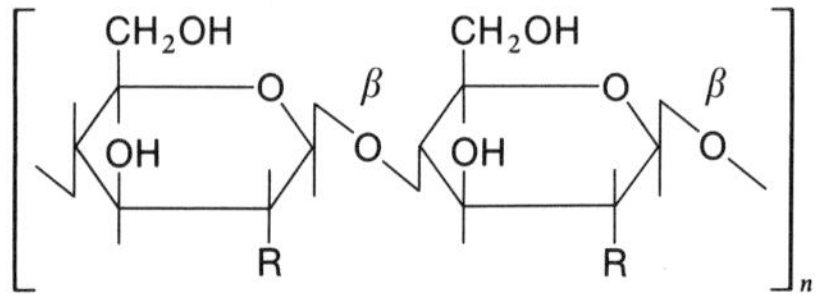

Cellulose, R = OH

Chitin, R = NH—C(=O)—CH_3

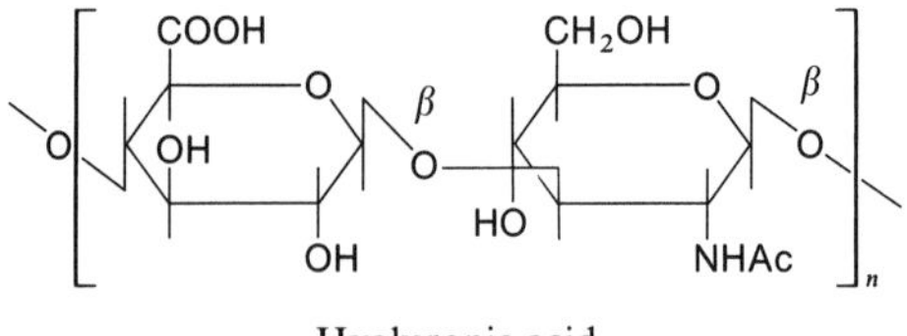

Hyaluronic acid

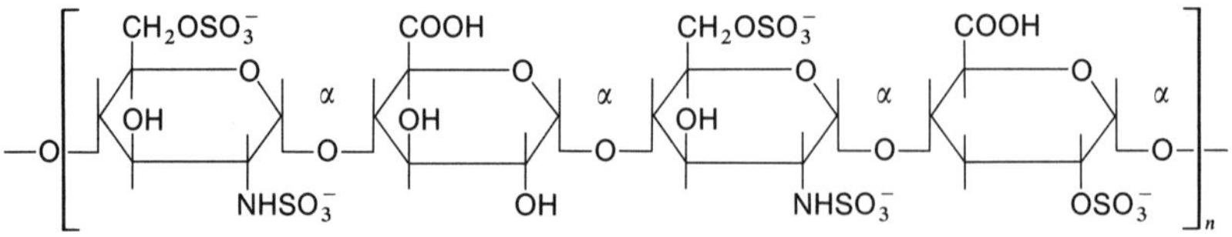

Heparin

Figure 4-1

Primary structure of several typical polysaccharide chains. The numbering scheme for hexose sugars is shown in the amylose structure. Note that α linkages start from the anomeric configuration of a sugar in which the C-1 bond to oxygen is on the opposite side of the sugar plane from the C-6 bond to oxygen. In β linkages, the C-1 and C-6 bonds to oxygen are on the same side of the sugar plane.

Cellulose and amylose are both homopolymers of glucose. The only difference in their primary structures is the stereochemistry at position-1 of the glucose. In the free sugar, two anomeric forms are in equilibrium:

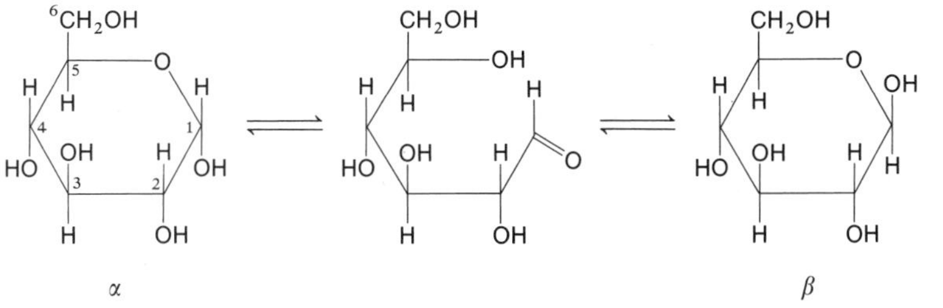

In polysaccharides, the hemiacetal at position-1 is replaced by an acetal, which is much less labile. As a result, the configuration at this position becomes locked in. Amylose has α-1,4 linkages between glucose, whereas in cellulose the linkages are β-1,4.

Other linear polysaccharides have a disaccharide as the fundamental repeating unit. One example (shown in Fig. 4-1) is hyaluronic acid, in which *N*-acetylglucosamine and glucuronic acid alternate in β-1,4 and α-1,3 linkages. Such alternating polymers generally form gels efficiently, and they frequently are found composing the matrix in which fibrous components of cell walls or other cellular support structures are imbedded. In more complex polysaccharides, the sequence of residues is not as regular. For example, the alginic acids of brown seaweed may consist of alternating blocks of different homopolysaccharides. Other matrix materials are sulfonated at various positions. The structure of heparin (Fig. 4-1) is a repeating sulfonated tetrasaccharide.

The primary structure of some polysaccharides is complicated by branching. In this respect, they are totally unlike proteins and nucleic acids, which always have linear or circular unbranched backbones. Glycogen, a storage form of glucose, consists mostly of linear α-1,4 polymerized regions. However, these are joined together by occasional α-1,6 branches, creating a complex network (Fig. 4-2). Some very complex polysaccharides make up the cell walls of higher plant forms. Pectins are acidic polysaccharides containing galacturonic acid, rhamnose, varying degrees of branching, and varying degrees of esterification of the galacturonate carboxyl groups. Hemicelluloses are similarly complicated polysaccharides. Some progress has been made in learning how all of these substances are organized in the plant cell wall.

Levels of structure in polysaccharides

The conformations of polysaccharides are rich and varied, and it is clear that the various levels of structure we have described for proteins or nucleic acids are useful concepts here as well. The primary structures of polysaccharides tend to be very regular, involving large blocks in which a single residue or a single sequence is repeated. In this respect they resemble such structural proteins as collagen.

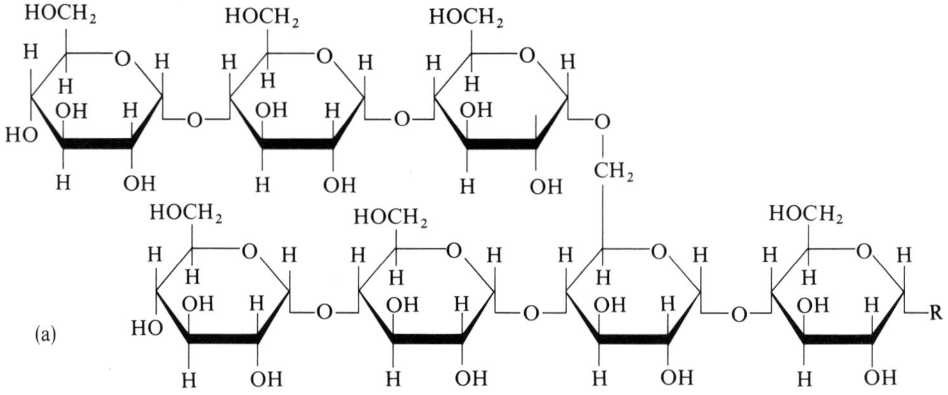

(a)

Figure 4-2

Structure of glycogen. (**a**) The covalent structure of two cross-branches. (**b**) A diagrammatic view of a larger section of the glycogen polymer, with the boxed region corresponding to the detailed structure of part a.

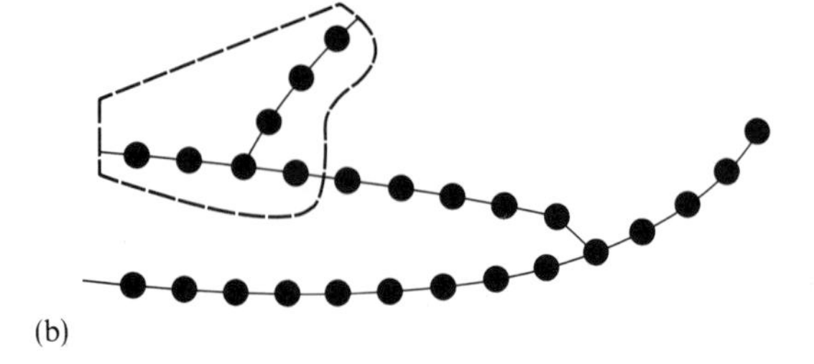

(b)

Fibrous polysaccharides, such as chitin and cellulose, form very rigid secondary structures. Steric hindrance between bulky sugar groups greatly limits the number of accessible conformations. The fibers are well ordered enough to give very detailed x-ray fiber diffraction patterns, but few of these patterns have been successfully analyzed to give unique structural models. Some polysaccharide fibers are composed of ribbons that contain two chains stabilized by interresidue hydrogen bonding. Other polysaccharide secondary structures are helices. For example, some algal cells use xylan (a β-1,3 polymer of the sugar xylose) in their walls instead of cellulose or chitin. Xylan forms a triple-stranded helix in which all three strands are parallel (Fig. 4-3a). The helix has six residues per turn and is stabilized by a central core of interstrand hydrogen bonds between sugar hydroxyl groups (Fig. 4-3b). Thus the general pattern of organization of xylan is very much like that of the collagen structure.

Figure 4-3

Three-dimensional structure of two polysaccharide helices as determined by x-ray fiber diffraction. (**a**) The β-1,3-xylan from blue-green algae. This is a triple-strand helix; one strand is shown in color. (**b**) A schematic representation of the hydrogen bonding that links the three strands. (See Atkins et al., *Proc. Roy. Soc. Ser. B* 173:209 [1969].) (**c**) The sodium salt of hyaluronic acid. This is a single-strand helical structure with four disaccharides per turn; just slightly more than one turn is shown. In the fiber are interstrand and intrastrand hydrogen bonds. Many coordinated Na^+ ions are present. Na^+–oxygen contacts are shown in detail for some of these. For clarity, one strand of the polymer is shown in color. (For more details, see Guss et al., *J. Mol. Biol.* 95:359 [1975].) [Parts a and b after drawings provided by Dr. E. Atkins. Part c after a drawing provided by Dr. W. Winter.]

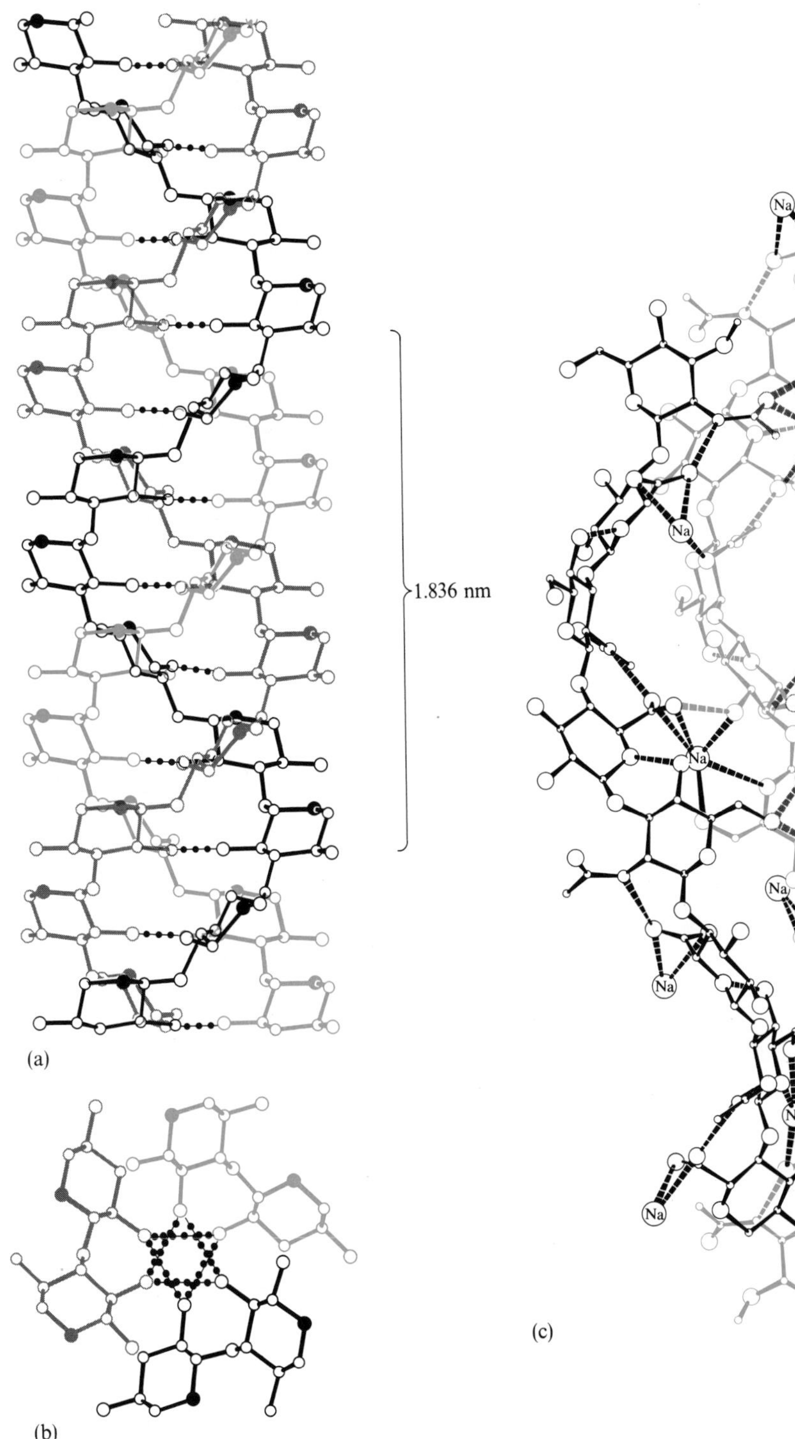
1.836 nm
(a)
(b)

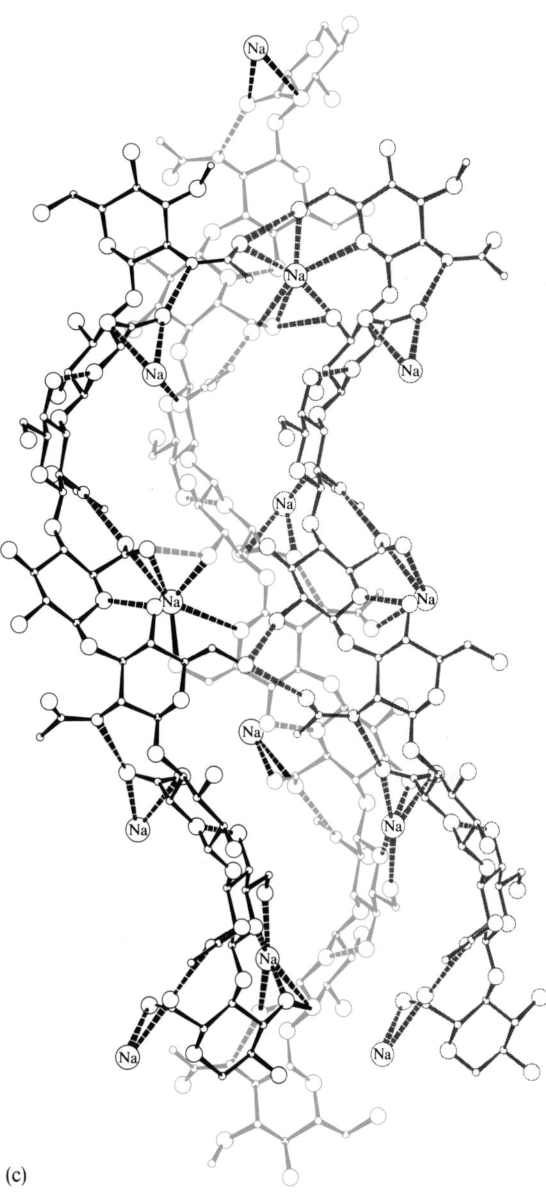
Na
(c)

Matrix-forming polysaccharides appear to form much more flexible structures. Carrageenans are double helices with each single strand having three disaccharide residues per turn. Hyaluronic acid forms a left-handed single helix with four disaccharide residues per turn. One strand is shown in color in Figure 4-3c. Intrastrand hydrogen bonds stabilize the helix. In the hyaluronic acid fiber, salt bridges and hydrogen bonds link adjacent helices. Irregularities in such a structure, caused either by insertion of occasional different saccharides or by branching, clearly will have a large effect on the overall configurations possible for a polysaccharide. Presumably these irregularities are responsible for the flexible mechanical properties of the fibers or gels it forms.

Tertiary structure in polysaccharides involves folding of the helical secondary structures. Quaternary structure is formed by association of individual helical or folded polysaccharides. For example, cellulose is most probably a two-strand ribbon. This can be folded into regular patterns. An ordered array of folded cellulose ribbons forms a quaternary structure. Cases are known in which this can extend over the whole surface of a plant cell. Other polysaccharides form gel structures. The gels produced by carrageenans appear to be a network of polysaccharide strands linked together by frequent stretches of double helix.

Questions about polysaccharide structure and function

It is worth listing some of the questions one would like to be able to answer about polysaccharides and some of the obstacles in studying these materials. How do particular modifications of the basic covalent structure alter secondary and tertiary structures? Are these modifications dispersed in any sort of regular way, leading to elements of ordered structure that have thus far eluded our attention? How is the synthesis of polysaccharides orchestrated in the cell in such a way as to produce desired macroscopic shapes and mechanical properties? For example, are particular chemical modifications introduced at random (so that only the number of each type is controlled) or are their spatial positions carefully regulated? What specific interactions occur between different polysaccharide molecules, and between polysaccharides and other cell constituents? Is the cell wall just a large association with many random features, incorporated molecules, and defects, or are particular short-range and long-range ordered features present? On a purely physical level, how can the mechanical properties of polysaccharide materials be explained quantitatively in terms of their chemical structure?

These are all interesting questions, but it probably will take much hard work to answer most of them. There are no biological assays yet established that can be used to determine whether a given polysaccharide sample is structurally intact, in the same way that assays are useful for proteins or nucleic acids. Most samples of polysaccharide appear to be heterogeneous in length, and many in composition as well. This seriously complicates the application of most of the physical techniques, which were designed to handle homogeneous systems. Spectroscopic techniques that look mostly at local structure are not bothered by heterogeneity as much, but the characteristics of polysaccharide residues, by and large, are not particularly favorable for most types of spectroscopy. Genetics is a powerful tool for rendering complex protein or nucleic acid systems more amenable to physical study. This tool probably will not be as useful for polysaccharides because the phenotypic expression of genetic effects on these substances is indirect. At present, it would seem that x-ray diffraction and electron microscopy of polysaccharides in intact cells (or at stages as close to this as possible) are likely to be among the most fruitful approaches. Improved preparative and purification procedures probably are a prerequisite to most high-resolution physical studies of polysaccharides in solution.

4-2 ASSOCIATIONS FORMED AMONG DIFFERENT MACROMOLECULE TYPES

Saccharide combinations with peptides, proteins, or lipids in bacterial cell walls

A number of complex polymeric structures in bacterial cell walls have been found to contain oligosaccharides or polysaccharides in combination with other materials. All known bacteria except mycoplasmas have a peptidoglycan network as the shape-maintaining structure. Also called murein, glycopeptide, or mucopeptide, this material contains long strands of polysaccharide with the repeating unit β-1,4-*N*-acetylglucosamine-β-1,4-*N*-acetylmuramic acid (Fig. 4-4a). These strands are cross-linked by short oligopeptide chains. The sequence of the peptides and the mode of cross-linking differ from one bacterial species to another.

Bacteria with different types of cell walls show different staining responses to the Gram procedure (treatment with crystal violet and iodine, followed by destaining with an organic solvent). Gram-negative bacteria are completely decolorized. Their surfaces consist of an outer wall that appears to be a fairly typical 80 Å lipid bilayer, a 20 Å to 30 Å peptidoglycan layer, a 75 Å gap (periplasmic space), and then an inner wall, which is another typical 80 Å lipid bilayer. The peptidoglycan of a typical Gram-negative bacterium such as *Escherichia coli* (Fig. 4-4b) has a low degree of cross-linking. The peptidoglycan layer contains about 3×10^6 repeating units and accounts for 5% to 10% of the dry weight of the cell wall. On the outer surface of the peptidoglycan, lipoprotein molecules are attached. These connect the peptidoglycan to the outer wall above it. The lipoproteins are attached by their C-termini to amino acids of the peptidoglycan. Their N-termini are modified to contain three fatty acid chains.

Gram-positive bacteria remain stained deep blue-black after the Gram procedure. They contain an outer cell wall anywhere from 100 Å to 500 Å thick and a single 80 Å lipid-bilayer inner wall. The thick outer wall is mostly peptidoglycan, which accounts for 40% to 90% of the dry weight of the cell wall. This peptidoglycan is highly cross-linked, as shown by the example of *Staphylococcus aureus* (Fig. 4-4b).

S. aureus

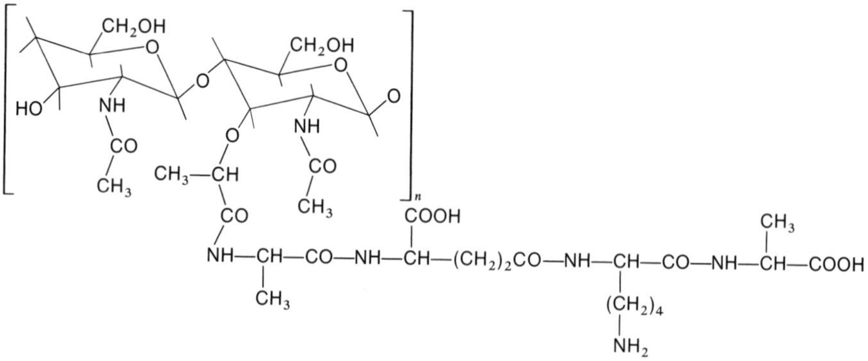

GlcNAc-MurNAc
|
L-Ala-D-isoGln-L-Lys-D-Ala

E. coli

GlcNAc-MurNAc
|
L-Ala-D-isoGlu-Dpm- -Ala

(a)

Figure 4-4

A section of the peptidoglycan cell-wall structure of two bacteria: *Escherichia coli*, a typical Gram-negative organism, and *Staphylococcus aureus*, a typical Gram-positive organism. (**a**) Chemical structure of the peptidoglycan chain. This consists of the repeating disaccharide unit *N*-acetylglucosamine–*N*-acetylmuramic acid with short peptide chains attached to the latter. Note that these contain D as well as L amino acids. In the *E. coli* structure Dpm is (*meso*)-diaminopimelic acid. (**b**) Schematic view of the cell wall, which consists of sheets of cross-linked peptidoglycans and lipid bilayers. In *E. coli* the lipoprotein is covalently attached to the peptidoglycan. The schematic structure shown for this protein is a prediction from the known sequence made by A. D. McLachlan (*J. Mol. Biol.* 122:493 [1978]). The structure shown for the lipoteichoic acid component of the *S. aureus* cell wall is purely a guess. For further details, see Braun and Hantke (1974), Stanier et al. (1976), and DiRienzo et al. (1978). [Drawing by Irving Geis.]

Mur
Glc
Glc
Mur
Glc
Mur
Mur
Glc
Mur
Polysaccharides
Lipid
bilayer
Gram negative: E. Coli
Mur
Glc
Mur
Glc
Mur
Glc
Mur
Glc
Mur
Glc
1 2 3 4
1 2 3 4
Gram positive: S. Aureus
Glc
Mur
Glc
Mur
Glc
Mur
Glc
Mur
Glc
Mur
1 2 3 4
1 2 3 4
(Gly)5
(Gly)5
(Gly)5
(Gly)5
(Gly)5
Nε
Nε
Nε
Nε
Lipoprotein
dimer
Periplasmic
space
Lipoteichoic
acid

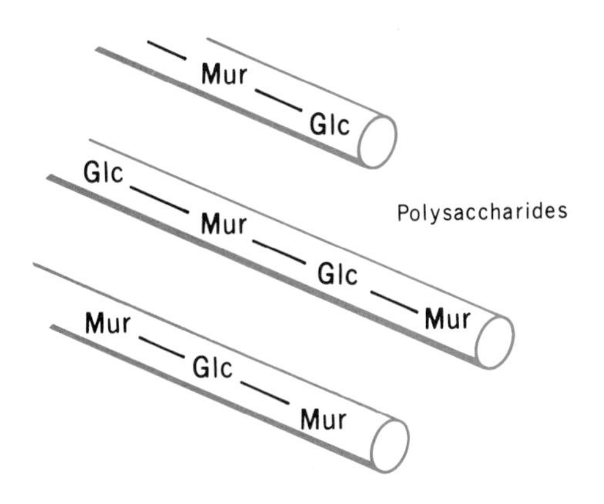

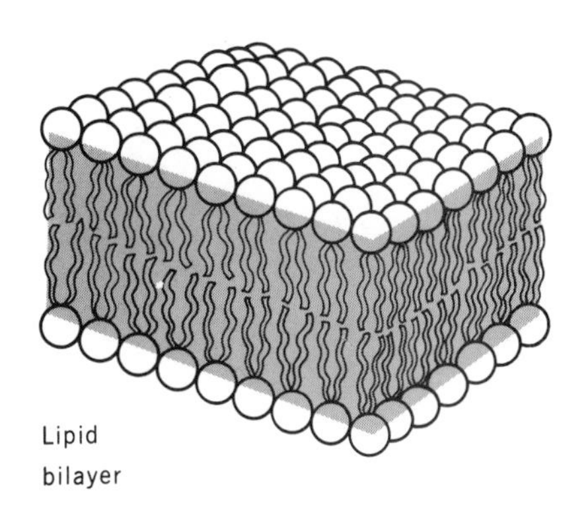

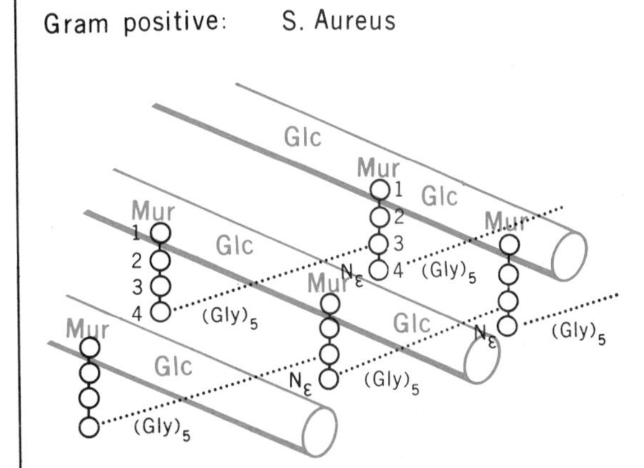

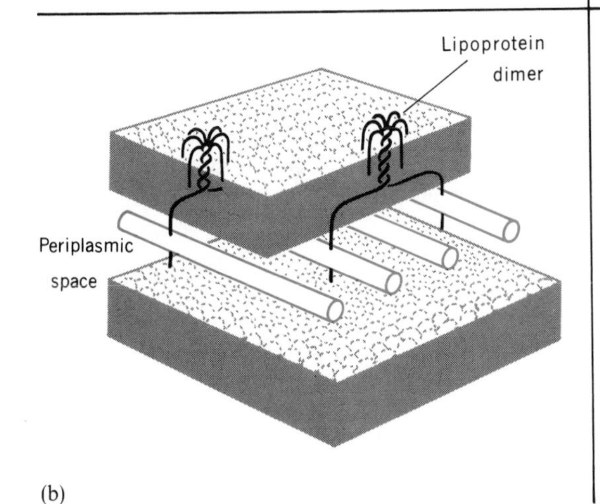

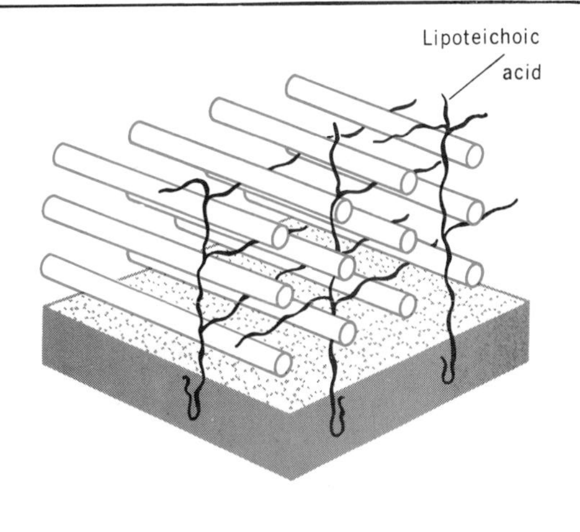

(b)

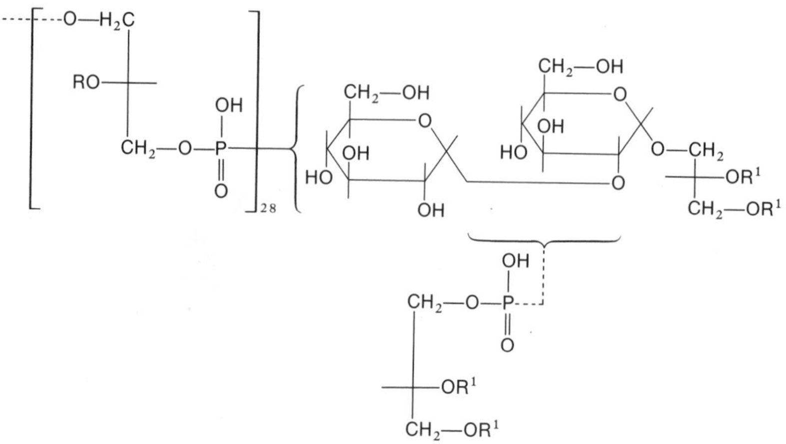

Figure 4-5

Chemical structure of the lipoteichoic acid from *S. faecilis.* R is diglucosyl or hydrogen; R^1 is fatty acid. The exact point of attachment of the lower phospholipid to the saccharide is not known. [After V. Braun and K. Hantke, *Ann. Rev. Biochem.* 43:89 (1974).]

In many Gram-positive bacteria, teichoic acids are attached to the peptidoglycan. Teichoic acids are substances such as polyglycerol phosphate. They are known to be linked in some instances by a phosphodiester bond to the 6-hydroxyl group of the muramic acid residues. One bacterial structure studied in detail comes from *Streptococcus faecalis.* This turns out to be a lipoteichoic acid containing a lipopolysaccharide linked to teichoic acid (Fig. 4-5). The teichoic acid residues probably do not play any role in maintaining the rigidity of the cell wall. They may be important in the binding and uptake of divalent cations.

The lipopolysaccharide attached to teichoic acid in *S. faecalis* is a very simple one. Much more complex lipopolysaccharides with molecular weights over 10,000 are found in the outer cell wall of Gram-negative bacteria. A tentative model for the lipopolysaccharide of *Salmonella typhimurium* is a complex branched structure containing scores of saccharides and numerous fatty acids. We know almost nothing about the probable conformation or other physical properties of such a structure. Even this model may be an oversimplification because—in *E. coli,* for example—there is evidence that proteins may be covalently attached to the cell-surface lipopolysaccharide.

Glycoproteins in animal-cell surfaces

Animal cells do not need a cell wall either for rigidity, as do plants, or for protection against osmotic shock, as do bacteria. They contain a lipid-bilayer cell membrane with many proteins. Quite a few of these are glycoproteins—proteins containing attached oligosaccharides. The sugar content can vary from just a few residues in a large protein to a substantial weight fraction of the protein. There can be one site of attachment or many. For example, bovine ribonuclease B differs from ribonuclease A (discussed in Chapter 2) by the addition of the oligosaccharide

$$\text{(mannose)}_6\text{—(}N\text{-acetylglucosamine)}_2\text{—}$$

attached specifically to Asn^{34}. Human chorionic gonadotropin contains the disaccharide

$$\text{galactose-}\beta\text{(1-3)-}N\text{-acetylgalactosamine—}$$

attached to serine residues 116, 121, and 123.

In contrast, glycophorin (the major glycoprotein of the human red blood cell membrane) has saccharides attached to one asparagine, seven serine, and eight threonine residues (shown schematically in Fig. 2-16). Furthermore, the structure of each of these saccharides is complex (Fig. 4-6). Sugars make up about 60% of the

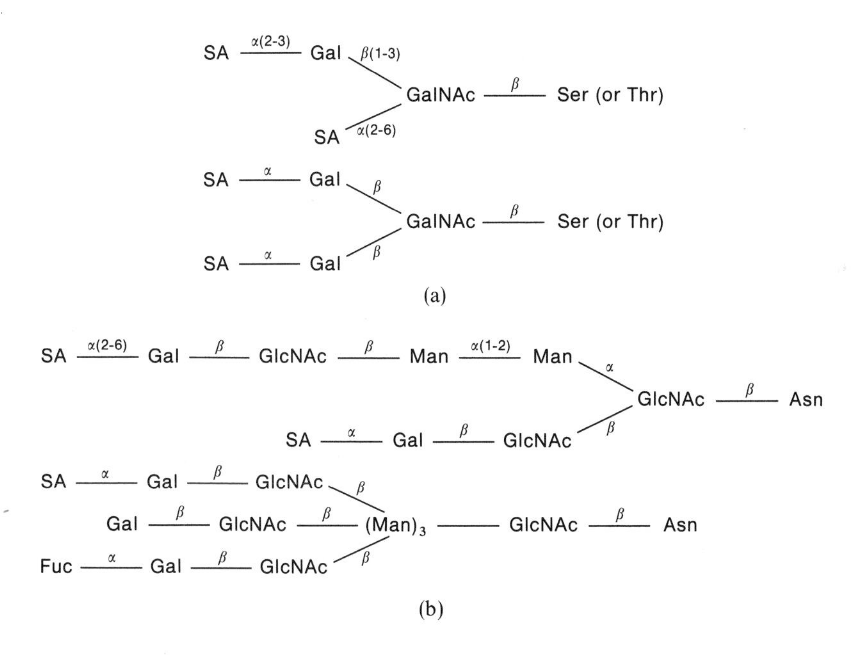

Figure 4-6

Possible structures of the saccharide portions of human erythrocyte glycophorin. (See Fig. 2-16 for the proposed attachment sites of these sugars.) **(a)** Two alternative structural models for the serine-linked and threonine-linked saccharides. **(b)** Two alternative structural models for the asparagine-linked saccharides. Gal = galactose; SA = sialic acid; GalNAc = *N*-acetylgalactosamine; GlcNAc = *N*-acetylglucosamine; Fuc = fucose; Man = mannose. [After V. T. Marchesi et al., *Ann. Rev. Biochem.* 45:667 (1976).]

Figure 4-7

Typical saccharide portions of glycoproteins. Structural elements common to many proteins are shown in color. (**a**) Threonine-linked or serine-linked saccharides. (**b**) Asparagine-linked saccharides. NANA = *N*-acetylneuraminic acid; Xyl = xylose; other abbreviations given in legend for Figure 4-6. [After R. Kornfeld and S. Kornfeld, *Ann. Rev. Biochem.* 45:217 (1976).]

total mass of glycophorin. Another example of a glycoprotein with a high sugar content is found in a protein isolated from Antarctic fish. This protein has the remarkable property of lowering the freezing point of water much more than would be expected from normal colligative properties of solutions. The protein has a tripeptide sequence, Thr–Ala–Ala, repeated 31 times. The disaccharide galactose-β(1-3)-*N*-acetylgalactosamine is attached to each threonine.

A wide variety of different saccharide structures are known. Certain features are quite common but not universal (Fig. 4-7). Saccharides linked to serine or threonine often contain the disaccharide

galactose-β(1-3)-*N*-acetylgalactosamine

Saccharides attached to asparagine very commonly contain the trisaccharide

mannose-β(1-4)-*N*-acetylglucosamine-β(1-4)-*N*-acetylglucosamine

The Asn-linked sugar portions are termed simple if they contain only these two types of sugars, and complex if they contain other monomer types—such as galactose, fucose, and sialic acid.

The effect of sugar residues on the properties of a protein is essentially unknown. At least some glycoproteins are very resistant to thermal denaturation. Most proteins with large numbers of sugars attached behave anomalously in techniques that measure size and shape.

The exact biological function of the sugar moieties of glycoproteins also is not completely clear. Sugars are potent antigens that apparently place an individual stamp on the cells of an organism so that immune defense mechanisms can distinguish these from foreign organisms. Sugars are probably also involved in cell–cell recognition and may serve as part of cell surface receptors for proteins, for elements of the immune system, and maybe even for smaller molecules. The rigidity of saccharide structures, their water solubility, and the great variety of their possible configurations would seem to make them ideal for these functions. Not all glycoproteins occur at the cell surface. Many are found in serum, are secreted into the digestive tract, and so forth. One speculation is that the sugar may correspond to a ticket that permits transport of the protein or that prevents destruction by cells until the sugar is removed.

Gal —β(1-3)— GalNAc ——— Ser(Thr)

Antifreeze glycoprotein, human IgA_1

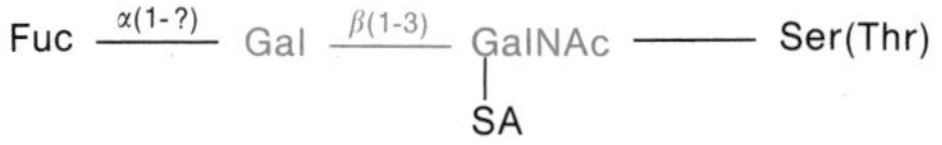

Fuc —α(1-?)— Gal —β(1-3)— GalNAc ——— Ser(Thr)
| SA

Some submaxillary mucins

NANA —α(2-3)— Gal —β(1-3)— GalNAc ——— Ser(Thr)
| α(2-6) NANA

Fetuin

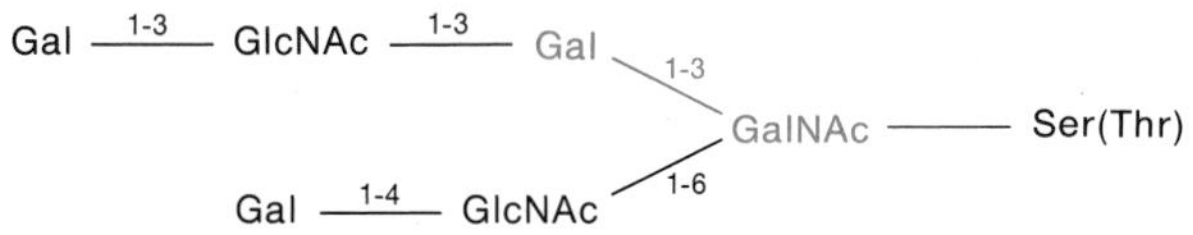

Gal —1-3— GlcNAc —1-3— Gal —1-3— GalNAc ——— Ser(Thr)
Gal —1-4— GlcNAc —1-6— GalNAc

Core region of human blood group substance

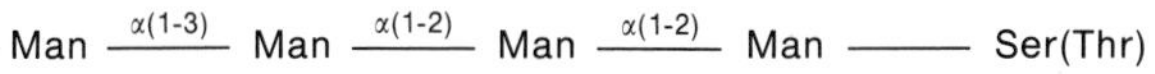

Man —α(1-3)— Man —α(1-2)— Man —α(1-2)— Man ——— Ser(Thr)

S. cervisiae mannan

(a)

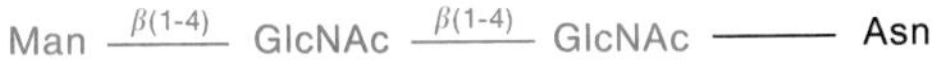

Man —β(1-4)— GlcNAc —β(1-4)— GlcNAc ——— Asn

Bovine RNase B

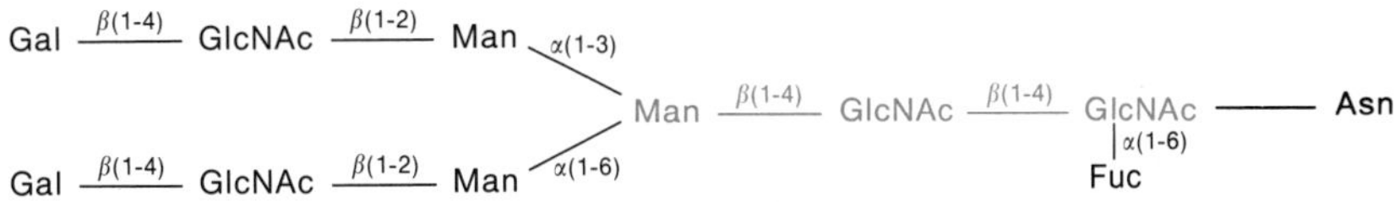

Gal —β(1-4)— GlcNAc —β(1-2)— Man —α(1-3)— Man —β(1-4)— GlcNAc —β(1-4)— GlcNAc ——— Asn
Gal —β(1-4)— GlcNAc —β(1-2)— Man —α(1-6)— Man
GlcNAc | α(1-6) Fuc

Bovine IgG

Man —α(1-2)— Man —α(1-3)— Man —β(1-4)— GlcNAc —β(1-4)— GlcNAc ——— Asn
Man —1-6— Man —α(1-6)— Man
Man | 1-2 Man

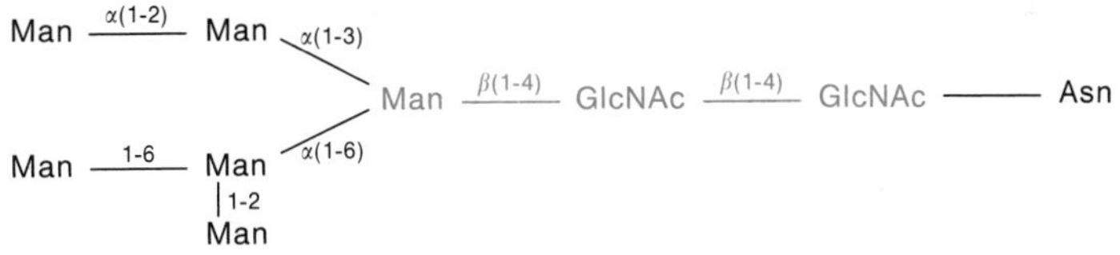

A. oryza α-Amylase, hen ovalbumin

Man —α(1-2)— Man —α(1-2)— Man —β(1-4)— GlcNAc —β(1-4)— GlcNAc ——— Asn
Man | α(1-6) Fuc; GlcNAc | β Xyl

Pineapple bromelain

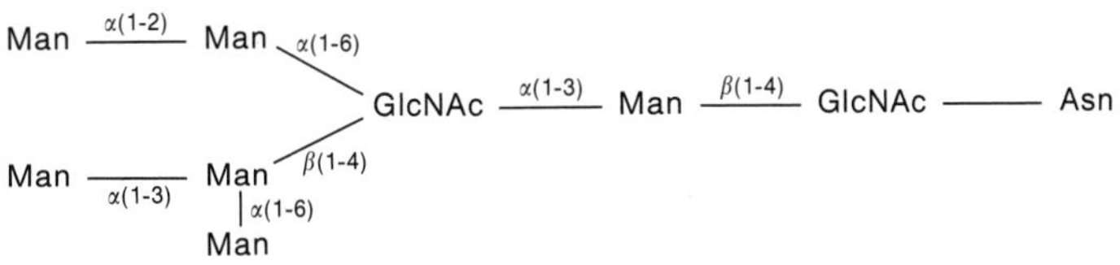

Man —α(1-2)— Man —α(1-6)— GlcNAc —α(1-3)— Man —β(1-4)— GlcNAc ——— Asn
Man —α(1-3)— Man —β(1-4)— GlcNAc
Man | α(1-6) Man

Human IgE

(b)

Nucleoproteins: noncovalent complexes between proteins and nucleic acids

Complexes between proteins and nucleic acids are at the core of cell processes that maintain and use genetic information. Nucleic acid–protein interactions take on a variety of forms—from simple cases, in which RNA or DNA are enzyme substrates, to complex ordered assemblies of proteins with one or more nucleic acids. In the latter class, three types of systems that have been extensively studied are viruses, chromatin, and ribosomes (Table 4-1).

Table 4-1
Properties of some nucleoprotein assemblies

Property	Nucleosome (chromatin)	*E. coli* 70S ribosomes	Tobacco mosaic virus	Phage P22	Adenovirus
Molecular weight	2.4×10^5	2.6×10^6	4.0×10^7	$\sim 5.5 \times 10^7$	$\sim 1.6 \times 10^8$
Percentage protein (by weight)	45	33	95	50	87
Number and kind of nucleic acids	1 DNA	3 RNAs	1 RNA	1 DNA	1 DNA
Number of proteins	9	57	2,130	$\sim$500	>2,500
Number of different proteins	5	53	1	6	$\sim$12

Most proteins present in nucleoprotein assemblies are quite small, having molecular weights of 10,000 to 25,000. A common feature is that many proteins associate with a single nucleic acid, and perhaps even with a single stretch of nucleic acid helix. They tend to be basic proteins. This is sensible because positive protein groups can partially neutralize the polyanionic nucleic acid and form tight complexes. However, not all proteins found tightly associated with nucleic acids are basic.

The general physical questions one must answer about nucleoproteins are similar in spirit to those raised about other structures formed by mixtures of macromolecules. One would like to know the extent of protein–protein and nucleic acid–nucleic acid contact, as compared to heterogeneous contact. Extreme cases of heterogeneous contacts would be a nucleic acid extended along a multiprotein core, or proteins assembled along a polynucleotide core.

What specific interactions exist between nucleotide residues and amino acid residues? If there is a simple code for mutual recognition, it has thus far eluded discovery. Two general features that may be operative in many cases are close interaction between either lysines or arginines and phosphates, and intercalation of planar aromatic amino acid residues between the bases of a polynucleotide chain. Model building suggests that the natural twist of β sheets may allow them to fit very nicely into one of the grooves of double-stranded DNA. Figure 4-8 shows an example of such a structure. Other model building suggests that α helices with consecutive arginines or lysines may be able to lie in the major or minor groove of the DNA double helix and neutralize phosphates in both sites of the groove.

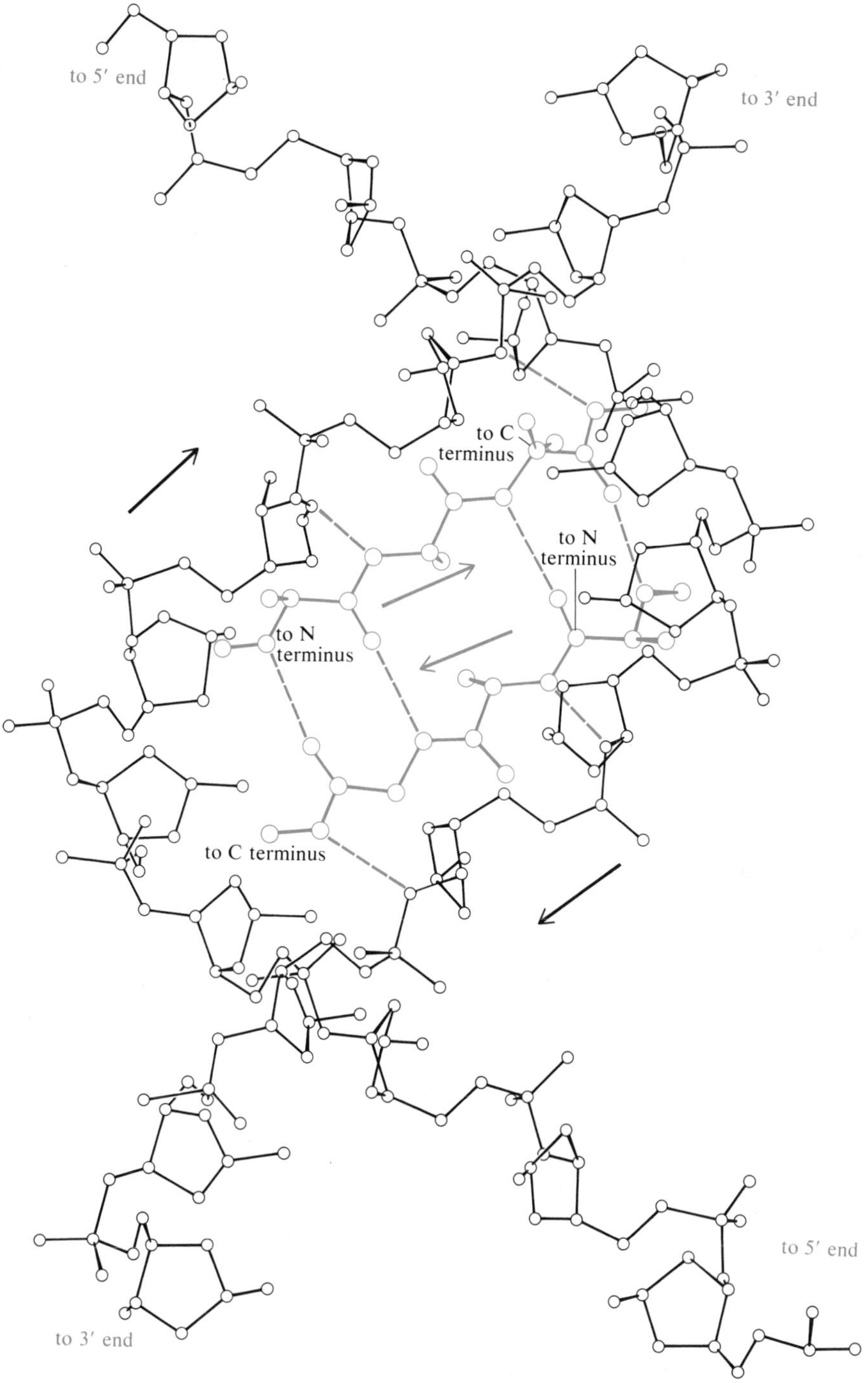

Figure 4-8

A possible model for the interaction between an antiparallel pleated sheet and double-stranded DNA. The sheet lies in the narrow groove; the twofold rotational symmetry axes of the sheet and of the underlying DNA coincide. Hydrogen bonds are formed between peptide NH groups and deoxyribose O-3′ atoms. [After G. M. Church et al., *Proc. Natl. Acad. Sci. USA* 74:1458 (1977).]

Hydrogen bonding between amino acids and bases may be able to produce sequence-specific interactions between proteins and single-stranded nucleic acids. For example, in DMSO–water mixtures, carboxylate anions (models for Asp and Glu side chains) form a complex with guanine that is 30 times stronger than cytosine-guanine base pairing under these conditions. Only guanine among the five usual nucleic acid bases can form this interaction.

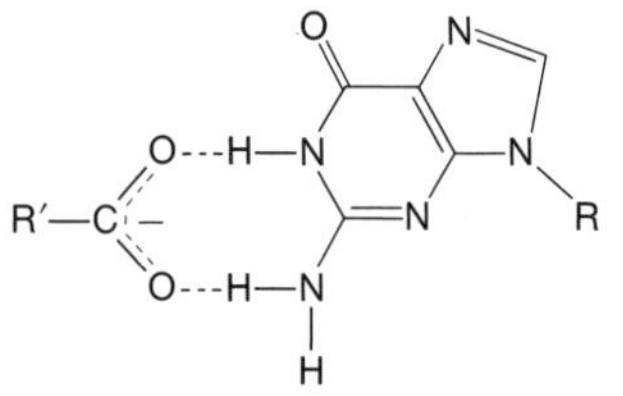

Such an interaction may explain how acidic proteins can form tight complexes with nucleic acids.

Are physical and chemical properties of individual proteins or nucleic acids changed by their association into nucleoprotein assembly? The general answer is yes, but in most cases it is not known whether this is due simply to conformational changes induced in one component by the other, or is due to more direct effects. For example, are there any unique properties of amino acids and nucleotides in direct contact? Most nucleic acids exist in association with proteins, but the reverse is not true. It is therefore of interest to know whether proteins that participate in nucleic acid association have any general properties different from those of normal soluble proteins. There are hints that many proteins that bind to nucleic acids may be less globular than the average small protein, but it is not yet clear how general this phenomenon is.

External protein and internal nucleic acid in viruses

Some viruses appear to consist of protein arranged about a nucleic acid core. Tobacco mosaic virus is an example of a rodlike virus composed of one single-strand RNA and many copies of a single protein. By itself, the protein forms an assembled structure virtually identical to that of the native virus, as described in Chapter 2. Therefore, although it is on the outside, it seems that the protein is really the organizing core of the structure, and that all the RNA does is determine the length. In the virus, the structure of the RNA is extremely extended. Base stacking is eliminated and replaced with base–protein interactions. Thus, during virus assembly, the RNA must change drastically from its structure when free in solution.

Many bacterial viruses (bacteriophage) consist of a protein head or capsid, into which DNA or RNA is inserted. Stages in the assembly of one such virus are shown in Figure 4-9. In some cases, the capsid is assembled independently. Then DNA is

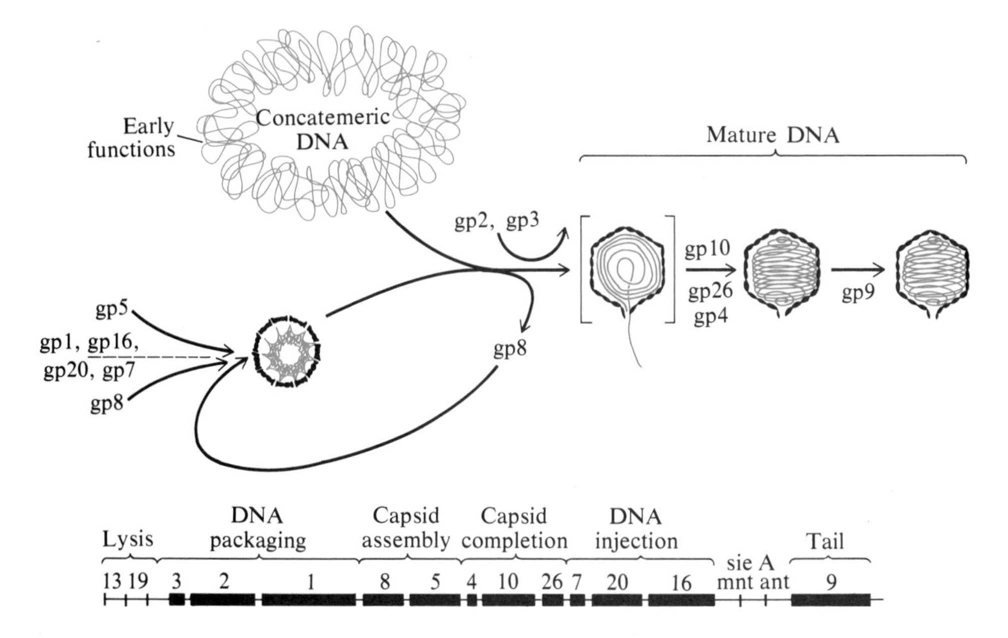

Figure 4-9

A schematic assembly pathway for P22, a bacteriophage that infects *S. typhimurium*. Individual gene products of the phage are denoted by gp. The first principal assembly stage is the formation of a single-shelled spherical procapsid composed principally of gp5 and gp8. When a section of concatemeric DNA enters, gp8 (a scaffolding protein) leaves. The DNA is cut to the proper length once inside the capsid; then the tail assembly is added. [After J. King et al., *Cell* 15:551 (1978).]

inserted with almost no change in capsid properties. Other cases are known in which insertion of DNA is accompanied by marked changes in the structure or composition of the capsid. In still other cases, the size of the nucleic acid within the capsid seems to determine the capsid's size. Finally, in more complex viruses, a protein core is found in addition to a protein coat.

Nature appears to use a variety of mechanisms to direct the assembly of viral nucleoproteins. All of these structures have extensive protein–protein contact, as judged not only from the appearance of the structure, but also by the ability of the protein assembly to form in the absence of nucleic acid. Except in the rodlike viruses, there also appears to be extensive nucleic acid self-contact, perhaps assisted by some noncoat proteins. These viruses thus keep protein and nucleic acid components largely separate, consistent with the protein function as a package for transport, protection, and injection of nucleic acid. Some enzymes involved in the assembly of viruses are sensitive to the state of viral organization. For example, proteins that cleave long precursor DNAs will not work on isolated nucleic acid, but require the presence of a filled capsid.

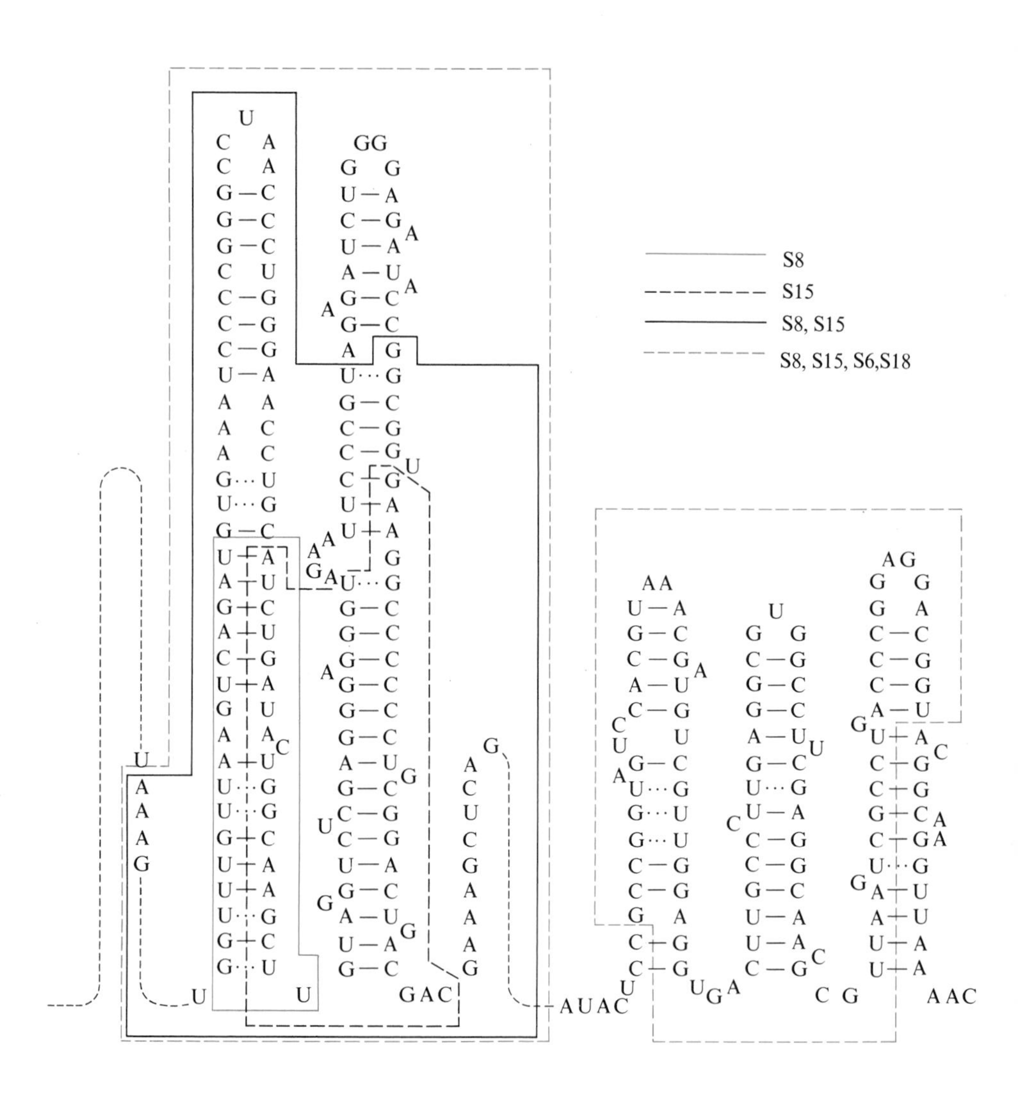

Figure 4-10

A possible secondary structure for a small portion of the *E. coli* 16S rRNA. Shown are regions protected against nuclease digestion (1) when proteins S8 and S15 are added singly to the RNA; (2) when both are added together; or (3) when S8, S15, S6, and S18 are present simultaneously. [After R. Zimmermann et al., *Nucleic Acids Research* 2:279 (1975).]

Intimate protein–RNA contact in ribosomes

The overall pattern of protein–nucleic acid organization of ribosomes appears to be quite different from that of viruses. There is extensive, and perhaps intimate, protein–nucleic acid contact over much of the structure. For example, many individual ribosomal proteins will bind to specific sites on the ribosomal RNAs (rRNAs). When bound, they protect large stretches of rRNA against enzymatic degradation. There are cooperative effects when more than one protein binds simultaneously (Fig. 4-10).

At present, it does not appear that ribosomal proteins can organize themselves into any ordered macrostructure in the absence of the rRNA. Indeed, only a few specific pairwise interactions of ribosomal proteins in solution have been found. Cross-linking studies on native ribosomes suggest that there is much protein–protein proximity, but it is not known how many protein pairs are in direct contact, nor whether such contacts are extensive enough to afford a significant stabilization to the assembled structure. (Fig. 1-7 shows a schematic picture of the protein–nucleic acid arrangement in the 70S ribosome and various 50S particles.)

In free ribosomal RNAs, there is much intrastrand secondary structure and, presumably, tertiary structure as well. Most available evidence suggests that, although assembly of the rRNA into ribosomes results in significant compaction of the molecule, there is little net change in secondary structure. Many observations lead to the speculation that there may be little additional RNA–RNA interaction in the ribosome, relative to that among free RNAs. If so, then RNA–protein interactions must be the major source of the stability of the final globular ribonsomal subunits.

There is now considerable evidence that the conformations of ribosomal subunits are easily perturbed. Binding of certain small proteins, tRNAs, or antibiotics appears to change the properties of the particle substantially. Mutations in proteins on one subunit can sometimes alter functions known to be carried out by the other subunit. These kinds of effects generally are not seen in smaller systems. If they serve biological functions, this could explain why nature has constructed such complex assemblies for certain tasks.

The properties of individual ribosomal components are markedly affected by the state of assembly. For example, lysine groups of one particular protein are chemically unreactive in the free state, but become quite reactive in the ribosome. The protein colicin E_3 causes a specific cleavage of the 16S rRNA of the 30S subunit only in the intact 70S particle, but not in the free 30S subunit or in free RNA. Various enzymatic or binding activities are associated with the ribosome or its subunits. Almost none of these activities have been found in isolated components. It will be interesting to learn whether individual proteins carry out these functions (but require the assembly to place them in an active conformation), or whether multiple components actually have a concerted role in some of the catalytic steps in protein synthesis.

Chromatin: a hierarchy of organized structures

Models and ideas of chromatin structure have been evolving rapidly within the last few years. Many levels of this structure must be considered. A complex of DNA with five types of histones forms a regular ordered structure. The fundamental unit of this structure is the nucleosome, apparently containing (1) two copies each of histones H2a, H2b, H3, and H4; (2) one copy of H1; and (3) about 200 base pairs of double-strand DNA. Nucleosomes are arranged along the DNA linearly, like beads on a string, perhaps with some free DNA between each nucleosome.

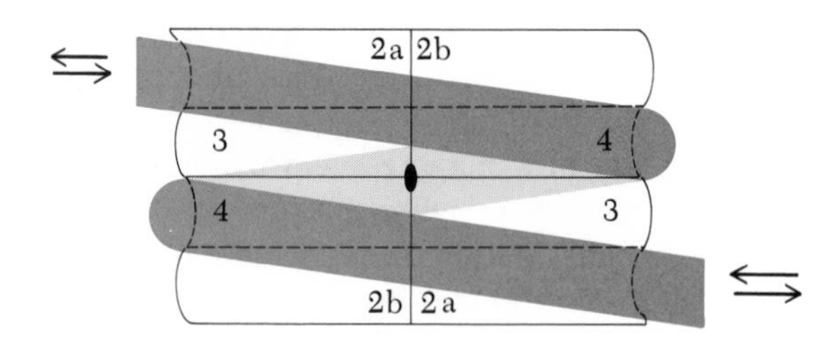

Figure 4-11

Schematic illustration of the repeating subunit structure of eukaryotic chromatin. The structure shown for one subunit (the nucleosome core) is speculative but plausible. This subunit contains two each of the four histones shown. A possible location of a C_2 axis is illustrated. This is a likely symmetry arrangement for the structure, in view of the protein composition and the known presence of pseudo-C_2 axes in the DNA. The actual arrangement of proteins is consistent with known protein–protein cross-links or subassemblies, but it is by no means a unique representation of available data. As shown, the DNA has one and a half toroidal supercoils (see Chapter 24).

Current models favor the idea of a nucleosome core particle containing (1) about 140 base pairs of DNA, and (2) all the histones except H1. The composition implies a twofold symmetry of the core particle. There is considerable evidence that the histones are at the center of the core. Besides cross-linking data, it is found that certain pairs of histones form stable dimeric complexes in solution; tetrameric protein complexes also have been observed. Presumably the DNA is wrapped around these protein tetramers in a regular way. Enough constraints are available about the organization of the DNA to propose detailed models such as the one shown in Figure 4-11, but insufficient data exist to establish one unique model. Furthermore, evidence exists that the nucleosomes containing actively transcribing genes have a more open or accessible structure than those containing nonexpressed genes.

In the core particle, DNA is compacted so that its linear extension is only one-seventh that of the B-form duplex. Whether DNA–DNA interactions participate in the compaction is not known. What is clear is that under some circumstances DNA can be induced into seemingly analogous condensed states without the presence of histones.

Natural chromatin contains a variable amount of a great many other proteins in addition to the histones and the DNA. These are called nonhistone proteins, an apt admission of our general ignorance about them. The amounts and types are tissue-dependent and span a broad range of possibilities. Some of these proteins must play a role in gene expression. Others may be involved in higher orders of packaging. Whether they modulate the structure or modulate the location of individual nucleosomes awaits further study.

Considerable evidence exists that the string of nucleosomes is folded in vivo (and, in concentrated gels, in vitro) into helical solenoidlike arrays. These are further organized into more complex structures. In metaphase cells, chromatin is packaged into specific large organized structures called chromosomes. The details of the packaging mechanism and the final structure are important problems for the next generation of biophysical chemists. It should be noted that prokaryotes apparently do not have chromatin. However, they too organize their DNA into compact structures, and we need to learn much more about these also.

4-3 LIPIDS IN BIOLOGICAL MEMBRANES

Every living cell is surrounded by an outer membrane, sometimes known as the plasma membrane. This structure serves as a barrier between the cell constituents and the external environment. It can act as a screening device, allowing (and sometimes even assisting) the penetration of some molecules but not others. In addition, the outer membrane can contain receptors that enable a cell to communicate with the external medium and with other cells.

Membrane structures are not simply confined to the outer edge of the cell, however. For example, in eukaryotes there is a membrane that encloses the nucleus, separating it from the surrounding cytoplasm. Eukaryotic cells also have an extensive system of membraneous cables, known as the endoplasmic reticulum. There are additional examples of membranes that occur in specialized situations, such as the myelin sheath around an axon and the so-called stacked disks of retinal rod cells.

The essential structural framework of membranes is provided by lipid amphiphiles. The amphiphiles have two main components: a polar group, and a larger nonpolar section. This type of molecule readily forms bilayers, which are themselves membraneous vesicles. In biological membranes, a number of proteins are situated within these bilayers.

Lipid components of membranes

From the standpoint of chemical composition, the distinctive feature of membranes is the high concentration of lipids. These molecules are easily extracted into organic solvents and have a sparse solubility in water. The major membrane lipids are particular amphiphiles (which all have long hydrocarbon tails), and cholesterol or cholesterol esters. The characteristics of these molecules, acting together in mixtures, in large part determine the structure and properties of biological membranes.

Table 4-2
Some naturally occurring fatty acids

Carbon atoms	Structure	Systematic name	Common name
Saturated fatty acids			
12	$CH_3(CH_2)_{10}COOH$	Dodecanoic acid	Lauric acid
14	$CH_3(CH_2)_{12}COOH$	Tetradecanoic acid	Myristic acid
16	$CH_3(CH_2)_{14}COOH$	Hexadecanoic acid	Palmitic acid
18	$CH_3(CH_2)_{16}COOH$	Octadecanoic acid	Stearic acid
20	$CH_3(CH_2)_{18}COOH$	Eicosanoic acid	Arachidic acid
24	$CH_3(CH_2)_{22}COOH$	Tetracosanoic acid	Lignoceric acid
Unsaturated fatty acids (all double bonds *cis* except where otherwise indicated)			
16	$CH_3(CH_2)_5CH{=}CH(CH_2)_7COOH$	9-Hexadecenoic acid	Palmitoleic acid
18	$CH_3(CH_2)_7CH{=}CH(CH_2)_7COOH$	*cis*-9-Octadecenoic acid	Oleic acid
18	$CH_3(CH_2)_7CH\underset{(trans)}{=}CH(CH_2)_7COOH$	*trans*-9-Octadecenoic acid	Elaidic acid
18	$CH_3(CH_2)_4CH{=}CHCH_2CH{=}CH(CH_2)_7COOH$	*cis*,*cis*-9,12-Octadecadienoic acid	Linoleic acid
18	$CH_3CH_2CH{=}CHCH_2CH{=}CHCH_2CH{=}CH(CH_2)_7COOH$	9,12,15-Octadecatrienoic acid	α-Linolenic acid
20	$CH_3(CH_2)_4CH{=}CHCH_2CH{=}CHCH_2CH{=}CHCH_2CH{=}CH(CH_2)_3COOH$	5,8,11,14-Eicosatetraenoic acid	Arachidonic acid
19	$CH_3(CH_2)_5HC{-}CH(CH_2)_7COOH$ (HC and CH bridged by $C H_2$)		Lactobacillic acid
19	$CH_3(CH_2)_7\underset{\mid \atop CH_3}{C}H(CH_2)_8COOH$		Tuberculostearic acid
24	$CH_3(CH_2)_{21}\underset{\mid \atop OH}{C}HCOOH$		Cerebronic acid

SOURCE: After A. L. Lehninger, *Biochemistry* (New York: Worth, 1975), p. 281; and A. White, P. Handler, E. L. Smith, R. L. Hill, and I. R. Lehman, *Principles of Biochemistry* (New York: McGraw-Hill, 1978), pp. 39–40.

In most cases, the amphiphiles are phospholipids (esters of glycerol phosphate), which have the following general structure:

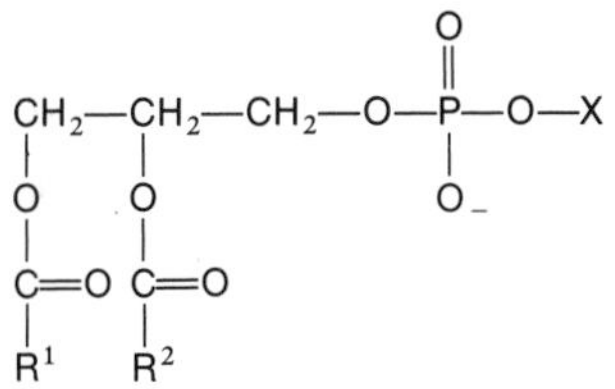

```
                              O
                              ‖
CH2—CH2—CH2—O—P—O—X
 |      |                     |
 O     O                     O−
 |      |
C=O  C=O
 |      |
R1    R2
```

where two fatty acids (R^1COOH and R^2COOH) are esterified to two of the glycerol hydroxyl groups, and a polar group X is coupled to the phosphate. The alkyl functions R^1 and R^2 of the fatty acids are long-chain hydrocarbons, which can be saturated or unsaturated. Some alkyl functions are listed in Table 4-2. The group X, in combination with the phosphate moiety, forms the polar head group, which may or may not bear a net charge, depending on the charge on X. Some X groups are shown in Table 4-3.

Table 4-3
Alcohol forms of polar X groups found in major phosphoglycerides

Phosphoglyceride	Alcohol component
Phosphatidyl ethanolamine	$HOCH_2CH_2NH_2$
Phosphatidyl choline	$HOCH_2CH_2N(CH_3)_3$
Phosphatidyl serine	$HOCH_2CHNH_2COOH$
Phosphatidyl inositol	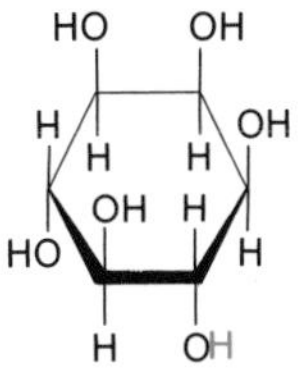 (inositol ring structure: HO, OH, H, OH, H, H, OH, H, HO, H, H, OH)
Phosphatidyl glycerol	$HOCH_2CHOHCH_2OH$
Phosphatidyl-3′-*O*-aminoacyl glycerol	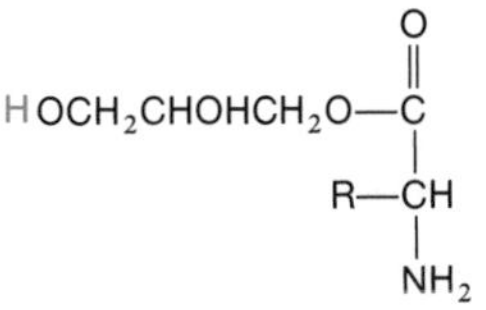 $HOCH_2CHOHCH_2O—C(=O)—CH(R)—NH_2$
Cardiolipin	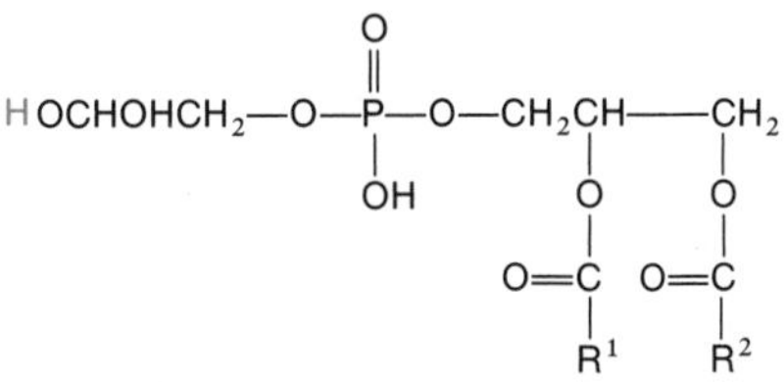 $HOCHOHCH_2—O—P(=O)(OH)—O—CH_2CH(—O—C(=O)—R^1)—CH_2(—O—C(=O)—R^2)$

§ The site of condensation with phosphate is indicated by the colored hydrogen atom.

The phospholipids are named in accordance with the substituents R^i and X. For example, if $R^1COOH = R^2COOH$ = palmitic acid, and X = choline, then the phospholipid is dipalmitoyl-phosphatidyl choline. The choline phospholipids are sometimes called lecithins.

In some phospholipids, the linkage of the long aliphatic moiety to the glycerol phosphate group does not involve an ester linkage, although the general amphiphile structure is similar and involves the same separation of hydrocarbon and polar groups. Here are some examples:

$$CH_3{-}(CH_2)_{12}{-}CH{=}CH{-}CH(OH){-}CH(NH{-}C(=O){-}R){-}CH_2{-}O{-}P(=O)(O^-){-}O{-}CH_2{-}CH_2{-}{}^+N(CH_3)_3$$

Sphingomyelin

$$R^1{-}CH{=}CH{-}CH(OH){-}CH(NH{-}C(=O){-}R^2){-}CH_2{-}O{-}X$$

X = a saccharide

Glycolipid

$$H_2C(O{-}C(=O){-}R^1){-}CH(O{-}CH{=}CH{-}R^2){-}CH_2{-}O{-}P(=O)(O^-){-}O{-}X$$

Plasmalogen

In the case of glycolipids, the group X is a carbohydrate monomer or oligomer. For sphingomyelin and the glycolipids, the backbone is derived not from glycerol, but instead from sphingosine:

$$CH_3{-}(CH_2)_{12}{-}CH{=}CH{-}CH(OH){-}CH(NH_3^+){-}CH_2OH$$

Sphingosine

Attachment of the fatty acid chain is via an amide instead of an ester linkage. On the other hand, the plasmalogens use the glycerol backbone, but one of the linkages from the glycerol moiety to the aliphatic chain is through an α,β-unsaturated ether.

The second major component of membranes is cholesterol and its derivatives.

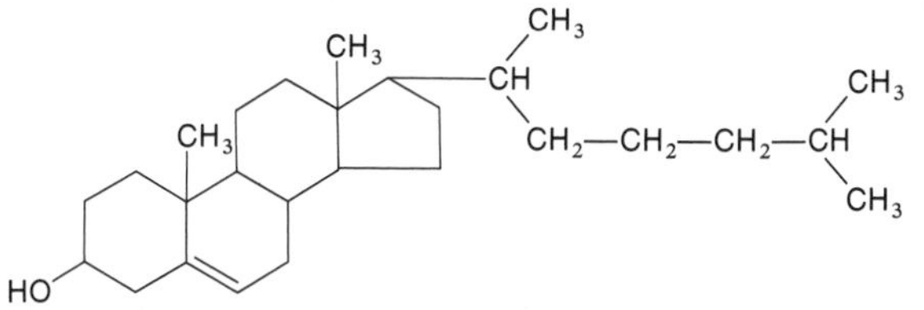

Cholesterol

The hydroxyl group is sometimes coupled with a fatty acid through an ester linkage. Cholesterol and cholesterol esters clearly are very weak amphiphiles.

Table 4-4 gives some examples of the lipid composition of membranes from different sources. Although there is a wide variation in the amounts of specific components, double-chain phospholipid amphiphiles dominate the composition. It is

Table 4-4
Lipid compositions of some biological membranes

Lipid	Human erythrocyte	Human myelin	Beef-heart mitochondria	*E. coli*
Phosphatidic acid	1.5	0.5	0	0
Phosphatidyl choline	19	10	39	0
Phosphatidyl ethanolamine	18	20	27	65
Phosphatidyl glycerol	0	0	0	18
Phosphatidyl inositol	1	1	7	0
Phosphatidyl serine	8.5	8.5	0.5	0
Cardiolipin	0	0	22.5	12
Sphingomyelin	17.5	8.5	0	0
Glycolipids	10	26	0	0
Cholesterol	25	26	3	0

NOTE: Values in the table are percentages by weight of the total lipid.
SOURCE: C. Tanford, *The Hydrophobic Effect* (New York: Wiley, 1973), p. 97.

not surprising, therefore, that "synthetic" membranes, made from a single pure double-chain phospholipid amphiphile, have many of the basic characteristics of natural biological membranes.

Pure lipids in bilayers

The crucial physical characteristic of the double-chain amphiphile is its propensity to form bilayers in aqueous dispersions, in the form of single vesicles or multilamellar structures. These are schematically illustrated in Figure 4-12, which shows that the amphiphiles arrange themselves with polar head groups exposed to the aqueous phase and the hydrocarbon moieties clumped together generating their own nonaqueous environment. The bilayer acts as a membrane, separating components encapsulated on the inside of the vesicle from those on the outside. Clearly the thickness of the bilayer will depend on the length and stiffness of the hydrocarbon moieties of the lipids.

In contrast, single-chain amphiphiles tend not to form bilayer structures in aqueous mixtures, but rather to form globular micelles. A typical micelle might contain 100 molecules; its structure is illustrated in Figure 4-12. Thus, the formation

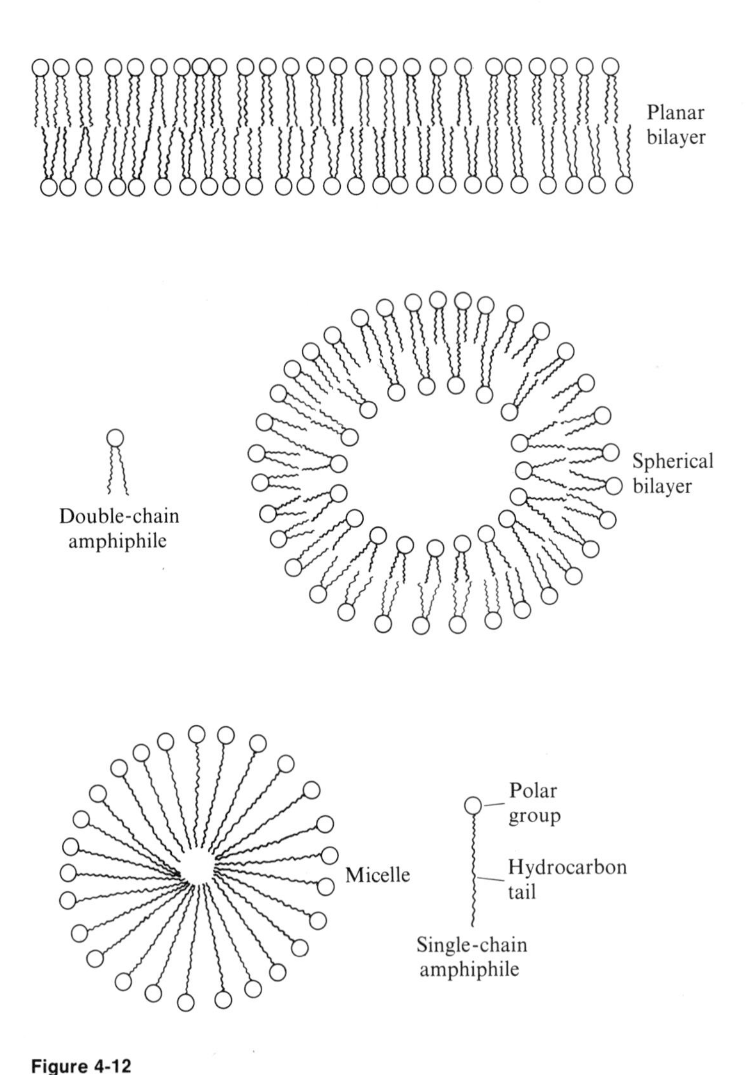

Figure 4-12

Schematic illustration of amphiphile structures. The planar bilayer and spherical bilayer (vesicle) are formed from double-chain amphiphiles. The micelle is formed from a single-chain amphiphile.

of membranelike bilayers is a specific property of the double-chain amphiphiles. In order to appreciate the underlying reasons for micelle and bilayer formation, and to appreciate better the characteristics of these aggregates, one must consider a number of detailed thermodynamic and structural aspects of micelles and bilayers (Chapter 25).

A flat bilayer composed of only a single type of lipid will have two solvent–bilayer interfaces (surfaces). If the solution on both sides is the same, one can expect that the properties and behavior of the two surfaces will be identical. However, if one side is exposed to a different solution than the other, the bilayer may develop asymmetric properties.

A spherical bilayer of sufficiently large diameter should behave just like a flat bilayer. Small-diameter bilayer vesicles may be able to show asymmetric behavior even if the solvent contained within the vesicle is the same as the external solution. This behavior arises because the curvature of the inner surface of the bilayer is different from the outer layer. Thus lipid packing arrangements must differ somewhat in the two surfaces, and the resulting properties can reflect this difference.

Mixtures of lipids in bilayers

When a bilayer is composed of more than one kind of lipid, more complex structure becomes possible. Figure 4-13 illustrates schematically a few cases of a spherical bilayer composed of two lipid types. The simplest imaginable structure is a completely homogeneous vesicle (Fig. 4-13a). Both surfaces have the same lipid composition, and lipid types are randomly distributed in each phase.

However, because of bilayer curvature or the presence of different solutions inside and outside, it is possible that lipids might segregate preferentially into two surfaces (Fig. 4-13b). Such an asymmetric distribution of lipids could be the result of a difference in the thermodynamic stability of each lipid type in a monolayer of a given curvature and type of solvent. It could also result kinetically in a biological cell, from an anisotropy in the transport of lipid through the bilayer: for example, specific delivery to the inside surface and removal from the exterior surface. In Chapter 25 we show that the rate of spontaneous motion of a lipid from the inner to the outer surface of a bilayer is very slow. Therefore a cell has the potential to maintain a metastable distribution of lipids by continuous insertion and removal.

In lipid bilayers, there is an additional possibility of asymmetry through lateral phase separations (Fig. 4-13c,d). These are two-dimensional analogs of solution–solute phase transitions, such as precipitation or crystallization. A phase separation can occur, in principle, either on one surface or on both surfaces of a vesicle. The phases do not have to be pure lipid components, just as metal alloy phases can consist of specific or variable mixtures of metal atoms.

To distinguish these possibilities, it is important to have techniques that can distinguish lipids on the inner and outer surfaces. It is important to be able to distinguish between patches of membrane with homogeneous or with heterogeneous lipid composition. Such techniques are in the process of development (Chapter 25).

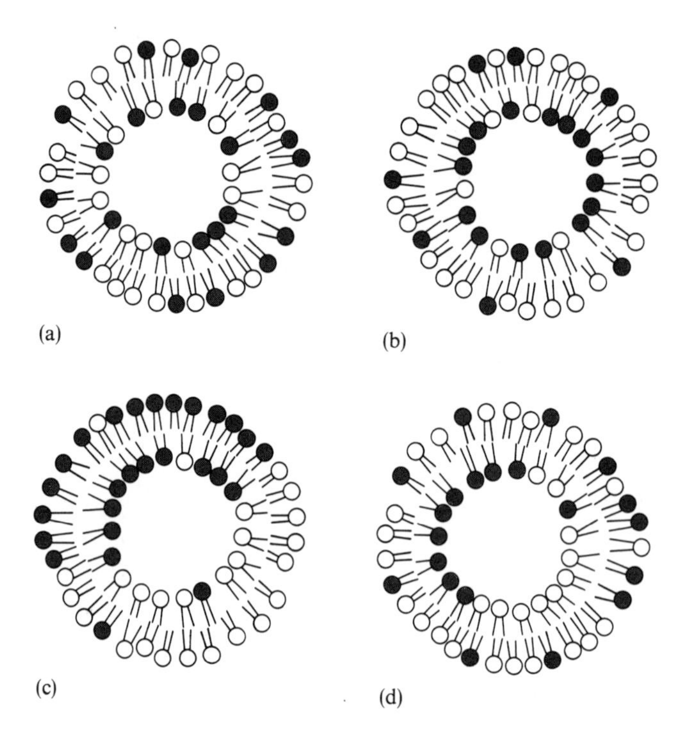

Figure 4-13

Schematic illustration of possible structures of a spherical bilayer composed of two different types of lipids. (**a**) A homogeneous bilayer. (**b**) Asymmetric composition of the two surfaces. (**c**) Lateral phase separation of both surfaces. (**d**) Lateral phase separation of the inner surface only.

Lipid phase transitions

The behavior of lipid molecules in bilayers is strongly influenced by the temperature. At sufficiently low temperatures, the lipid molecules are solidlike. They show little translational motion, and in many ways their properties resemble those of hydrocarbon crystals. As the temperature is raised, the properties of bilayers often show abrupt changes at one or more temperatures. The sharpness of the changes is reminiscent of the phase changes that occur when a solid melts. Above the transition temperature, the bilayer behaves more like a fluid system. Lipid molecules are then capable of rapid translational motion (lateral diffusion). The actual temperatures at which transitions occur, and the number of transitions, depend on the lipid composition. A high concentration of unsaturated fatty acids leads to more fluid bilayers and to lower temperatures for transitions. Some phase transitions involve lateral phase separations as described above. Others represent a solidification of the entire membrane.

4-4 PROTEINS IN BIOLOGICAL MEMBRANES

Protein, lipid, and carbohydrate components

There is much evidence to suggest that the phospholipid bilayer is a basic structure of biological membranes. However, biological membranes are considerably more complex than a simple bilayer. This is due to the presence of components other than lipids. In particular, there is generally a substantial amount of protein in biological membranes.

Table 4-5 gives a summary of the compositions of some cell membranes. Each membrane preparation contains protein, lipid, and carbohydrate. In all cases, the carbohydrate is a minor component, present generally in an amount of 10% or less by weight. This carbohydrate material is attached to the two main components—lipids and proteins—so that it is actually present as glycolipids and glycoproteins.

Table 4-5
Composition of cell membranes

Membrane	Protein (%)	Lipid (%)	Carbohydrate (%)	Ratio of protein to lipid
Myelin	18	79	3	0.23
Plasma membranes				
blood platelets	33–42	58–51	7.5	0.7
mouse liver cells	46	54	2–4	0.85
human erythrocyte	49	43	8	1.1
amoeba	54	42	4	1.3
rat liver cells	58	42	(5–10)§	1.4
L cells	60	40	(5–10)§	1.5
HeLa cells	60	40	2.4	1.5
nuclear membrane of rat liver cells	59	35	2.9	1.6
retinal rods, bovine	51	49	4	1.0
mitochondrial outer membrane	52	48	(2–4)§	1.1
Sarcoplasmic reticulum	67	33	—	2.0
Chloroplast lamellae, spinach	70	30	(6)§	2.3
Mitochondrial inner membrane	76	24	(1–2)§	3.2
Gram-positive bacteria	75	25	(10)	3.0
Halobacterium purple membrane	75	25	—	3.0
Mycoplasma	58	37	1.5	1.6

§ Deduced from the analyses.
NOTE: Values in the table are percentages or ratios by weights.
SOURCE: After G. Guidotti, *Ann. Rev. Biochem.* 41:731 (1972).

The protein and lipid material are each present in quite variable amounts, depending on the cell membrane. For example, in myelin the ratio of protein to lipid is about 0.2, whereas in the mitochondrial inner membrane, the ratio is about 15-fold higher. This kind of variability no doubt reflects the variations in the functions of different membranes. Myelin, for example, presumably functions mainly as an insulator, and has no apparent need for a large array of proteins. On the other hand, the inner membrane of the mitochondria is involved with enzymatic and transport processes, so that it can reasonably be expected to require a battery of proteins.

Types and arrangements of proteins

S. J. Singer has emphasized the concept that membrane proteins fall into two broad categories, depending upon how they are bound to the membrane. These are shown in Figure 4-14. *Peripheral* proteins are those that can be removed from membranes by relatively mild treatments, such as an increase in ionic strength. These proteins are generally stable in aqueous solutions and contain no tightly bound lipid material. Examples of this kind of protein are cytochrome *c* from mitochondrial membranes, and spectrin (a protein component of erythrocyte membranes).

In contrast, *integral* proteins are more difficult to remove from membranes—

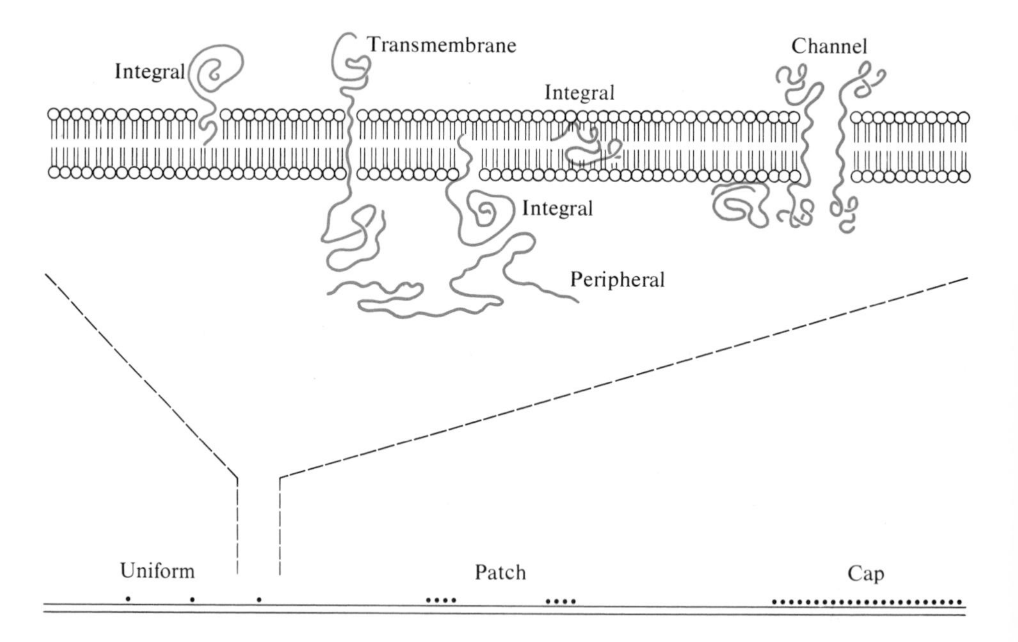

Figure 4-14

Types and arrangements of membrane proteins. The upper part of the figure is a greatly expanded portion of the lower part.

requiring treatment, for example, with organic solvents to be released from the membrane. These proteins are often isolated with bound lipid, without which they tend to aggregate or precipitate when in an aqueous environment. Finally, integral proteins are typically very heterogenous with respect to size, and in many membranes they comprise over 70% of the total protein.

The very different properties and behavior of integral and peripheral membrane proteins are reflected in such basic parameters as their amino acid compositions. As shown in Chapter 2, from this measurement alone, it is usually possible to correctly classify an integral membrane protein.

Most peripheral proteins can be expected to be located on the surfaces of the membrane. In principle, they can be restricted to one surface, or can occur on both surfaces. It is reasonable to expect that in many cases such proteins associate with the membrane by contacts with integral membrane proteins rather than by direct contact with lipids.

Surface location of proteins

Integral proteins can be localized on one or both membrane surfaces or can actually penetrate through the membrane. These latter are called transmembrane proteins. Integral proteins appear to have a region rich in hydrophobic residues, which permits direct contact with lipids. Some integral proteins localized to one surface have a hydrophobic tail, which anchors them in the membrane. If this tail is clipped enzymatically, the tail remains in the membrane whereas the head moves to the solution and behaves fairly much like a normal soluble protein. Transmembrane proteins can be expected to have a hydrophobic sequence someplace in the middle of their primary structure. An example is glycophorin, a transmembrane protein from erythrocytes. The amino acid sequence of this protein is shown in Figure 2-16.

Sometimes, rather simple experiments allow the surface location of membrane proteins to be established. Reagents are known that cannot pass through the lipid bilayer. A simple example is antibodies elicited against a specific membrane protein. Starting with a sealed spherical vesicle or intact cell, one asks whether the reagent can bind to the protein when added from the external solution. Then one can disrupt the vesicle and see if the reagent can now bind to proteins previously sequestered on the internal surface. Alternatively, procedures exist for forming a vesicle in the presence of a blocked reagent and then washing the sealed vesicles so any external reagent is removed. Then the blocked reagent is activated and can attack proteins on the internal surface.

Surface-labeling techniques cannot easily distinguish a protein that is distributed on both surfaces from a transmembrane protein. To do this, it is necessary to show whether the very same polypeptide chain is simultaneously accessible from both surfaces.

Lateral arrangement of proteins

A detailed description of membrane proteins requires not only a statement of which surface they are on, but also how they are distributed within a surface. Proteins can float individually in the membrane or can be associated with other membrane proteins. They can form symmetric oligomers, helical structures, or sheetlike structures. Very little is currently known about the quaternary structure of membrane proteins. One area in which this question may be of particular interest is the possibility of channels through the membrane, caused by single proteins with internal holes or by associations of proteins that produce a central cavity. The presence of a channel must be suspected wherever a membrane shows high permeability to ionic molecules, because pure lipid bilayers are extremely good insulators. However, very little is known at the molecular level about the formation or control of channels.

On a molecular level, the immediate neighbors of a membrane protein can be determined with techniques similar to those described for quaternary structures of normal proteins (Chapters 2, 8, and 14). However, on a much more macroscopic level, it is important to learn whether each type of membrane protein is distributed uniformly or shows any evidence of lateral phase separations. Electron microscopy is particularly useful for examining long-range organization of membrane proteins. In various systems, individual proteins may be widely spaced, or may be associated into patches or even caps that can cover a significant fraction of the surface of a cell (Fig. 4-14b).

Electron microscopy of freeze-etched membranes

During the past decade, considerable advances have been made in the study of membrane structure by electron microscopy. This has been possible through the development of freeze-fracture and freeze-etching techniques. The basic idea is to freeze rapidly a specimen to extremely low temperatures, with the hope of producing as little disturbance of the specimen as possible. The frozen specimen is then fractured with a knife, so as to expose an internal face. The fractured specimen is then prepared for electron microscopy.

Figure 4-15 is a schematic illustration of the basic procedure used. The figure is based on the methodology of Daniel Branton and others. As shown in the figure, the specimen is rapidly frozen in liquid freon. With a cold knife, the frozen sample is fractured in vacuo (to prevent frosting of the knife). Etching is achieved by evaporating away some of the ice in vacuo at about $-100°C$. After this, the exposed surface is shadowed with platinum and carbon. The deposited film is then removed to yield a replica of the surface that can be examined by electron microscopy.

It is now known that fracturing occurs mainly along the interface of the two hydrocarbon layers in the bilayer (Fig. 4-16). Therefore, it is this interior region that is etched, shadowed, and replicated. An example of a fractured membrane (from *Acholeplasma laidlawii*) is shown in Figure 4-17. The photograph shows a relatively

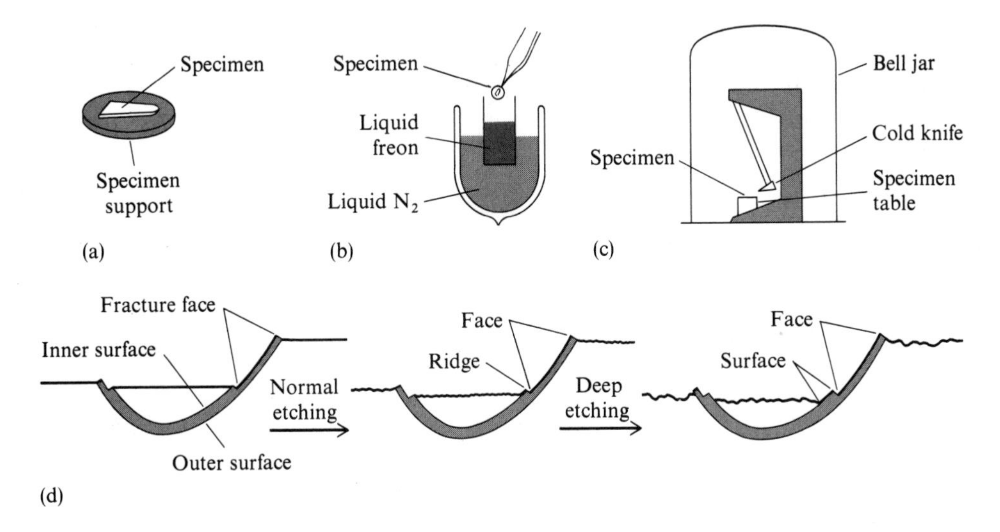

Figure 4-15

Freeze-etching procedure. Specimen is (**a**) placed on support, (**b**) frozen, and then (**c**) placed in a vacuum chamber to be fractured with a cold knife. The etching process (**d**) removes frozen ice by sublimation, thus exposing surface features. [After D. Branton, *Proc. Natl. Acad. Sci. USA* 55:1048 (1966).]

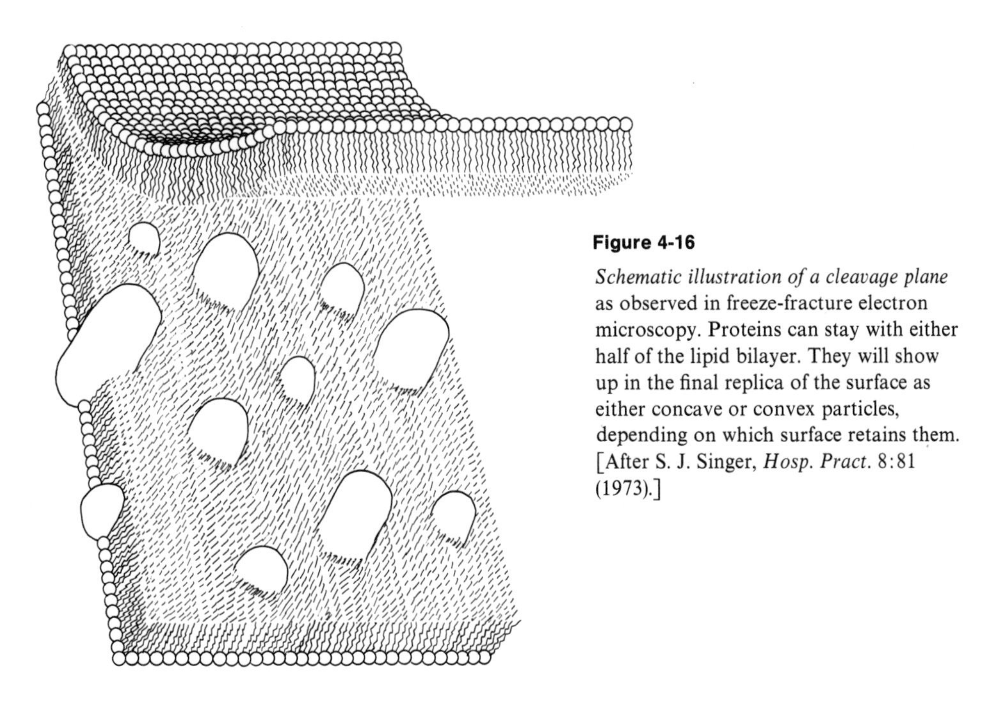

Figure 4-16

Schematic illustration of a cleavage plane as observed in freeze-fracture electron microscopy. Proteins can stay with either half of the lipid bilayer. They will show up in the final replica of the surface as either concave or convex particles, depending on which surface retains them. [After S. J. Singer, *Hosp. Pract.* 8:81 (1973).]

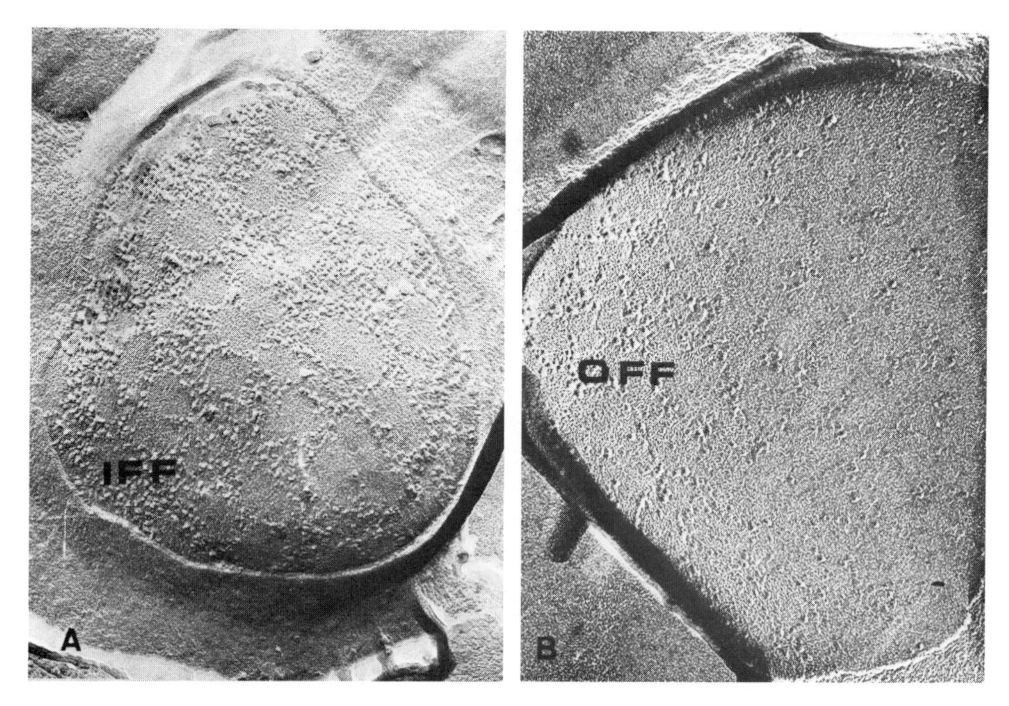

Figure 4-17

Fracture faces of B membranes from *Acholeplasma laidlawii.* OFF = outer (concave) fracture face; IFF = inner (convex) fracture face. [From A. J. Verkleij et al., *Biochim. Biophys. Acta* 288:326 (1972).]

smooth fracture face that is speckled with many fairly randomly distributed particles. These particles are almost certainly proteins and protein aggregates. The pattern shown in Figure 4-17 is fairly typical for membranes at the cells' growth temperature, where the lipids are—in most membranes—in their fluid state (above the lipid membrane transition temperature). Membranes that are freeze-fractured from an inital state that is below the transition temperature show ridges in the fracture face and a clustering of the particles into large arrays.

The ridges are also seen in pure lipid bilayer vesicles (liposomes) that have been freeze-fractured from a state below the transition temperature, whereas only smooth faces are found when fracturing is done from above the transition zone. Thus, the freeze-fractured liposomes show many of the basic features of fractured natural membranes. This is one of many pieces of evidence supporting the concept that biological membranes make use of the basic bilayer structure of the liposome. The photograph in Figure 4-17 (and many others like it) shows that in natural membranes many proteins are scattered throughout the basic bilayer matrix of phospholipids.

The distribution of particular macromolecules as seen by electron microscopy

Electron microscopy can also be used to visualize individual membrane *surface* components. To accomplish this it is necessary to have a means of "staining" the specific molecule of interest so that it will be visible in an electron micrograph. This can be done by using an electron-dense tag attached to a reagent that specifically labels the surface component under investigation.

An example of this kind of approach is provided by work of Singer and colleagues. They examined the distribution of the Rh_0(D) antigen on the surface of the human erythrocyte membrane. O,Rh-positive cells were first reacted with human antibodies directed against the specific antigen. With the antibodies attached to the specific surface antigens, the cells were lysed and the membranes flattened out on an electron microscope grid. The flattened membranes were then treated with *goat* IgG antibodies directed against the *human* IgG antibodies attached to the Rh_0(D) antigens on the membrane. (Thus, the bound human antibodies acted as antigens for the goat antibodies.) The goat antibodies had been previously coupled to ferritin, an iron-rich protein that is sufficiently electron dense so as to be easily visualized in the electron microscope. Thus, ferritin, attached to the goat antibodies, provides a specific stain by which the Rh_0(D) surface antigens are visualized.

The results obtained are displayed in Figure 4-18, which shows the distribution of ferritin (black dots) on the outer surface of the membrane. The ferritin dots are in circled clusters of two to eight. Each cluster represents a single Rh_0(D) antigen; the individual dots give an indication of the number of goat antibodies reacting with the human IgG attached to this antigen. Figure 4-18 shows that the Rh_0(D) surface antigen is randomly distributed over the membrane outer surface. This distribution is consistent with that found from freeze-fracture studies shown in Figure 4-17, where protein particles in *Acholeplasma laidlawii* are shown distributed in an irregular fashion over an inner cleavage plane of the membrane.

General picture of structure

It is instructive to consider a working model of biological membrane structure. The schematic illustration in Figure 4-19 is adapted from the fluid-mosaic model of S. J. Singer. The phospholid bilayer is envisioned as a fluid matrix 60 Å to 100 Å thick in which various integral proteins are implanted. Some proteins protrude through both surfaces of the membrane others through only one side. Consequently, the membrane is asymmetric, a feature for which there is abundant evidence. The proteins are viewed as containing both hydrophobic and hydrophilic sections, enabling them to be thermodynamically accomodated by the corresponding parts of the phospholipid bilayer. Over the long range, a given protein can be randomly distributed, as suggested

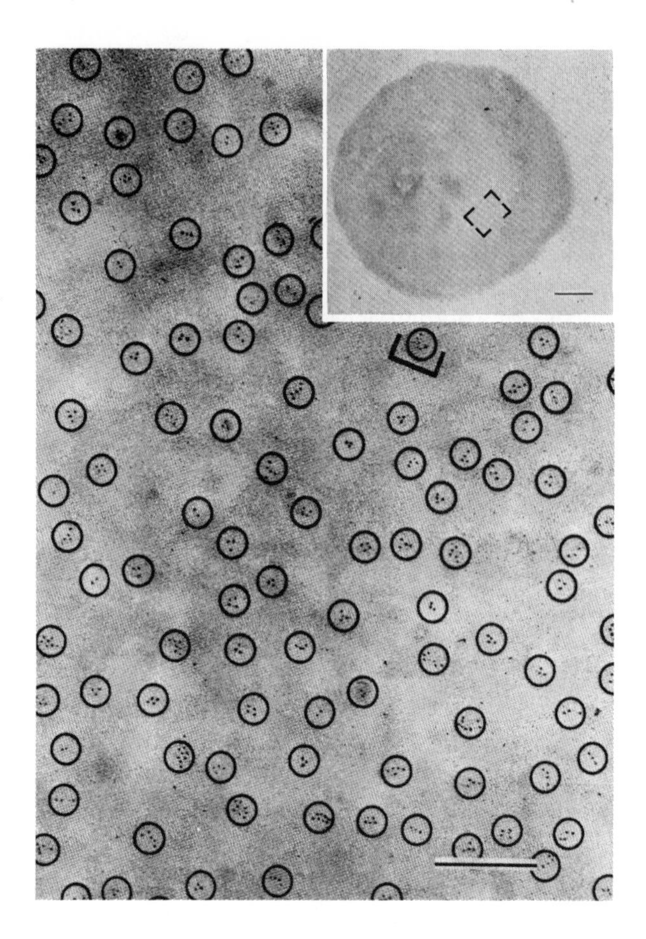

Figure 4-18

Distribution of $Rh_0(D)$ *antigens* on the surface of a human erythrocyte. Each circled cluster of black dots corresponds to one $Rh_0(D)$ molecule. The black dots are ferritin molecules attached to goat antibodies; the goat antibodies have reacted with human antibodies that are bound to the $Rh_0(D)$ molecules. Scale is 0.2 μm; inset scale is 1 μm. [From G. L. Nicholson et al., *Proc. Natl. Acad. Sci. USA* 68:1416 (1971).]

by the distribution of the $Rh_0(D)$ antigen shown in Figure 4-18. Over the short range, a specific distribution may exist for a given protein.

The membrane has dynamic aspects so that lateral movement of proteins in the plane of the bilayer can occur. There are numerous experimental demonstrations of this kind of motion. Also, it is possible that a given protein may rotate about axes either perpendicular or parallel to the bilayer plane. Such motion might be essential for carrying out many of the functions associated with membranes. However, definitive studies of these aspects are still in the early stages.

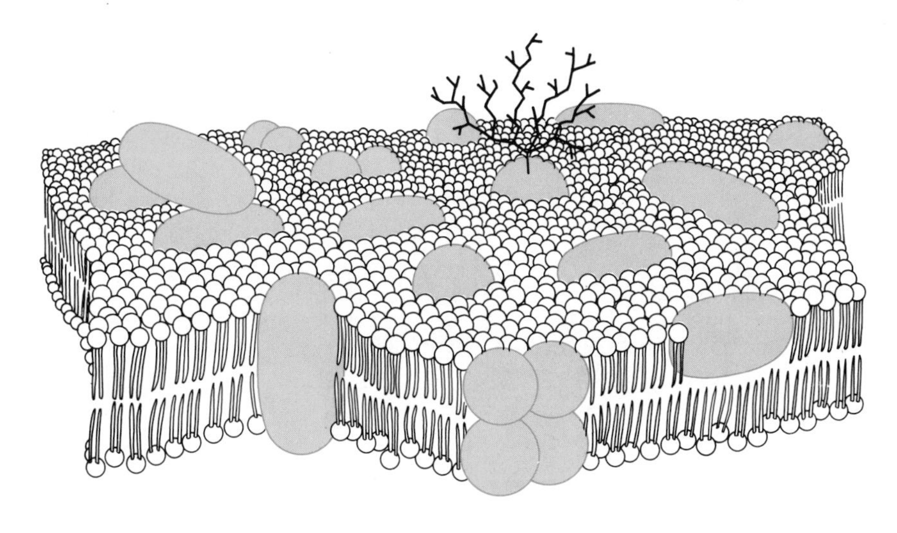

Figure 4-19

A general model for the structure of biological membranes. [After J. M. Clark, Jr., and R. L. Switzer, *Experimental Biochemistry*, 2nd ed. (San Francisco: W. H. Freeman and Company. Copyright © 1977).]

The erythrocyte membrane

The cell membrane of the red blood cell is the most complex natural membrane that is known in considerable detail. The major polypeptides that have been found in this membrane are summarized in Figure 4-20a.

The peripheral proteins are about 40% of the total. These are components 1, 2, 4.1, 4.2, 5, and 6. (The numbers correspond to bands seen in electrophoresis.) These components are all on the inside surface of the membrane, as shown by enzymatic and immunochemical accessibility tests. Bands 1 and 2 are a high-molecular-weight fibrous protein called spectrin. The individual polypeptide chains have molecular weights between 225,000 and 250,000. Components 1 and 2 appear to be closely associated with each other because they are easily cross-linked.

Band 5 is actin or an actinlike protein. It has a molecular weight of 45,000 and can associate into homopolymers. Many investigators feel that the spectrin and actin components assemble together to form a fibrous lattice that underlies the entire cell membrane (shown schematically in Fig. 4-20b). The lattice, if it exists, could play the role of determining the shape of the red blood cell, and also possibly of organizing the locations of some of the other membrane proteins.

Band 6 is glyceraldehyde 3-phosphate dehydrogenase. It is placed near band 3 (an integral protein) in Figure 4-20a, but some recent results suggest that band 6 is really just associated with spectrin. Little is known about bands 4.1 and 4.2.

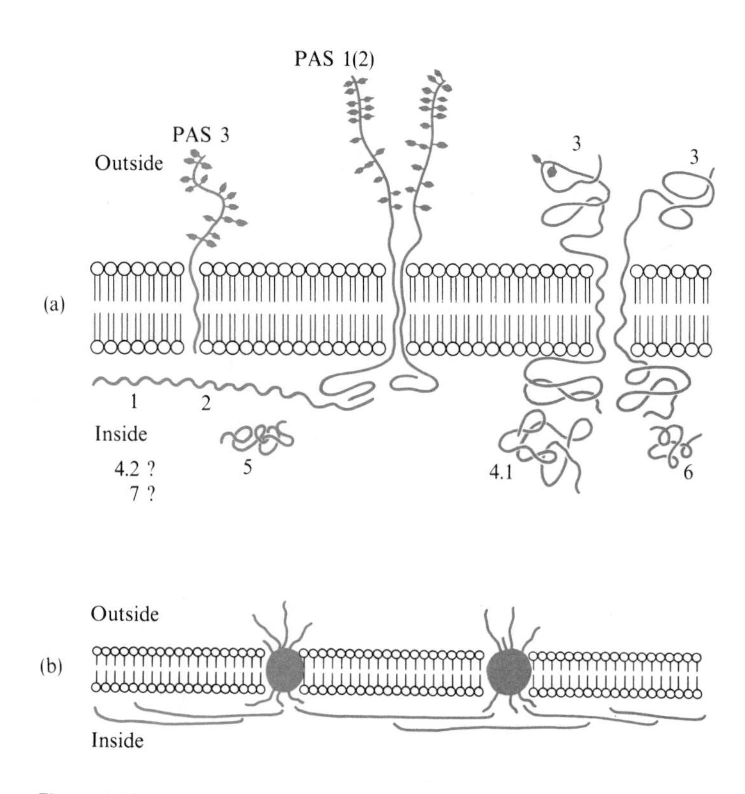

Figure 4-20

Tentative model of the structure of the erythrocyte membrane. (**a**) The major polypeptide chains; current information about each is summarized in the text. Sugars are shown as colored diamonds. (**b**) Possible structure of an internal fibrous protein network. [After V. T. Marchesi et al., *Ann Rev. Biochem.* 45:667 (1976).]

Band 3 and PAS-1 are the most prominent integral membrane proteins. Band 3 is a glycoprotein with about 5% to 8% carbohydrate. It is shown as a dimer, because it can be cross-linked under fairly mild conditions. Chemical modification data suggest that band 3 is a transmembrane protein. It may play a role in the transport of anions or glucose.

PAS-1 is glycophorin, which has already been discussed extensively in this chapter and in Chapter 2. It is shown as a dimer, because it appears to behave as a dimer in detergent solutions such as sodium dodecyl sulfate. Glycophorin may be responsible for the bulk of the globular structures seen in the interior of the erythrocyte membrane when examined by freeze-fracture microscopy. There is some evidence that the part of glycophorin inside the cell interacts directly with the spectrin network (as shown in Figure 4-20b).

PAS-3 is the last major membrane component. It is a glycoprotein, with a peptide moiety consisting of a chain of roughly 25,000 mol wt. It may be a semiintegral protein, somewhat more easily extractable than PAS-1 and band 3.

The principal lipid components of the erythrocyte membrane are listed in Table 4-4. Recent evidence suggests that these are not symmetrically distributed in the bilayer. About 70% of the total phosphatidyl choline and 80% to 85% of the total sphingomyelin are located on the outer surface of the bilayer. About 80% to 90% of the phosphatidyl serine and phosphatidyl ethanolamine is located in the inner surface of the bilayer. The best current evidence suggests that cholesterol, the remaining major lipid component, is distributed about equally between the two surfaces.

4-5 PROTEIN–LIPID INTERACTIONS

Contacts between proteins and lipids in bilayers

Critical features of membrane structure and properties must depend on specific interactions between proteins and lipids. From general knowledge about noncovalent interactions, one can say quite confidently that lipid hydrophobic tails can interact favorably with nonpolar amino acid side chains. Polar lipid heads can interact with polar moieties of proteins. However, important details about such interactions remain to be understood. Here are a few questions just beginning to be explored.

Does a membrane protein perturb the lipid phase adjacent to it? Such perturbation could involve specific attraction of certain lipid components, enhanced mobilization or immobilization of nearby lipids, or a change in bulk membrane properties (such as bilayer thickness, curvature, and permeability). Some evidence suggests that lipids near proteins behave differently from the other lipids. This boundary lipid is probably quite analogous to the water molecules found tighly bound to soluble proteins (see Chapter 10).

Do particular lipids affect the properties of membrane proteins? For example, will a membrane protein function in any bilayer, or is a particular lipid composition required? It is quite clear that the fluidity of the membrane is important for some enzyme functions. This is easy to understand for reactions in which substrates or enzymes must diffuse toward each other in the membrane. There is not much information yet on whether proteins can strongly differentiate between two bilayers that have the same transition temperature but different lipid compositions. However, some proteins do appear to undergo substantial changes when the lipid transition temperature is crossed.

Do lipids penetrate the membrane protein or lie on its surface? The answer is really not very clear at present. We badly need model systems in which a protein interacts with just a few lipids in the absence of the complexities imposed by a membrane. Attractive possibilities include the integral membrane enzyme C-55 isoprenoid alcohol phosphokinase and the peripheral enzyme pyruvate oxidase (Gennis and Jonas, 1977).

Only the first glimpses of the shapes of integral membrane proteins are now available (Chapter 14). Several three-dimensional structures of such proteins must be determined before it is safe to make too many generalizations. One reasonable speculation is that parts of the membrane protein that are within the bilayer probably have extensive α-helical or β-sheet structure. This is an efficient way to satisfy the need to pair up hydrogen-bond donors and acceptors.

Lipoproteins

Protein–lipid interactions are found in lipoproteins as well as in membranes. Lipoproteins are micellelike assemblies. Attention will be restricted here to plasma lipoproteins, which are noncovalent assemblies of various lipids and peptides. Some covalent lipid–protein conjugates are known to exist, as we showed earlier, at the bacterial cell surface. Four major classes of lipoproteins are found in the plasma. These are (in order of increasing density) chylomicrons, very-low-density lipoprotein (VLDL), low-density lipoprotein (LDL), and high-density lipoprotein (HDL).

Relatively little is known about the first class, but the properties of the other three are beginning to be elucidated; Table 4-6 summarizes what is known. A major function of plasma lipoproteins is lipid transport, and LDL is known to be the principle vehicle for carrying cholesterol. As isolated from humans or other organisms, the lipoproteins are large globular particles with some heterogeneity in protein and lipid content. In all cases, there appears to be a small number of major protein components, and the amino acid sequences of most of these from VLDL and HDL are known. The proteins from these two classes have an unusually high percentage of α helix, which appears to be promoted or stabilized by the presence of the lipid, because it is lower in purified apoproteins than it is in native lipoprotein.

As expected, lower protein densities are found with higher lipid contents. Larger particle sizes are also found with higher lipid contents. Each lipoprotein class seems to favor one type of lipid, although many types of lipids are repesented to some extent in each class. More than half the lipids in VLDL are triglycerides, in LDL they are cholesterol esters, whereas in HDL they are phospholipids. Note that two components (triglycerides and cholesterol esters) are almost totally nonpolar and water insoluble. They require the other lipids and proteins to form a stable suspension that can be transported in the blood. Most of the peptide chains in VLDL are quite small. Those in HDL are more typical of average proteins, except that protein apo-A-II is a most unusual symmetrical dimer of two identical chains joined by a single disulfide bond. The protein chains in LDL are not yet well characterized, but it is quite likely that they are larger than the other apolipoproteins.

Table 4-6
Properties of plasma lipoproteins

Property	Very low density	Low density	High density
Density (g/cm^3)	0.95–1.006	1.019–1.063	1.063–1.21
Molecular weight	5×10^6	2×10^6	1.7×10^5 to 3.6×10^5
Percentage protein (by weight)	10	25	50
Major protein components	Apo-C-I (57 amino acids) Apo-C-II (78 amino acids) Apo-C-III (79 amino acids) Apo-E (33,000 mol wt) Apo-B	Apo-B (up to 275,000 mol wt)	Apo-A-I (245 amino acids) Apo-A-II (dimer of two 77-amino acid chains)
Protein secondary structure	73% α helix 27% other	25% α helix 37% β sheet 37% disordered	70% α helix 11% β sheet 19% disordered
Apoprotein secondary structure	56% α helix	—	52% α helix 10% β sheet 38% disordered
Lipid content (as percentage of total lipoprotein weight)			
triglycerides	55	6–8	Small
phospholipid	20	21–22	>20
cholesterol	10	7–8	Small
cholesterol esters	5	36–41	15

SOURCE: From results summarized by J. D. Morresett, R. L. Jackson, and A. M. Gotto, Jr., *Ann. Rev. Biochem.* 44:183 (1975); and L. C. Smith, H. J. Pownall, and A. M. Gotto, Jr., *Ann. Rev. Biochem.* 47:751 (1978).

Major points of interest about native lipoproteins are (1) the mode of organization of lipid and protein components into the globular particles, and (2) the details of the protein–lipid interactions. Very little is known. A micellar arrangment seems plausible for most lipoproteins, with nonpolar lipids internal and polar parts of proteins and other lipids external. However, this leaves many questions unanswered. Is there an ordered arrangement of protein subunits? Some studies on LDL suggest there may be. Are protein subunits in contact with each other, or are they all floating independently in a sea of lipid? Proteins apo-A-I and apo-A-II of HDL, when present in equimolar amounts in the absence of natural lipids, form a 1:1 dimer. This interaction apparently occurs also in the presence of lipid, because in HDL the two proteins are readily cross-linked together by 1,5-difluoro-2,4-dinitrobenzene.

It has been shown that at least some of the peptides in lipoproteins can interact with enormous numbers of lipids. For example, apo-C-III can bind between 18 and 80 phosphatidylcholine molecules. Studies on lipid binding by protein fragments, and on protection of the intact protein against proteolysis by lipid, suggest that most

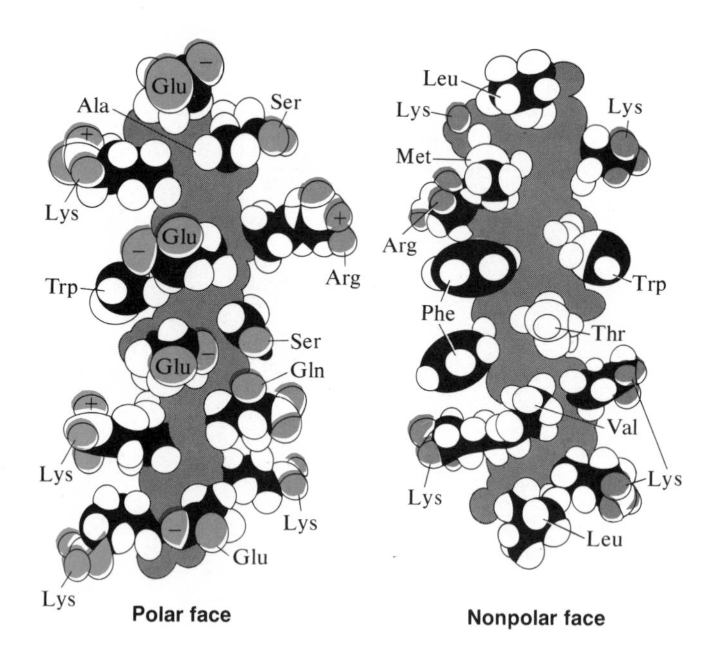

Figure 4-21

A schematic α-helical structure for a region of the peptide portion of apolipoprotein C-I. Charged groups are shown in color. [After J. D. Morrisett et al., *Ann. Rev. Biochem.* 44:183 (1975).]

lipid is bound to the C-terminal half. Similar fragment studies with apo-A-II show that principal phospholipid binding occurs in the two-thirds of the protein closest to the C-terminus. The observation that lipid increases the α-helical content leads to an interesting hypothesis for the detailed nature of the binding. The easiest way to rationalize this observation is to say that lipid binds to α-helical sections of the protein.

When the amino acid sequence of possible lipoprotein binding regions are arranged in an α helix, a suggestive observation can be made (Fig. 4-21). For several stretches of the apo-C-I sequence, the α helix (viewed as a cylinder) has a nonpolar face and a polar face. The striking feature of the polar face is that negatively charged side chains lie at its center, whereas positively charged groups are positioned at the edges of the face. It is tempting to speculate that phospholipids bind to the α helices so that the negative phosphate is held close to the positive edge, and a few of the methylene groups on the fatty acid nearest the phosphate are in contact with the adjacent nonpolar face of the peptide. The remaining part of the fatty acid is thought to extend away from the peptide to interact with other lipid molecules.

It should be possible to design experiments to examine the mode of interaction between lipids and peptides experimentally. This will have important bearing, not only on our understanding of lipoprotein structure, but also on the much more general problem of membrane protein structure and function.

Unknown macromolecular associations

From four basic types of macromolecules—lipids, polysaccharides, proteins, and nucleic acids—six types of covalent pairwise assemblies, and six more noncovalent assemblies, are possible in principle. Table 4-7 shows that only 8 of the 12 potential

Table 4-7
Frequency of heterologous associations of macromolecules found thus far in nature

Type	Covalent	Noncovalent
Protein–nucleic acid	Occasional	Very frequent
Protein–lipid	Occasional	Very frequent
Protein–saccharide	Very frequent	Occasional
Nucleic acid–saccharide	Occasional	——
Lipid–saccharide	Fairly common	——
Lipid–nucleic acid	——	——

associations have been observed thus far in ordered structures that have long-term stability in the cell. It is not known whether the others are nonexistent, or whether they simply have not been found yet. What is most suspicious is the lack of any known ordered complex between lipids and nucleic acids. One possible source of such a complex is the rough endoplasmic reticulum, which consists of membrane and associated ribosomes. It will be interesting to learn whether, in this structure, any direct interaction occurs between lipid and RNA.

Summary

The properties of some of the less-well-studied biological polymers and assemblies are informative, yet raise many interesting questions.

Polysaccharides exist with both linear and branched chain structures. Sequences in many cases are homopolymers or alternating copolymers. Secondary structures are single-stranded or multiple-stranded helices. These helices organize into fiber or gel (tertiary and quaternary) structures. Saccharides are found also as components of other biological assemblies, such as the cell-surface peptidoglycan of bacteria, and the glycoproteins of eukaryotes.

Protein–nucleic acid complexes perform very important roles. The overall basic organization of these complexes appears to be quite variable. In simple viruses, the protein component appears to be major structural determinant, and it is organized on the exterior of the particle. Chromatin has a protein core with the DNA coiled on its surface. In ribosomes, there appears to be a pattern of intimate protein–RNA contacts. The detailed mechanisms by which general or specific protein–nucleic acid recognition occurs are unknown.

Lipids are amphiphiles; they form micelles if they have single fatty acid chains, and bilayers or bilayer vesicles if they have two fatty acid chains. The properties and compositions of the two surfaces of a bilayer are not necessarily identical. Natural membranes contain extensive amounts of protein in addition to lipid. Peripheral membrane proteins are easily removed, whereas integral membrane proteins are anchored tightly to the bilayer, presumably by a hydrophobic section of peptide chain. Some integral proteins are restricted to one surfce of the membrane; others pass through the membrane. Lipid bilayers undergo phase transitions from a solidlike to a liquidlike state. Natural membranes show similar transitions and also have the possibility of lateral phase separation into domains with different protein or lipid concentrations. What distinguishes membranes from the other biological structures is their dynamic behavior. Lateral translational motion of proteins or lipids is quite rapid. Rotational motion of components within a membrane surface also is fast. However, flip-flops of components from one surface to the other are extremely slow.

Not much is known about the details of protein–lipid interactions. Serum glycoproteins, which are micellelike structures composed of a number of different peptides and lipids, may serve as useful model systems, and they are interesting in their own right because of their role in lipd transport.

Problems

4-1. Suppose you could isolate a micelle containing a single protein that normally exists as a transmembrane protein. How would you expect the lipid and protein to be arranged in this micelle?

4-2. Cross-linking frequently has been used to identify proteins located adjacent to each other in cell membranes. However, because of the rapid lateral mobility of some membrane proteins, there is a real risk that proteins initially located far apart might become trapped together due to a random collision that allows a cross-link to form. How would you test for this possibility or try to prevent it in a cross-linking experiment?

4-3. Why would you expect an oligomeric protein to be involved in the specific recognition of a palindromic nucleic acid sequence?

4-4. Why do irregular sequences and chain branching enhance the possibility of gel formation over other possible polysaccharide tertiary and quaternary structures?

4-5. Some nucleoprotein complexes are stabilized by increasing the salt concentration, whereas others are destabilized. How are the structures of these two kinds of complexes likely to differ? Consider the types of proteins that may be involved, possible detailed protein–nucleic acid contacts, and the number of small cations bound to the complexes are compared with the separate components.

References

GENERAL

Gennis, R. B., and A. Jonas. 1977. Protein–lipid interactions. *Ann. Rev. Biophys. Bioeng.* 6:195.

Kirkwood, S. 1974. Unusual polysaccharides. *Ann. Rev. Biochem.* 43:401.

Rogers, H. J., and H. R. Perkins, 1968. *Cell Walls and Membranes*. London: Spon.

Singer, S. J. 1974. The molecular organization of membranes. *Ann. Rev. Biochem.* 43:805.

Von Hippel, P. H., and J. D. McGhee. 1972. DNA–protein interactions. *Ann. Rev. Biochem.* 41:231.

SPECIFIC

Braun, V., and K. Hantke. 1974. Biochemistry of bacterial cell envelopes. *Ann. Rev. Biochem.* 43:89.

Casjens, S., and J. King. 1975. Virus assembly. *Ann. Rev. Biochem.* 44:555.

Champoux, J. J. 1978. Proteins that affect DNA conformation. *Ann. Rev. Biochem.* 47:449.

DiRienzo, J. M., K. Nakamura, and M. Inouye. 1978. The outer membrane proteins of Gram-negative bacteria: Biosynthesis, assembly and functions. *Ann. Rev. Biochem.* 47:481.

Elgin, S. C. R., and H. Weintraub. 1975. Chromosomal proteins and chromatin structure. *Ann. Rev. Biochem.* 44:900.

Kornfeld, R., and S. Kornfeld. 1976. Comparative aspects of glycoprotein structure. *Ann. Rev. Biochem.* 45:217.

Lancelot, G., and C. Helene. 1977. Selective recognition of nucleic acids by proteins: The specificity of guanine interactions with carboxylate ions. *Proc. Natl. Acad. Sci. USA* 74:4872.

Marchesi, V. T., H. Furthmayr, and M. Tomita. 1976. The red cell membrane. *Ann. Rev. Biochem.* 45:667.

Morrisett, J. D., R. L. Jackson, and A. M. Gotto, Jr. 1975. Lipoproteins: Structure and function. *Ann. Rev. Biochem.* 44:183.

Rothman, J. E., and J. Lenard. 1977. Membrane asymmetry. *Science* 195:743.

Stanier, R. Y., F. A. Adelberg, and J. L. Ingraham. 1976. *The Microbial World*, 4th ed. Englewood Cliffs, N.J.: Prentice-Hall. [See Chapter 11 for an excellent treatment of cell walls.]

Tanford, C. 1973. *The Hydrophobic Effect*. New York: Wiley.

Thomas, J. O. 1978. Chromatin structure. In *Biochemistry of Nucleic Acids*, vol. 2, ed. B. F. C. Clark (Baltimore University Park Press), p. 181.

Warrant, R. W., and S. H. Kim. 1978. α-Helix–double helix interaction shown in the structure of a protamine-transfer RNA complex and a nucleoprotamine model. *Nature* 271:130.

Wittmann, H. G. 1976. Structure, function and evolution of ribosomes. *Eur. J. Biochem.* 61:1.

5

Conformational analysis and forces that determine protein structure

5-1 BASIC PROBLEMS OF PROTEIN STRUCTURE

One of the most fascinating and challenging problems of biochemistry is that of the physical basis for the markedly organized and specific structures of proteins. This problem takes on particular significance because the biological activity of these macromolecules is sensitive to variations in three-dimensional conformation. Because biological polymers generally can be unfolded (for example, by heat or urea treatment) and later refolded to the original state, it is reasonable to assume that, in the main, the conformations adopted by various biopolymers are the thermodynamically preferred ones. Some support for this assumption was obtained in a series of experiments by C. B. Anfinsen and colleagues (1961). These investigators demonstrated that, even when all four disulfide bonds of ribonuclease are broken and the protein is completely denatured by urea, the correct native structure is readopted upon removal of urea and reoxidation of the disulfide linkages.[§]

In this chapter, we discuss the various factors that determine protein conformation. We first consider inherent geometric factors such as the essentially fixed bond lengths and bond angles of the chain. Next we examine the restrictions on available conformations imposed by steric interactions. We also take up the refinement of this analysis with the use of more realistic potential functions. Finally, we look at other factors that are of extreme importance in determining protein structures.

[§] In spite of the results of this experiment, however, one must consider that native structures may represent local free energy minima, and that kinetic barriers may prevent attainment of conformations with even lower free energies.

These include the well-characterized hydrogen bond, as well as the less-understood "hydrophobic" interaction. If all of the above-mentioned factors were well enough understood, and if sufficient mathematical tools were available, then it should be possible to predict, for example, the three-dimensional structure of a protein from its amino acid sequence. This goal has not yet been achieved but, in view of recent progress, it is not completely unrealistic. (Further aspects of protein folding are discussed in Chapter 21.)

5-2 POLYPEPTIDE CHAIN GEOMETRIES

Figure 5-1 displays the α-L-polypeptide chain in its all-*trans* (planar zig-zag) form. Residues are indexed serially from 1 to n, starting from the N-terminus. Note that

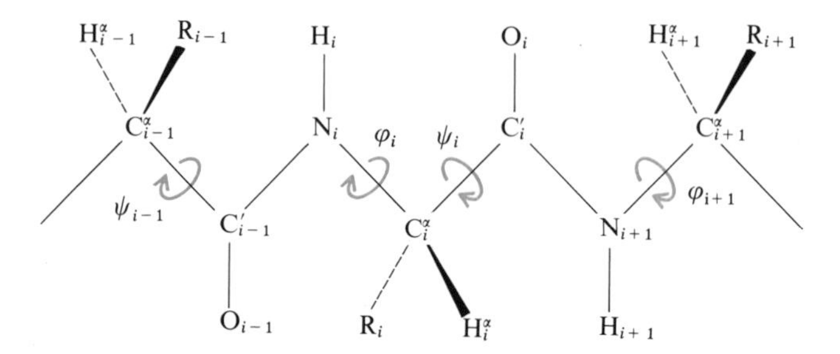

Figure 5-1
The α-L-polypeptide chain in the all-*trans* form.

similar atoms within the same residue are also distinguished; for example, the ith alpha-carbon is denoted C^{α}_i while the ith carbonyl carbon is C'_i. The individual amino acid units differ according to their R-group side chains. The various bond lengths and bond angles that characterize the chain backbone may be regarded as essentially fixed. Table 5-1 gives the generally accepted values for these geometric parameters.

An important aspect of polypeptide structure is that the amide bond usually occurs in the planar *trans* conformation. This can be rationalized to be a result of a resonance form that imparts double-bond character to the amide group.

The partial double-bond character shortens the amide bond by about 0.1 Å. As a result of the planar *trans* form of the amide group, the distance between successive alpha-carbon atoms is fixed at 3.8 Å. This means that the conformation of the chain as a whole is determined by the rotations about the N–C^{α} and C^{α}–C′ bonds.

Table 5-1
Polypeptide chain geometries

Bond	Bond length (Å)	Bonds	Bond angle
C^{α}—C'	1.53	C^{α}—C'—N	113°
C'—N	1.32	C'—N—C^{α}	123°
N—C^{α}	1.47	N—C^{α}—C'	110°
C'=O	1.24		
N—H	1.00		
C^{α} – C^{β}	1.54		
C^{α}—H^{α}	1.07		

SOURCE: Data from V. Sasisekharan in *Collagen*, ed. N. Ramanathan (New York: Interscience, 1962), p. 39.

Internal rotation angles

In order to specify the three-dimensional conformation of a macromolecule, it is necessary to specify the internal rotation angles, or torsion angles. We follow the scheme of the IUPAC–IUB Commission on Biochemical Nomenclature (1970). Consider a four-atom system A–B–C–D:

A B ϕ C D

Given fixed bond lengths and bond angles, the torsion angle ϕ describes the relationship between the C–D and A–B bonds. This angle is defined as the angle between the plane specified by bonds B–C and C–D and the plane specified by bonds A–B and B–C.

The torsion angle is, by convention, assigned a unique numerical value as follows. Figure 5-2a,b shows Newman and perspective projections of our four-atom system. A positive value of ϕ is assigned to the *right-hand rotation* necessary to bring the front atom (A or D) into an eclipsed position with the rear atom (D or A). For example, in the first and third illustrations of Figure 5-2a, right-hand rotation (viewed B to C) about B–C through the angle ϕ_1 results in atom A eclipsing atom D. Conversely, in the second and fourth illustrations (viewed C to B), D is the front atom, and a right-hand rotation about C–B through the angle ϕ_2 results in D eclipsing A. Thus, whether we look from B to C or from C to B, the same positive value for ϕ is obtained.

Alternatively, the same value of ϕ is obtained if we look from B to C and consider ϕ as the right-hand rotation of bond C–D (about B–C) that is required to bring atom D (the rear atom) out of eclipse by A (the front atom) to the designated position. We obviously can perform the equivalent operation by looking from C to B. Negative values of the torsion angle are *left-hand rotations* (Fig. 5-2b).

The value $\phi = 0°$ corresponds to the eclipsed conformation in which the bonds

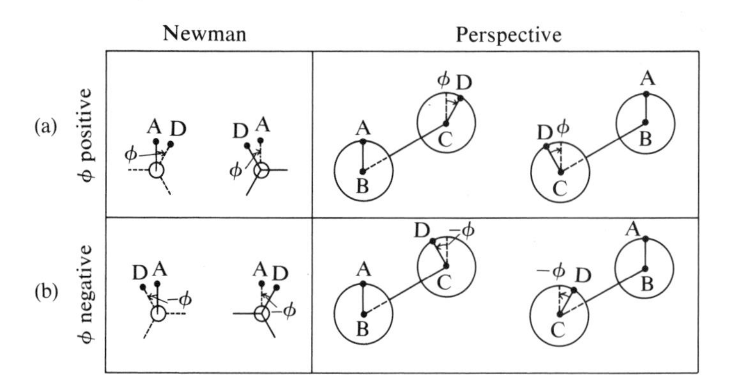

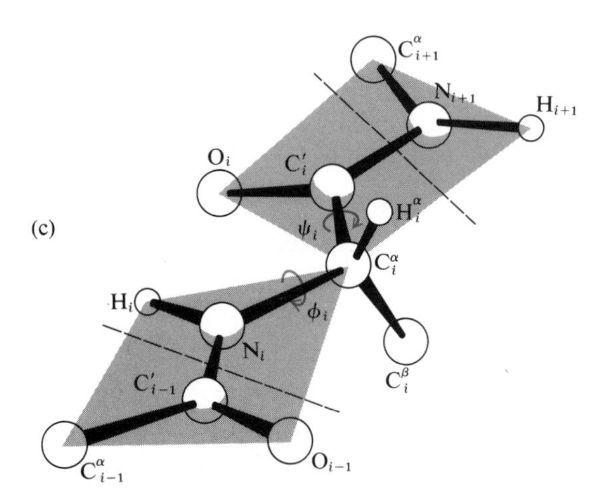

Figure 5-2

Torsion angles. (**a**) Newman and perspective diagrams illustrating positive values of the torsion angle ϕ for the four-atom (ABCD) system. (**b**) Similar diagrams illustrating negative values of ϕ. (**c**) Perspective drawing of the α-L-polypeptide chain, illustrating the torsion angles ϕ and ψ. The chain is drawn in the planar zig-zag (all-*trans*) form, for which $\phi = \psi = 180°$. The dashed lines delineate the extent of the ith amino acid residue. [After IUPAC–IUB Commission on Biochemical Nomenclature, *Biochemistry* 9:3471 (1970).]

A–B and C–D are *cis*. When these bonds are *trans*, as in

$$\text{A}\diagup\text{B—C}\diagup\text{D}$$

the torsion angle $\phi = 180°$ (or, equivalently, $\phi = -180°$). It is convenient to specify the values of ϕ as falling in the range $-180° \leqslant \phi \leqslant +180°$. In this way, the value of ϕ makes clear the relationship between enantiomeric conformations.

For the polypeptide backbone, with its fixed *trans* conformation for the amide bond, the important torsion angles are those for the N_i–C_i^α and the C_i^α–C_i' bonds. These angles are designated ϕ_i and ψ_i, respectively. For an α-L-polypeptide in the planar *trans* conformation (Fig. 5-2c), $\phi_i = \psi_i = 180°$ for all *i*.

(It should be noted that much of the earlier literature in the field is based on the convention that the reference state, $\phi_i = 0°$ and $\psi_i = 0°$, be assigned to the planar *trans* arrangement. Therefore, in reading this literature, 180° must be added to—or subtracted from—the numerical values of ϕ and of ψ in order for these values to correspond to the established convention used here. See Box 18-1 for further discussion, particularly with respect to conventions used in treating polymer chain statistics.)

The conformation of the polypeptide chain backbone as a whole is specified by enumerating the value of each ϕ_i and ψ_i. A helical form is generated when every $\phi_i = \phi_j$ and every $\psi_i = \psi_j$. For the right-handed alpha helix, $\phi = -57°$ and $\psi = -47°$. The ϕ, ψ coordinates of a variety of ordered forms of polypeptides are given in Table 5-2. These are all periodic, regular forms in which each rotation angle is the same for each residue. (Some of these structures are illustrated in Chapter 2.) Note that the pleated-sheet conformations are closest to the planar zig-zag structure shown in Figure 5-1.

Table 5-2
Approximate torsion angles for some regular structures

Structure	ϕ	ψ
Right-handed α helix [α-poly(L-alanine)]	−57°	−47°
Left-handed α helix	+57°	+47°
Parallel-chain pleated sheet	−119°	+113°
Antiparallel-chain pleated sheet [β-poly(L-alanine)]	−139°	+135°
Polyglycine II	−80°	+150°
Collagen (triple helix)	−51°, −76°, −45°	+153°, +127°, +148°
Poly(L-proline) I	−83°	+158°
Poly(L-proline) II	−78°	+149°

NOTE: For a fully extended chain, $\phi = \psi = +180°$. The imide bond is *cis* in polyproline I; it is *trans* in polyproline II.
SOURCE: After IUPAC–IUB Commission on Biochemical Nomenclature, *Biochemistry* 9:3471 (1970) [published simultaneously in *J. Biol. Chem.* 245:6489 (1970) and *J. Mol. Biol.* 52:1 (1970)].

In contrast to the periodic forms listed in Table 5-2, globular proteins (or random coils, of course) show ϕ, ψ coordinates varying considerably from residue to residue.

Ramachandran or steric contour diagrams

In theory, a chain molecule can adopt an essentially infinite variety of backbone conformations, each corresponding to a unique set of values for the various backbone rotation angles. However, many of these hypothetical conformations can be excluded from consideration on the basis of unfavorable steric overlaps. G. N. Ramachandran and colleagues (1963) were among the first to investigate this problem.

In studying steric interactions, we look first at overlaps between atoms whose distance of separation depends on a single rotation angle. For example, the distances C_i^α to O_i, C_i^α to N_{i+1}, or C_i^α to C_{i+1}^α depend solely upon the value of ψ_i. By determining the steric overlaps that occur as we vary ψ_i, we can exclude certain values of ψ_i. However, by examination of atomic scale models, it is immediately apparent that additional values of ψ_i can be eliminated because of overlaps between atoms that are separated by ϕ_i and ψ_i—for example, O_i and O_{i-1}. Because of these overlaps, which depend simultaneously upon the values of both ϕ_i and ψ_i, rotations about N_i–C_i^α and C_i^α–C_i' are said to be interdependent.

One can readily determine those values of the pair ϕ_i, ψ_i that are sterically allowed, by investigating contacts between all atoms whose distance of separation depends solely on these two rotation angles. The allowed domain of ϕ_i and ψ_i is *not* further reduced by interactions that are also simultaneously dependent on the values of the rotation angles in adjacent units, such as ϕ_{i-1} and ψ_{i-1}. For example, the distance of separation of O_i and R_{i-1} depends on ϕ_i, ψ_i, and ψ_{i-1}, but no contact between O_i and R_{i-1} is possible when ϕ_i and ψ_i are confined to their previously determined allowed domains. Thus, with amide groups fixed in the *trans* form, rotations within a given residue are *interdependent*, but they are sterically independent of rotations within neighboring residues.

At this point it is important to distinguish between short-range and long-range interactions. Those occurring between atoms or groups that are neighboring or close to each other in the sequence are termed short-range. This is the type discussed in the preceding paragraph. Long-range interactions are those between groups that are far apart in the sequence, and they occur only when the chain folding brings such groups into close proximity. This section is concerned largely with short-range effects, which are capable of giving considerable insight into protein conformation.

How can we best represent or diagram those values of ϕ_i and ψ_i that are sterically allowed for a given amino acid residue? This is accomplished by a map, in which ψ is the ordinate and ϕ is the abscissa. Lines then are drawn to enscribe the ϕ, ψ coordinates in which no unfavorable steric contacts occur.

Figure 5-3 is a steric diagram for a glycyl residue (R = H) situated within a polypeptide chain; Figure 5-4 is a similar diagram for alanine ($R = H_3$). The dark zones show normal contact radii for the various atom pairs, whereas the light zones

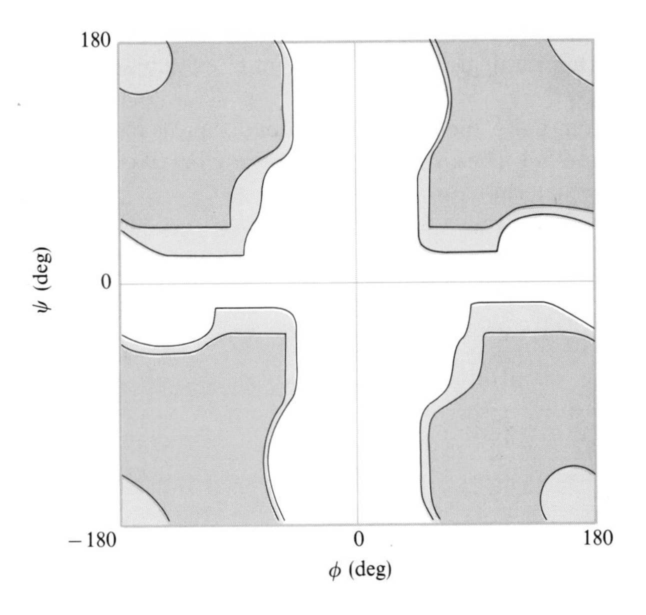

Figure 5-3

Steric contour diagram for a glycyl residue in a polypeptide chain. Dark zones show "normal" and light zones show "outer-limit" contours. [After G. N. Ramachandran et al., *Biophys. J.* 6:849 (1966).]

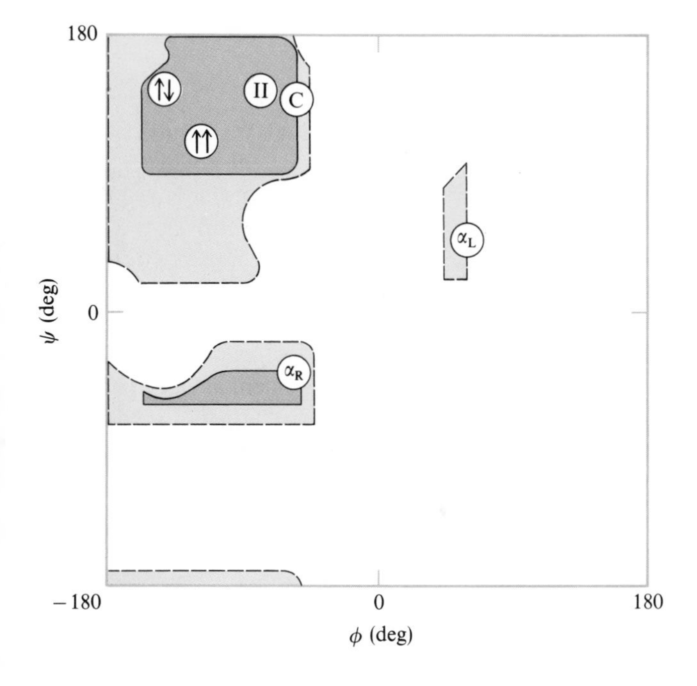

Figure 5-4

Steric contour diagram for an L*-alanyl residue in a polypeptide chain.* Dark zones show "normal" and light zones show "outer-limit" contours. Coordinates of right- and left-handed α helices (α), parallel (↑↑) and antiparallel (↑↓) pleated sheets, polyglycine II (II), and collagen (C) are denoted (see also Table 5-2). [After P. J. Flory, *Statistical Mechanics of Chain Molecules* (New York: Interscience, 1969), and G. N. Ramachandran et al., *J. Mol. Biol.* 7:95 (1963).]

show some of the shortest contact radii observed in crystal structures of relevent molecules; these somewhat shorter radii give the "outer limit" of the sterically allowed values.

Table 5-3 lists, for various atom pairs, the minimal contact distances important for the generation of Figures 5-3 and 5-4. These contact distances are based on values found in crystallographic structure determinations.

Table 5-3
Minimum contact distances for important polypeptide atom pairs

Atom pair	Normal (Å)	Outer limit (Å)
C—C	3.2	3.0
C—O	2.8	2.7
C—N	2.9	2.8
C—H	2.4	2.2
O—O	2.8	2.7
O—N	2.7	2.6
O—H	2.4	2.2
N—N	2.7	2.6
N—H	2.4	2.2
H—H	2.0	1.9

SOURCE: O—O distances from C. Ramakrishnan and G. N. Ramachandran, *Biophys. J.* 5:909 (1965); other data from G. N. Ramachandran et al., *J. Mol. Biol.* 7:95 (1963).

The steric diagram for glycine is centrosymmetric—that is, symmetric with respect to any line that passes through the center ($\phi = 0°$, $\psi = 0°$). This is a consequence of the symmetry of the glycine residue. As might be anticipated, the *cis* configuration about either bond ($\phi = 0°$ or $\psi = 0°$) and configurations in the neighborhood of the vertical lines ($\phi = 0°$ or $\psi = 0°$) are sterically forbidden. In the case of alanine, the steric diagram is asymmetric, and the sterically allowed domain is significantly smaller. This smaller domain is due to the additional unfavorable contacts that involve the $-CH_3$ side chain. Steric diagrams for residues with longer side chains but that are not branched at the β-carbon—such as leucine, R = $-CH_2CH(CH_3)_2$—are essentially the same as Figure 5-4. A side chain that is branched at the β-carbon—such as valine, R = $CH(CH_3)_2$—does reduce somewhat the sterically permitted domain below the areas of Figure 5-4.

5-3 ESTIMATES OF POTENTIAL ENERGY

The steric diagrams of Figures 5-3 and 5-4 give an approximate indication of the allowed conformations of amino acid residues. They do not indicate, however, which parts of the sterically allowed domain are preferred—that is, they do not show gradations in energy within the sterically permissible regions. In order to obtain an estimate of relative preference, it is necessary to compute the potential energy as a function of ϕ and ψ.

Methods for computing rotational potential functions are only semiempirical; therefore, energies computed by such methods are not to be viewed too literally. Nevertheless, when these methods are used in conjunction with configuration statistics (see Chapter 18), there usually is reasonable accord between calculated and observed chain dimensions; on the other hand, the use of "hard sphere" diagrams (such as Figs. 5-3 and 5-4) does not always give good agreement with experiment.

"Nonbonded" interactions

Consider first the attractive and repulsive interactions between atoms whose distance of separation is a function of ϕ and ψ. These nonbonded interactions give rise to an energy $E_{kl}(\phi_i, \psi_i)$, which is generally expressed as

$$E_{kl}(\phi_i, \psi_i) = a_{kl}/r_{kl}^m - c_{kl}/r_{kl}^6 \tag{5-1}$$

where the parameters a_{kl} and c_{kl} are characteristic of the atom pair or atomic group (e.g., CH_3) k and l. A typical plot of E_{kl} versus r_{kl} is given in Figure 5-5. The plot shows that there is no interaction at large distances but that, as the atoms approach each other, an attractive interaction sets in. This attractive interaction is soon overcome by a strong repulsion as the atoms begin to penetrate each other's atomic radii.

The sixth-power attractive term is the familiar London dispersion energy. The distribution of electron density around any nucleus undergoes rapid fluctuations that give rise to a separation between the center of electronic charge and the center

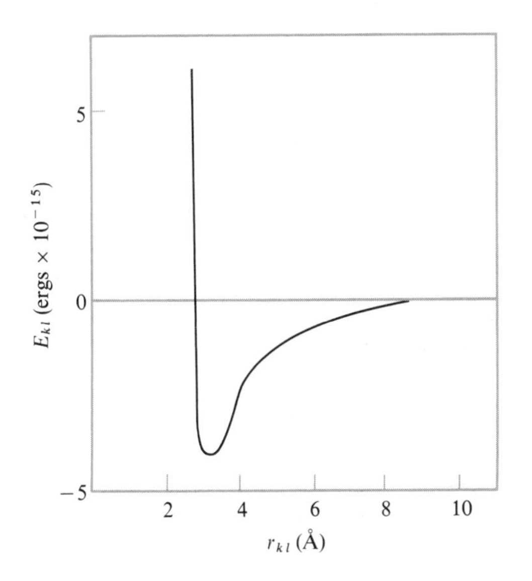

Figure 5-5

Nonbonded interaction energy E_{kl} between atoms or groups k and l plotted as a function of their separation. The example shown is for a pair of hydrogen molecules. [After W. J. Moore, *Physical Chemistry*, 3d ed. (Englewood Cliffs, N.J.: Prentice-Hall, 1962), p. 715.]

of nuclear charge. This separation creates a transient dipole moment, which can resonate with that of a neighboring atom, so that the transient charges orient in an attractive configuration. The result is an induced-dipole–induced-dipole interaction, which accounts for the sixth-power dependence; the parameter c_{kl} is obtained from the atomic polarizabilities of atoms k and l, and the effective number of valence-shell electrons. (See Brant et al., 1967, and references therein.)

The repulsive term a_{kl}/r_{kl}^m in Equation 5-1 has no strict theoretical justification;[§] experimental results obtained with molecular beams suggest that m is in the range of 9 to 12. The parameter a_{kl} may be obtained by requiring the minimum in Figure 5-5 to occur at the sum of the atomic radii or by a variation of this procedure (see Brant et al., 1967). Some values for the atomic radii (r^0) that are used in polypeptide calculations are $r_H^0 = 1.2$ to 1.3 Å; $r_O^0 = 1.5$ to 1.6 Å; and $r_{CH_2}^0 = 1.95$ Å (the CH_2 group usually is treated as a single spherical entity).

Dipolar interactions

A second contribution to the energy producing rotations ϕ and ψ arises from the dipole interaction between adjacent amide groups. The amide-group dipole is approximately parallel to the N–H bond and points from N to H (Fig. 5-6). It has a

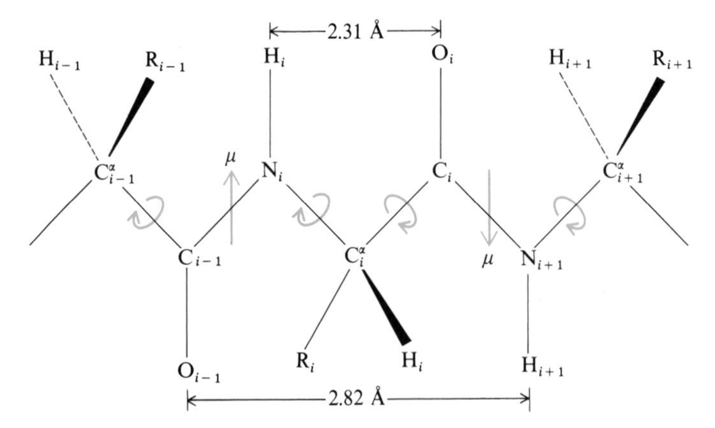

Figure 5-6

Representation of the amide-group dipole moment μ. In the fully extended chain, these moments alternate in direction along the chain. [After P. J. Flory, *Statistical Mechanics of Chain Molecules* (New York: Interscience, 1969), chap. VII.]

[§] An exponential repulsive potential also has been used. See, for example, Brant and Flory (1965).

magnitude of approximately 3.7 D. This is a relatively large moment; for example, the dipole moment of HCl is 1.03 D, that of CH_3Cl is 1.87 D, and that of HCN is 2.93 D. Therefore, we expect dipolar interactions to have a significant influence on polypeptide backbone conformation.

Clearly, the orientation of any two successive amide-group dipoles is determined by ϕ and ψ. The interaction energy E_d between two point dipoles $\boldsymbol{\mu}_A$ and $\boldsymbol{\mu}_B$ separated by the vector $\mathbf{r}$ is given by

$$E_d = \varepsilon^{-1}[\boldsymbol{\mu}_A \cdot \boldsymbol{\mu}_B / r^3 - 3(\boldsymbol{\mu}_A \cdot \mathbf{r})(\boldsymbol{\mu}_B \cdot \mathbf{r})/r^5] \tag{5-2}$$

where r is the scalar magnitude of $\mathbf{r}$, and ε is the dielectric constant. (See Box 5-1.) This expression is, of course, only approximate because it requires that the distance of separation of the amide dipoles be large compared to the separation of the partial charges of the dipole moment itself. To achieve a more accurate representation of the electrostatic interaction between adjacent amide dipoles, the energy may be computed by assigning partial charges to the appropriate atoms and then calculating the total electrostatic energy as a sum of charge–charge coulombic interactions. The energy then is computed from

$$E_d = \left(\sum_{ij} q_i q_j\right) \Big/ \varepsilon r_{ij} \tag{5-3}$$

where q_i and q_j are the partial charges on atoms i and j, and r_{ij} is their distance of separation. Partial charges must be chosen, of course, that will approximately reproduce the overall amide-group dipole moment as well as the accepted values for the individual N–H and C–O bond moments. These criteria are fulfilled by assigning partial charges of $-0.28e$ and $+0.28e$ to N and H, respectively, and $-0.39e$ and $+0.39e$ to O and C, respectively.

Choosing a value for the dielectric constant ε is not a simple task. Although the macroscopic dielectric constant of water is about 80, the local microscopic constant should be considerably less. One expects the lines of force between two *adjacent* dipoles to pass, in large part, through the local backbone of the polymer itself, and not through the surrounding bulk water. Therefore, the high-frequency dielectric constant of solid amides and polyamides, to which contributions from the orientations of permanent dipoles are small, might be a more appropriate measure of the local dielectric constant along a polypeptide backbone in solution. The appropriate data suggest that a value in the range of $\varepsilon = 2$ to 5 might be a realistic estimate of the effective dielectric constant between two successive amide groups in a polypeptide chain.

To a good approximation, dipolar interactions between group dipoles that are second neighbors may be ignored because the dipolar interaction energy falls off rapidly with increasing distance of separation of the dipoles (see Eqn. 5-2). Moreover, at larger distances, the lines of electrostatic force may pass through some of the bulk solvent, thus raising the effective dielectric constant.

Intrinsic torsional potential

In addition to contributions examined above, a small hindrance to rotation also occurs because of an intrinsic torsional potential associated with single bonds. That is, apart from nonbonded or electrostatic interactions, the bond itself presents a barrier to rotation. Thus, the ϕ and ψ rotations each have intrinsic rotational hindrance potentials that are probably threefold (that is, three minima occur). These minima

Box 5-1 DERIVATION OF INTERACTION ENERGY BETWEEN TWO DIPOLES

The mutual potential energy of two point dipoles is not conceptually difficult to calculate. If we designate the respective dipole moments as $\boldsymbol{\mu}_A$ and $\boldsymbol{\mu}_B$, then the interaction energy is simply the potential energy of $\boldsymbol{\mu}_B$ in the electric field generated by $\boldsymbol{\mu}_A$ (or vice versa)—that is, $-\mathbf{E}_A \cdot \boldsymbol{\mu}_B$, where $\mathbf{E}_A$ is the electric field generated by the dipole $\boldsymbol{\mu}_A$. Therefore, the problem reduces to calculating $\mathbf{E}_A$.

For this purpose, we first consider the more general case of the field produced by an arbitrary charge distribution, and we illustrate the concept of the multipole expansion. Imagine a distribution of charges about an origin; the ith one has a charge $z_i e$ and has coordinates $\tilde{x}_i, \tilde{y}_i, \tilde{z}_i$. The electric potential Φ at some distance ($\mathbf{r} = \hat{\mathbf{i}}\tilde{x} + \hat{\mathbf{j}}\tilde{y} + \hat{\mathbf{k}}\tilde{z}$) from the origin is given by

$$\Phi = \sum\{z_i e/\varepsilon[(\tilde{x} - \tilde{x}_i)^2 + (\tilde{y} - \tilde{y}_i)^2 + (\tilde{z} - \tilde{z}_i)^2]^{1/2}\}$$

This expression may be expanded in a Taylor series in $\tilde{x}_i, \tilde{y}_i, \tilde{z}_i$ about the origin. The result is

$$\begin{aligned}\Phi = (1/\varepsilon)\{&\textstyle\sum(z_i e/r) + \sum[z_i e\tilde{x}_i(\partial X^{-1/2}/\partial\tilde{x}_i)] + \sum[z_i e\tilde{y}_i(\partial X^{-1/2}/\partial\tilde{y}_i)] \\ &+ \textstyle\sum[z_i e\tilde{z}_i(\partial X^{-1/2}/\partial\tilde{z}_i)] + \frac{1}{2}\sum[z_i e\tilde{x}_i^2(\partial^2 X^{-1/2}/\partial\tilde{x}_i^2)] \\ &+ \textstyle\sum[z_i e\tilde{x}_i\tilde{y}_i(\partial^2 X^{-1/2}/\partial\tilde{x}_i\partial\tilde{y}_i)] + \cdots\}\end{aligned}$$

where $X = (\tilde{x} - \tilde{x}_i)^2 + (\tilde{y} - \tilde{y}_i)^2 + (\tilde{z} - \tilde{z}_i)^2$, and all the derivatives are to be evaluated at $\tilde{x}_i = \tilde{y}_i = \tilde{z}_i = 0$.

Note that the first term is simply the electrostatic potential at the point r for the situation in which all the charges $z_i e$ are concentrated at the origin (potential = [net charge]/εr. The following terms correct for the fact that the charges are distributed about the origin. The first three terms contain the $\tilde{x}$, $\tilde{y}$, and $\tilde{z}$ components of the dipole moment $\boldsymbol{\mu}$ of the distribution. That is,

$$\mu_{\tilde{x}} = \textstyle\sum z_i e\tilde{x}_i; \qquad \mu_{\tilde{y}} = \sum z_i e\tilde{y}_i; \qquad \mu_{\tilde{z}} = \sum z_i e\tilde{z}_i$$

The next nine terms involve the nine components of the quadrapole moment $\underset{\sim}{Q}$ of the distribution, where

$$Q_{\tilde{x}\tilde{x}} = \textstyle\sum z_i e\tilde{x}_i^2; \qquad Q_{\tilde{y}\tilde{y}} = \sum z_i e\tilde{y}_i^2; \qquad Q_{\tilde{z}\tilde{z}} = \sum z_i e\tilde{z}_i^2; \qquad Q_{\tilde{x}\tilde{y}} = \sum z_i e\tilde{x}_i\tilde{y}_i; \ldots$$

The nine components of the quadrapole moment make up the quadrapole moment tensor.

occur at rotation angles of (approximately) 60°, 180°, and 300°. The barrier heights between minima are believed to be fairly small, on the order of 1 kcal mole^{-1}. The torsional energy $E_{\mathrm{tor}}(\phi_i, \psi_i)$ thus can be represented as

$$E_{\mathrm{tor}}(\phi_i, \psi_i) = (E_\phi^0/2)(1 + \cos 3\phi) + (E_\psi^0/2)(1 + \cos 3\psi) \tag{5-4}$$

where E_ϕ^0 and E_ψ^0 are the barrier heights associated with ϕ and ψ rotations, respectively.

Using the relationships for the components of the dipole and quadrapole moments, and noting that $(\partial X^{-1/2}/\partial \tilde{x}_i)_{\tilde{x}_i=0} = -\partial r^{-1}/\partial \tilde{x}$ (and so on for derivatives with respect to $\tilde{y}_i$ and $\tilde{z}_i$), we are able to write the series expansion of Φ as

$$\begin{aligned}\Phi = (1/\varepsilon)[(\textstyle\sum z_i e)/r &- \mu_{\tilde{x}}(\partial r^{-1}/\partial \tilde{x}) - \mu_{\tilde{y}}(\partial r^{-1}/\partial \tilde{y}) - \mu_{\tilde{z}}(\partial r^{-1}/\partial \tilde{z}) \\ &+ \tfrac{1}{2}Q_{\tilde{x}\tilde{x}}(\partial^2 r^{-1}/\partial \tilde{x}^2) + \tfrac{1}{2}Q_{\tilde{y}\tilde{y}}(\partial^2 r^{-1}/\partial \tilde{y}^2) + \tfrac{1}{2}Q_{\tilde{z}\tilde{z}}(\partial^2 r^{-1}/\partial \tilde{z}^2) \\ &+ Q_{\tilde{x}\tilde{y}}(\partial^2 r^{-1}/\partial \tilde{x}\partial \tilde{y}) + \cdots]\end{aligned}$$

This is the multipole expansion of the potential Φ; the terms containing the components of the dipole moment make up the dipole potential, and the second-derivative terms are the quadrapole potential.

Assume now that the charge distribution contains equal numbers of opposite charges (so that $\sum z_i e = 0$) displaced a small distance from the origin to give a dipole moment $\boldsymbol{\mu}_A$ (for example, the amide dipole is equivalent to two equal and opposite unit charges displaced a distance of 3.7 Å from each other). In the expansion of Φ, the leading nonvanishing term is the dipole potential, which (for small displacements of charges) dominates over the quadrapole potential, which involves second-power terms in the displacements. Noting that $\partial r^{-1}/\partial \tilde{x} = -\tilde{x}/r^3$, we obtain

$$\Phi = (\mu_{A\tilde{x}}\tilde{x} + \mu_{A\tilde{y}}\tilde{y} + \mu_{A\tilde{z}}\tilde{z})/\varepsilon r^3$$

where $\mu_{A\tilde{x}}$, $\mu_{A\tilde{y}}$, and $\mu_{A\tilde{z}}$ are the components of $\boldsymbol{\mu}_A$. The field $\mathbf{E}_A$ is

$$\begin{aligned}\mathbf{E}_A &\doteq -\nabla\Phi \\ &= -\nabla\boldsymbol{\mu}_A \cdot \mathbf{r}/\varepsilon r^3 \\ &= -\boldsymbol{\mu}_A/\varepsilon r^3 - (\boldsymbol{\mu}_A \cdot \mathbf{r}/\varepsilon)\nabla(1/r^3)\end{aligned}$$

The potential energy of interaction with the dipole $\boldsymbol{\mu}_B = \hat{\mathbf{i}}\mu_{B\tilde{x}} + \hat{\mathbf{j}}\mu_{B\tilde{y}} + \hat{\mathbf{k}}\mu_{B\tilde{z}}$ located at the distance $\mathbf{r}$ from $\boldsymbol{\mu}_A$ is simply $-\mathbf{E}_A \cdot \boldsymbol{\mu}_B$, which after simplification can be written as

$$E_d = (\boldsymbol{\mu}_A \cdot \boldsymbol{\mu}_B/\varepsilon r^3) - 3(\boldsymbol{\mu}_A \cdot \mathbf{r})(\boldsymbol{\mu}_B \cdot \mathbf{r})/\varepsilon r^5$$

This is the result given in Equation 5-2.

Total energy as the sum of individual contributions

Our goal now is to calculate the total energy as a function of ϕ_i and ψ_i for the ith amino acid residue situated within a polypeptide chain. As ϕ_i and ψ_i are varied, the distance between all atoms whose separation depends solely on ϕ_i and/or ψ_i must be calculated. For each atom pair k and l, the nonbonded energy E_{kl} is computed according to Equation 5-1, and the sum over all atom pairs gives the total nonbonded interaction energy. In addition, the distance of separation between the two dipoles on either side of the ith α-carbon atom depends on ϕ_i and ψ_i, so that the dipolar energy must be calculated for every value of ϕ_i and ψ_i (Eqn. 5-2 or 5-3). Finally, the contribution of the torsional potential must also be computed for every value of ϕ_i and ψ_i (Eqn. 5-4). The total rotational potential $E(\phi_i, \psi_i)$ thus is given by

$$E(\phi_i, \psi_i) = \sum_{k,l} [E_{kl}(\phi_i, \psi_i) + E_{\mathrm{d}}(\phi_i, \psi_i) + E_{\mathrm{tor}}(\phi_i, \psi_i)] \tag{5-5}$$

where the summation extends over all atom pairs k and l whose distance of separation depends on ϕ_i and/or ψ_i. Equation 5-5 is, therefore, the basis for the calculation of conformational energies of polypeptide chain molecules.[§]

5-4 RESULTS OF POTENTIAL-ENERGY CALCULATIONS

Glycyl and L-alanyl residues

The results of energy estimates calculated according to Equation 5-5 are shown in Figure 5-7 for a glycyl residue and in Figure 5-8 for an alanyl residue, both situated within a polypeptide chain. The energy contour zones are drawn at intervals of 1 kcal mole^{-1} relative to the minima marked by X. In the case of glycyl (Fig. 5-7), two symmetrically disposed minima occur at about $\phi = -80°$, $\psi = +90°$, and $\phi = +80°$, $\psi = -90°$. There are four distinct regions on the glycyl map in which the residue energy is less than 5 kcal mole^{-1}. These regions are comparable to the "sterically allowed" regions shown in Figure 5-3.

In the case of the L-alanyl residue (Fig. 5-8), three distinct low-energy regions are designated by roman numerals. The lowest points of regions I and II are close to the coordinates of the right-handed and left-handed alpha helices, respectively (see Table 5-2).[§§] It is clear, however, that the minimum within region I is lower than that of region II. The energy map, therefore, provides a rational basis for the well-known preference of L-polypeptides for the right-handed α-helical conformation.

[§] This equation ignores special solvent interactions that depend on conformation.

[§§] Hydrogen-bonding interactions between peptide units separated by three pairs of ϕ, ψ angles play a significant role in stabilizing the α-helical conformations near the minima within regions I and II. These interactions have *not* been included in the calculations of Figures 5-7 and 5-8, and consequently these diagrams should be viewed as applicable to situations where the solvent or other factors render such interactions unlikely.

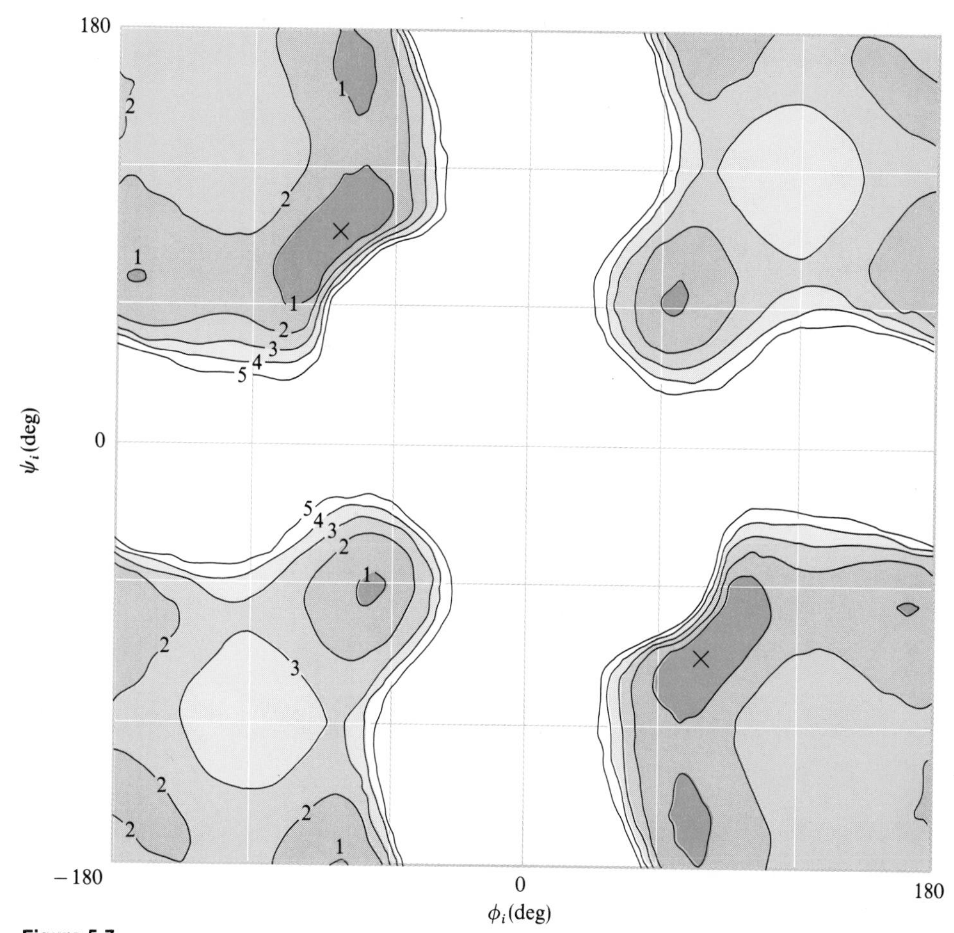

Figure 5-7

Energy contour diagram for a glycyl residue in a polypeptide chain calculated from Equation 5-5. Changes in shading show 1-kcal intervals; X indicates positions of lowest energy. [After D. A. Brant and P. J. Flory, *J. Mol. Biol.* 23:47 (1967).]

Importance of the dipolar interaction

Region III is the lowest-energy portion of the energy diagram of Figure 5-8, with the 2 kcal zone enscribing a relatively large area around the minumum (X). This region encompasses the extended forms of the residue, including the well-characterized β-pleated sheet structures (see Table 5-2). However, it should be mentioned that, energetically speaking, the chief reason for the preference of region III over region I arises from the amide dipolar interactions. If the dipolar interactions are omitted, region I becomes preferred over region III. The necessity for including the dipolar

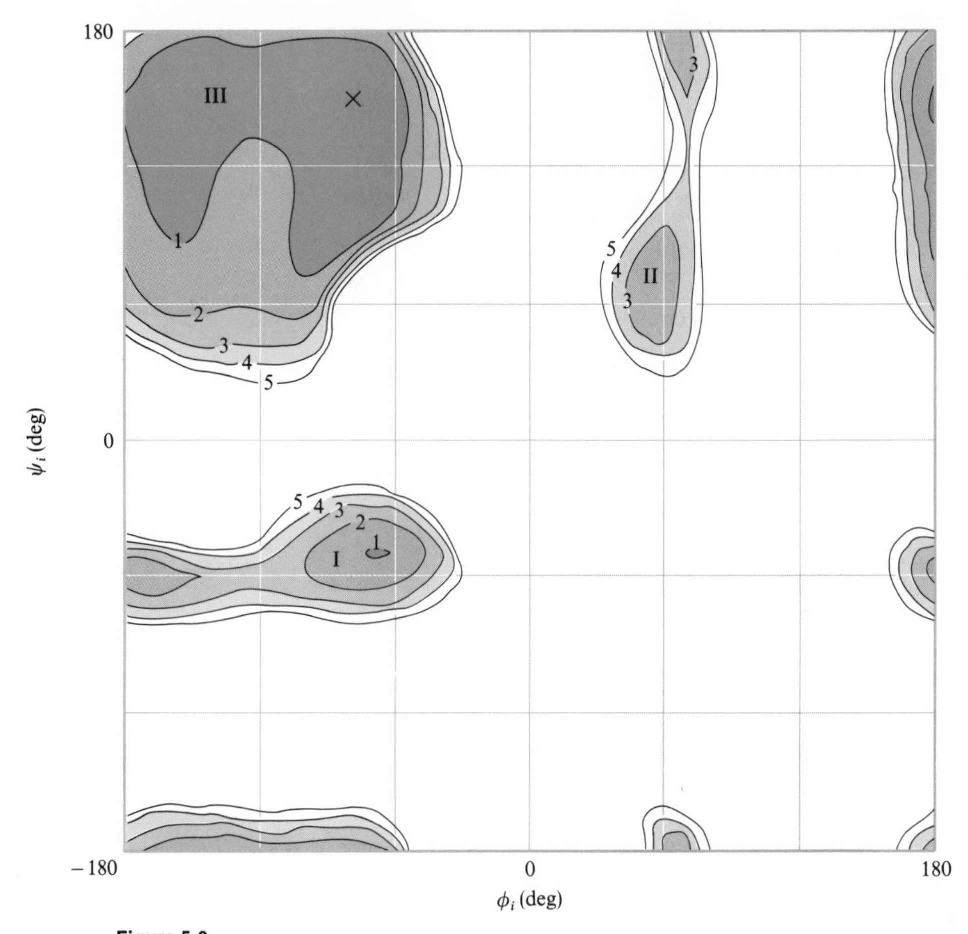

Figure 5-8

Energy contour diagram for an L*-alanyl residue in a polypeptide chain* calculated from Equation 5-5 [After D. A. Brant and P. J. Flory, *J. Mol. Biol.* 23:47 (1967).]

interactions in the calculations has experimental support from measurements of the random-coil dimensions of polyalanine-like chains. These measurements have shown that the backbone is biased toward extended conformations to a degree that is explained only by taking into account the electrical interactions of the backbone. (This point is discussed in greater detail in Chapter 18.)

Consequences of glycyl and L-alanyl conformational energies

The energy maps for glycine and alanine have some interesting implications. Because of their symmetrically disposed low-energy regions, glycine residues are naturally more "flexible" than alanine or alaninelike amino acids (those bearing side chains).

Moreover, the preferred regions of glycine, somewhat more than those of alanine, penetrate toward the center of the energy map where $\phi = 0°$ and $\psi = 0°$. These features enable glycine residues to assist a chain in adopting more compact or folded structures. This prediction is underscored by the fact that the root-mean-square end-to-end distance of a polyglycine random coil of 100 residues is about a factor of two smaller than that of a polyalanine random coil of 100 units. Moreover, introduction of a few glycyl residues into a polyalanine chain causes a disproportionately large drop in the chain dimensions (see Chapter 18 on conformational statistics).

Conformational energies of some other residues

L-Amino acid residues that are not branched at the β-carbon atom have energy maps essentially the same as that shown for a L-alanyl residue (mentioned above in connection with the steric diagrams). The disposition of the three regions is very similar for residues, such as valine, that are branched at the β-carbon atom, but the regions tend to be smaller. One residue that is completely distinct, however, is L-proline, and this special case must be treated separately. Moreover, the residue that precedes proline in a polypeptide chain must also be given special consideration.

Isolated L-prolyl residue

In most proteins, proline residues occur as isolated residues within the chain sequence—that is, sequences of two or more prolines are absent or rare. Because the proline residue—with its bulky pyrrolidine ring—restricts the domain available to the preceding residue, the conformational freedom of an isolated proline residue is considerably different from that of one imbedded within a sequence of prolines. For our purposes, however, the isolated proline is of greater relevance, and we restrict our attention to this particular situation.

Figure 5-9 is a schematic diagram of an α-L-polypeptide chain containing an isolated proline residue in the *trans* configuration.[§] Because of the rigid pyrrolidine ring, the rotation angle ϕ is fixed at about $-60°$. Therefore, the conformational energy of the prolyl residue depends only on the value of ψ.

Figure 5-10 plots conformational energy in terms of ψ for an isolated L-prolyl residue. There are two minima, at $\psi \cong -55°$ and $\psi \cong 145°$. The two minima in Figure 5-10 fall within regions I and III of Figure 5-8, and along the line $\phi = -60°$. The amide–imide dipolar interaction, which is difficult to estimate, has not been included in this calculation; at most it would lower the relative position of the minimum at $\psi \cong 145°$ by about 1 kcal mole^{-1}.

Of the two conformations available to the isolated prolyl residue, one corresponds to a relatively compact form, and the other to a relatively extended form. The difference in energy between the two states is very small, so either form can be readily adopted.

[§] The prolyl residue is also known to exist in polypeptides with the imide bond *cis*. In proteins, available x-ray data suggest that the *trans* form occurs with a much higher frequency than the *cis* form.

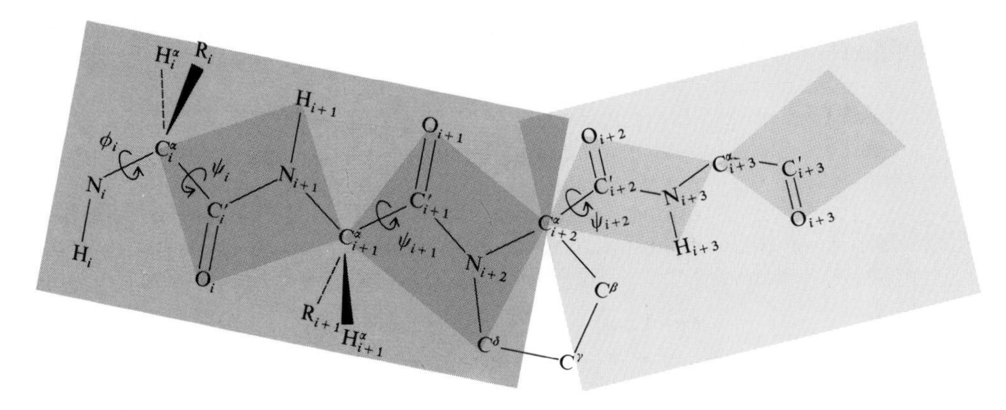

Figure 5-9

Schematic illustration of a polypeptide chain containing an isolated proline residue. [After P. R. Schimmel and P. J. Flory, *J. Mol. Biol.* 34:105 (1968).]

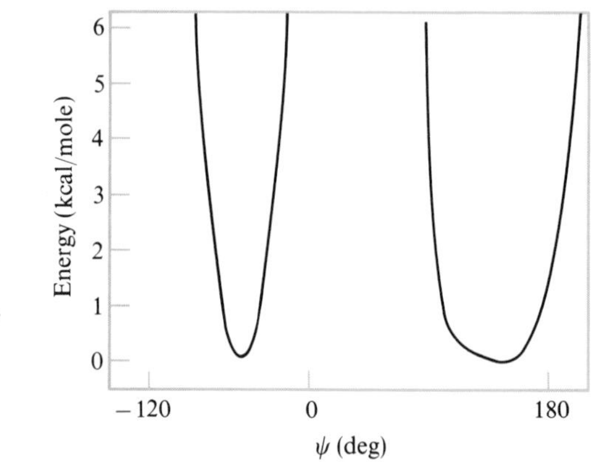

Figure 5-10

Plot of conformational energy versus ψ for an isolated proline within a polypeptide chain. [After P. R. Schimmel and P. J. Flory, *J. Mol. Biol.* 34:105 (1968).]

Certainly the relative ease with which the residue can adopt the compact $\psi \cong -55°$ conformation makes it a desirable candidate for accommodating turns and bends in a polypeptide chain.

Conformational energy of the residue preceding L-proline

We have mentioned that rotations ϕ_i, ψ_i of a residue within a polypeptide chain are independent of rotations within neighbors $i - 1$ and $i + 1$. The steric interactions of residue i with its predecessor $i - 1$ are, therefore, independent of the nature of the predecessor because only the carbonyl group (C═O) of the preceding residue is involved. However, in the case of the succeeding residue $(i + 1)$, it matters whether

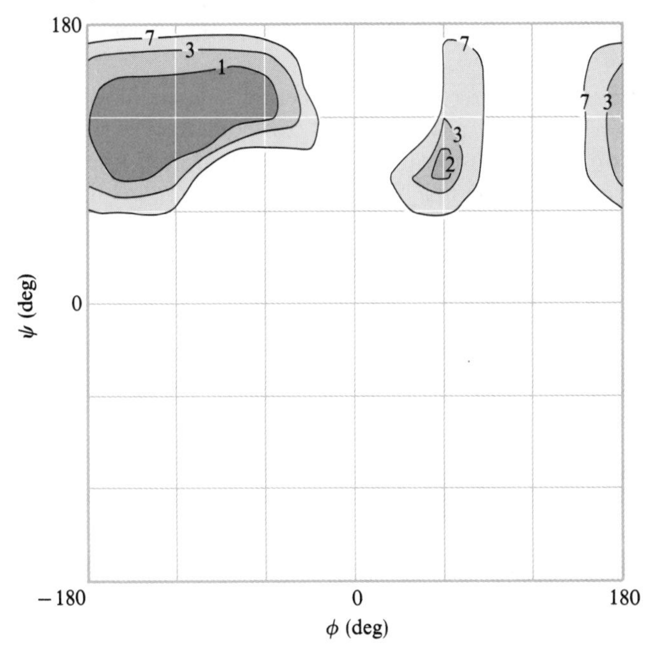

Figure 5-11

Energy contour diagram for an L*-alanyl residue that is succeeded by proline.* [After P. R. Schimmel and P. J. Flory, *J. Mol. Biol.* 34:105 (1968).]

or not the $(i + 1)$th residue is proline; the usual interactions of the ith residue with the amide hydrogen of the $(i + 1)$th are replaced by steric conflicts involving the atoms of the pyrrolidine ring, particularly the $-CH_2-$ group attached to the imide nitrogen. Therefore, the rotational freedom of a residue succeeded by proline may be expected to be more restricted than if it preceded one of the other residues.

Figure 5-11 is an energy diagram for an L-alanyl residue succeeded by a L-prolyl residue. Comparison of Figure 5-11 with Figure 5-8 reveals that the proline residue has significantly curtailed the domain accessible to alanine. Region I is completely inaccessible, with conformations throughout the entire range of $-180^\circ < \psi < 60^\circ$ now involving unfavorable steric overlaps. These added unfavorable interactions especially involve contacts between the $-CH_3$ side group of alanine with the $-CH_2$ group attached to the imido nitrogen.

When glycine precedes an L-proline residue, the general aspects of its energy map are not greatly altered from that shown in Figure 5-7. All four sterically allowed regions remain intact. The energy in the vicinity of $\psi = 0^\circ$ is raised owing to overlaps involving the pyrrolidine ring.

Conformational constraints on the residue preceding proline

Region I is not accessible for residues that bear a side chain and are succeeded by proline. This fact has an especially significant consequence: the right-handed α-helical conformation is no longer possible for those residues. On the other hand, even though its nitrogen cannot participate in a hydrogen bond, the proline residue can adopt the right-handed α-helical conformation (see Fig. 5-10). Therefore, a proline can occur at the beginning of an α-helical sequence, but it prevents the preceding residue from adopting the helical conformation if that residue is not a glycine. Thus, the fact that the proline nitrogen cannot participate in a hydrogen bond is not the primary cause for the ability of this residue to disrupt α-helical sequences, although this reason is often given as the basis for proline's helix-disrupting tendencies.

5-5 EXPERIMENTALLY OBSERVED VALUES FOR ROTATION ANGLES

In the preceding sections we have considered the energetically allowed conformations of amino acid residues situated within polypeptide chains. The energy functions that were used to calculate the energy diagrams are only semiempirical, of course. The question naturally arises as to whether or not the observed conformations of amino acid residues agree with the general predictions of these calculations. This question can best be answered by considering the ϕ, ψ rotation angles of residues in the various crystal structures of peptides, polypeptides, and proteins.

A word of caution must be voiced, however. Several factors can cause or magnify a discrepancy between the predicted and observed conformations. One arises from the fact that the diagrams given in Figures 5-3, 5-4, 5-7, 5-8, 5-10, and 5-11 are based on the assumption of fixed bond angles and bond lengths. A variation of a few degrees in a bond angle, which is not unreasonable, can cause some alterations in the shapes of the "allowed" regions; a small change in bond lengths can similarly alter the diagrams. In addition, the accuracy to which the ϕ, ψ coordinates are determined in a protein crystal, for example, is subject to an error of at least several degrees.

Lysozyme as an example

With these limitations in mind, we can pursue the question of how well the observed conformations agree with the energy diagrams. For an example, we consider the residues in crystalline lysozyme that occur in nonhelical sequences. (Residues within sequences experience contributions to their conformational energy that are not included in our calculations.[§])

[§] For example, hydrogen bonds occur between groups separated by three ϕ, ψ pairs in a α-helix. These bonds, however, are not necessarily favored in aqueous environment, where water molecules can compete for hydrogen-bonding sites within the protein. In a folded protein, of course, the chain itself can form pockets of nonpolar environments, which can allow internal hydrogen bonding within the protein.

Figure 5-12a is an energy contour map for a residue bearing a CH_2–R′ side chain at the alpha-carbon, and situated within a polypeptide chain; the individual conformations of the appropriate residues are represented by points. (The energy diagram is identical, of course, to the one for alanine given in Figure 5-8.)

It is clear that almost all of the residues fall within areas that are not excessively high in energy. In fact, 22 of the 61 residues plotted in Figure 5-12a fall within the 1 kcal $mole^{-1}$ zone surrounding the minimum X. In general, of course, the relative number of residues within a particular region depends on the energy as well as the effective area of that region, according to the principles of statistical mechanics. We might qualitatively expect, therefore, that the relative abundancies of the three regions should follow the order III > I > II. This is, in fact, what is observed. However, it is also observed that more residues appear to fall in the relatively high-energy saddle region between regions I and III than in region II. This apparent discrepancy can be rationalized, according to the limitations discussed above.

The degree to which certain residues do not conform to the calculated preferred conformations can also serve as an index of the extent and magnitude of forces that have not been accounted for. Steric or other kinds of interactions between residues that are far apart in the sequence, solvent–protein interactions, and forces of crystal packing can presumably be strong enough to encourage a residue to adopt a conformation that is at the fringes of or outside of an "allowed" domain. Hence, some of the discrepancies between calculated and observed conformations may not reflect any shortcoming in the calculations or in the crystallographic data, but rather reflect the magnitude of these compensatory forces.

Figure 5-12b is an energy contour map for a glycyl residue (compare Figure 5-7) situated within a polypeptide chain, with the glycine residue conformations of lysozyme represented by points. Only eleven residue conformations are plotted. Of these, two fall in excessively high-energy areas. The remaining residues appear to scatter throughout the "allowed" regions, and show no particular tendency to cluster within the very lowest (1 kcal) contours.

Other proteins, as well as various peptide structures, have been subjected to comparisons of the predicted and observed ϕ, ψ coordinates of the individual residues. By and large, the residues in these structures fall within the "allowed" domains of the energy maps. Thus, in trying to predict the conformation of a given protein it is reasonably safe to assume that the residues will adopt conformations within the predicted "allowed" regions. However, the latitude of choice of conformations within the permissible areas is broad enough to allow for a great number of polymer conformations. For this reason, attention must be turned to the rather specialized forces that aid in stabilizing the highly ordered conformations of native protein molecules. It is these forces that determine, from the sterically permitted conformations, the one (or ones) that is thermodynamically preferred by a significant margin.

A number of specialized forces and interactions occurring in proteins are major determinants of the native three-dimensional structure. These include the covalent cross-linking of chain segments via disulfide bridges, noncovalent interactions such as the hydrogen bond, ionic interactions, "hydrophobic" interactions, and other solvent–macromolecule interactions. These forces, taken together, are sufficiently strong and specific to bias the molecule toward a highly preferred conformation.

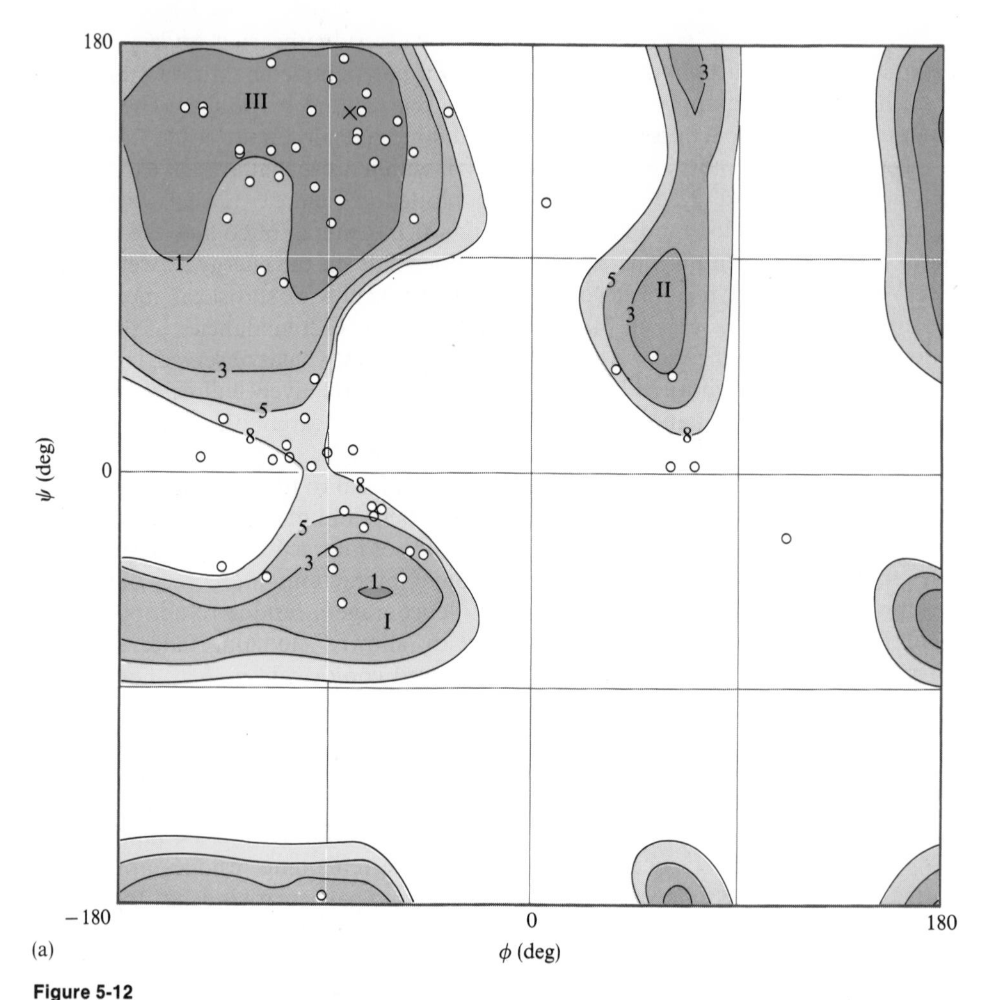

Figure 5-12

Observed residue coordinates plotted on conformational energy maps. Points show observed coordinates of residues in crystalline egg-white lysozyme occurring in nonhelical sequences. Conformational energy maps are calculated for residues situated within polypeptide chains. (**a**) L-Alanyl-like residues. (**b**) Glycyl residues. [After D. A. Brant and P. R. Schimmel, *Proc. Natl. Acad. Sci. USA* 58:428 (1967).]

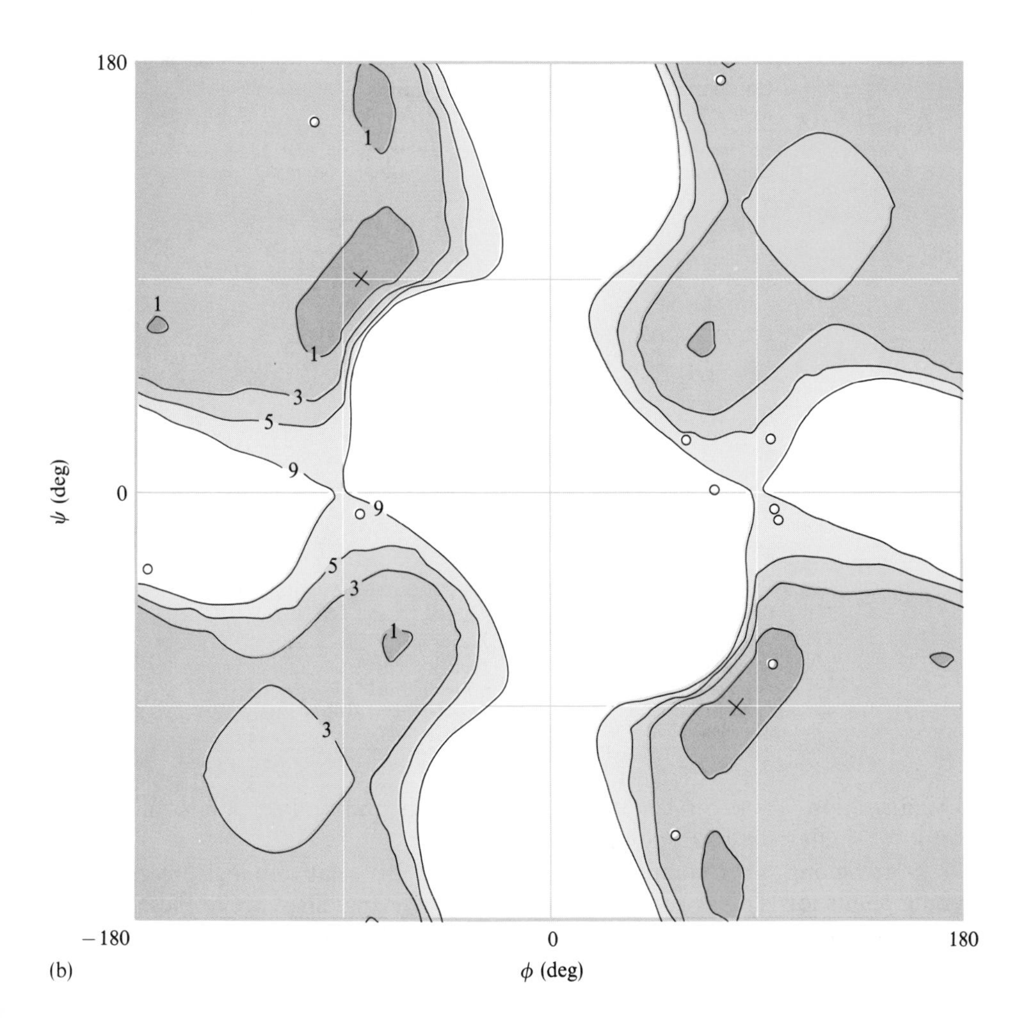

(b)

5-6 HYDROGEN BONDING

A hydrogen bond generally is said to exist between a donor molecule D–H and an acceptor A when there is evidence that the two molecules associate in a fashion specifically involving the hydrogen atom of the donor. It is not hard to marshall spectroscopic evidence for the existence of a hydrogen bond in particular situations where it might be suspected to form. For example, frequency shifts of infrared and Raman bands are readily interpreted in terms of hydrogen-bond formation. Another direct physical method that can be used for detecting hydrogen bonds is proton magnetic resonance (see Chapter 9). On the other hand, evidence for strong intermolecular associations also can be obtained by comparing rather ordinary macroscopic properties such as freezing and boiling points of pure liquids, and their heats

Table 5-4
Comparison of melting and boiling points for molecules of similar size

Compound	Melting point (K)	Boiling point (K)
H_2O	273	373
H_2S	190	211
$CH_3-\overset{O}{\overset{\|}{C}}-OH$	290	391
$CH_3-\underset{H}{\overset{O}{\overset{\|}{C}}}-CH_3$	178	330
CH_3CH_2OH	156	351
$CH_3CH_2CH_3$	83	231
CH_3-O-CH_3	135	249
CH_3NH_2	181	267
CH_3CH_3	101	185

SOURCE: Data from G. C. Pimentel and A. L. McClellan, *The Hydrogen Bond* (San Francisco: W. H. Freeman and Company, 1960), p. 36; and *Handbook of Chemistry and Physics*, 39th ed. (Cleveland, Ohio: Chemical Rubber Pub. Co., 1958).

of vaporization. Table 5-4 compares melting points and boiling points for several groups of similar sized substances.

Note among other comparisons, for example, the relatively high freezing and boiling points for H_2O as compared to H_2S. The striking difference in these macroscopic properties clearly points to a strong intermolecular association existing in water. Consider also the boiling points of three closely similar aromatic liquids: the boiling point of chlorobenzene is 132°, that of *o*-chlorophenol is 175°, and that of *p*-chlorophenol is 217°. Can you rationalize these differences?

It is experimentally observed that a hydrogen bond between D–H and A can form when the D and A atoms are strongly electronegative. In most cases, the three atoms also must be essentially colinear when forming the bond. But what is the physical basis for the hydrogen bond's strength? A simple and early view holds that the bond is essentially electrostatic in nature and forms as a result of attractive coulombic interactions. The small charge separations responsible for the coulombic interaction are envisioned as existing before bond formation. This view is supported by the fact that the strongest hydrogen bonds form when D and A are fluorine atoms and the next strongest when they are oxygen atoms, with nitrogen being somewhat weaker. However, this simple electrostatic viewpoint cannot account for all experimental observations—for example, the lack of any relationship between the dipole moment and the hydrogen-bond strength of a base. It is clear, therefore, that a more sophisticated quantum mechanical description is appropriate, and many have been offered. Nevertheless, the electrostatic model remains useful for making some qualitative predictions and rationalizations.

Water's competition for hydrogen-bonding sites on a protein

The hydrogen bonds of most importance for protein structure are those between the backbone amide nitrogens and carbonyl oxygens. These hydrogen bonds play a major role in stabilizing the various helical and ordered polypeptide conformations (such as the α-helix and β-pleated sheet structures), and they also play a prominent role in stabilizing ordered segments of folded protein molecules. Consequently it was assumed by many investigators that hydrogen bonds must play the dominant role in determining protein conformation.

It was eventually pointed out that water—the solvent for native proteins—can readily participate in hydrogen-bond formation. Consequently, the more pertinent question involves the free energy change for the process

$$\mathrm{NH}\cdots\mathrm{H_2O} + \mathrm{OH_2}\cdots\mathrm{O{=}C} \rightleftharpoons \mathrm{NH}\cdots\mathrm{O{=}C} + \text{released } \mathrm{H_2O} \tag{5-6}$$

It is not clear what the free energy change should be for this process, especially because the liberated water molecules probably participate in some kind of hydrogen bonding with other water molecules.

The general problem was carefully investigated by I. M. Klotz and J. S. Franzen (1962), using infrared spectroscopy. These investigators studied the association of *N*-methylacetamide (NMA), a good analog for the peptide backbone. The NMA molecule can form an amide hydrogen bond as follows:

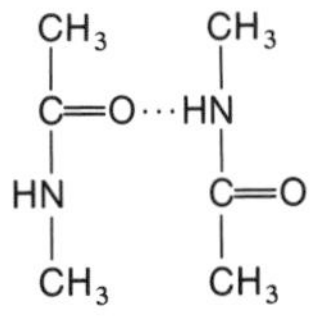

```
 CH3      CH3
 |        |
 C=O···HN
 |        |
HN        C=O
 |        |
 CH3      CH3
```

where the amide group is in the *trans* conformation. The equilibrium between monomer (M) and dimer (D) was studied, and the data were analyzed in terms of the degree of association (α), which is defined as

$$\alpha = 2(\mathrm{D})/[(\mathrm{M}) + 2(\mathrm{D})] = \text{fraction of molecules in dimeric state}$$

The parameter α was measured in three different solvents—carbon tetrachloride, dioxane, and water—so that the effect of solvent could be directly tested. Figure 5-13 shows the results as plots of α versus the *N*-methylacetamide (NMA) concentration. It is strikingly apparent that the association of NMA differs markedly in the three

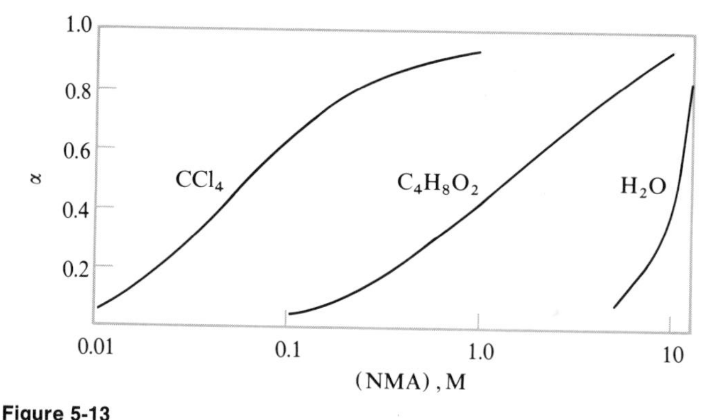

Figure 5-13

Association of N-methyl acetamide (NMA) in carbon tetrachloride, dioxane, and water. [After I. M. Klotz and J. S. Franzen, *J. Am. Chem. Soc.* 84:3461 (1962).]

solvents. The strongest association occurs in carbon tetrachloride, which cannot form any hydrogen bonds; a far weaker association occurs in water, an obviously strong competitor for the hydrogen-bonding sites on NMA. Dioxane, which has two oxygens that can accept a hydrogen bond, is intermediate between carbon tetrachloride and water.

The temperature dependence of the association in the three solvents also was measured in order to assess the thermodynamic parameters of the association (Table 5-5). It is clear that the association of NMA is thermodynamically preferred in carbon tetrachloride, but not in water.

These results clearly demonstrate the importance of the solvent in determining whether the change in state will be favorable. How do these results affect our understanding of the strength of hydrogen-bonding interactions in proteins? The interior of a protein is largely hydrocarbon in nature, so that the measurements carried out in carbon tetrachloride might be most relevant for a process in which a hydrogen

Table 5-5

Thermodynamics of amide hydrogen-bond formation by NMA at 25°C

Solvent	ΔH^0 (kcal mole^{-1})	ΔS^0 (cal deg^{-1} mole^{-1})	ΔG^0 (kcal mole^{-1})
CCl_4	−4.2	−11	−0.92
Dioxane	−0.8	−4	0.39
H_2O	0.0	−10	3.1

NOTE: The standard state is 1 mole liter^{-1}.

SOURCE: Data from I. M. Klotz and J. S. Franzen, *J. Am. Chem. Soc.* 84:3461 (1962).

bond forms and breaks within the interior of a protein. On the other hand, if the bond is to form near the polar surface, or if the unbonded state has access to water, the results obtained in water should be more appropriate. However, one must not overlook the fact that the specific water structure around the surface of the protein introduces further complications in interpretation. It is toward these complexities in water structure, and their importance for protein conformation, that we now direct our attention.

5-7 HYDROPHOBIC INTERACTIONS AND WATER STRUCTURE

Physical properties of water

It has long been appreciated that the solvent for biological systems—water—possesses some unusual properties. In fact, liquid water is now known to be so complicated that it stands out as a challenging problem of experimental and theoretical physical chemistry. An example of the peculiar properties of water is its well-known expansion when it freezes at 0°C, which is just the opposite of what is observed for most liquids. The expansion amounts to 1.6 cm^3 $mole^{-1}$. Furthermore, when ice is melted at 0°C and gradually warmed, the liquid continues to contract even after it has melted. This contraction continues right up to about 4°C, above which temperature the liquid expands. Therefore, a density maximum occurs around 4°C. In addition to these peculiar macroscopic physical properties, water has high melting and boiling points, as well as an unusually large heat capacity.

Infrared spectroscopy provides a powerful means of exploring the structure of molecules in liquids. However, the infrared spectrum of liquid water is incredibly complex, and at present most of the absorption bands cannot be quantitatively accounted for. Hence, detailed structural information has not been derived from infrared data.

Figure 5-14 is a diagrammatic representation of crystalline water. Each water molecule is tetrahedrally bonded and consequently participates in four hydrogen bonds. Two of these involve the bonding of the hydrogens to adjacent molecules, whereas the oxygen atom serves as a recipient for two hydrogen atoms. The question may be raised, then, whether hydrogen bonding *in the liquid* might not play a key role in determining the characteristics of the liquid state. In fact, this is generally assumed to be the case, although the precise explanation of how the hydrogen bonding influences the microscopic structure may vary from theory to theory.

A popular qualitative picture is that water molecules in solution form transient hydrogen-bonded clusters containing several to many water molecules (see H. S. Frank and W.-Y. Wen, 1957). These clusters can be visualized as microscopic icebergs, although the hydrogen-bonding pattern and geometries might differ from those in the crystal state; for example, all four bonding positions might not be occupied. Interspersed with these clusters of various sizes are unbonded water molecules, which possess more translational freedom, and which are packed more closely together

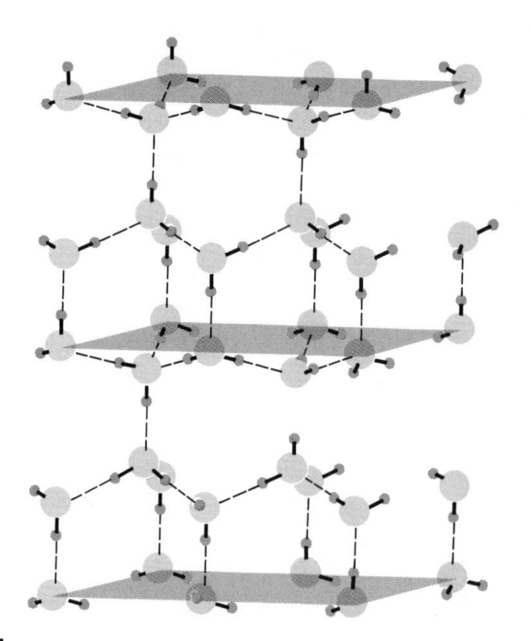

Figure 5-14

Diagrammatic model of crystalline water. Dashed lines indicate hydrogen bonds between water molecules. [After L. Pauling, *The Nature of the Chemical Bond*, 3d ed. (Ithaca, N.Y.: Cornell Univ. Press, 1960).]

because they are not constrained to keep their prescribed distances in a crystal "lattice."

With this qualitative picture, it is possible to rationalize some of the macroscopic properties. For example, the contraction of water upon melting at 0°C may be regarded as due to the liberation of some of the water molecules from the hydrogen-bonded lattice. These "free" water molecules can now pack more closely, thus giving rise to a contraction in going from the solid to the liquid state. As the temperature is increased from 0°C, more and more water molecules are liberated, and the liquid continues to contract. However, thermal agitation of the molecules also increases as the temperature rises, and this effect tends to cause the liquid to expand. Hence, the opposing effects (which encourages both contraction and expansion) give rise to the density maximum at about 4°C, above which temperature the liquid undergoes a net expansion with rising temperature.

Effect of solvent on protein structure

What relevance do these ideas have to the determination of native protein structure? An understanding of liquid structure and of solvent–protein interactions is necessary in order to comprehend the profound influence the solvent can exert on protein structure. To illustrate just how sensitive the structure can be to solvent composition, consider the fact that a relatively small molar proportion of urea or of guanidinium chloride (for example, 1 urea molecule per 5 to 10 H_2O molecules) is often sufficient to disrupt completely the native protein structure and cause a gross change in conformation, so that random, disorganized structures are adopted. On the other hand, it is also possible to produce a solvent-induced structural change to another unique, highly ordered state. For example, 2-chloroethanol can be used for this purpose. Proteins are soluble in this solvent which, although polar, has a much weaker hydrogen-bonding ability than water. When ribonuclease, for example, is dissolved in this solvent, it adopts another ordered structure that is believed to be a much more highly helical conformation than that found in water (see Doty, 1961). Similar results have been obtained for other proteins. These results can be rationalized, of course, by the fact that intramolecular hydrogen bonds have a much greater propensity to form in poor hydrogen-bonding solvents. Such findings underscore the significant influence that solvent exerts on conformation.

It is clear that water can influence protein conformation by competing for hydrogen-bonding sites on the macromolecule and by effectively solvating polar groups. These interactions in themselves, however, fail to account for the total influence of water on conformation. For example, urea also can form hydrogen bonds and can interact with polar groups, but it acts as a denaturant when present in sufficient proportions. To gain further insight into the role of water, therefore, we turn our attention to a consideration of the interactions between nonpolar side groups with aqueous environments.

Apolar molecules in water

Many years ago, it was observed that apolar molecules form very stable crystalline hydrates, in which a cagelike polyhedron of water molecules forms a cavity that encloses the apolar substance. These polyhedra can join together to form a large lattice. Figure 5-15a shows a model of a pentagonal dodecahedron in which the individual balls represent water molecules. The cavity inside this dodecahedron is about 5 Å in diameter, large enough to enclose an appropriate apolar molecule. Figure 5-15b shows a cluster of pentagonal dodecahedrons.

These hydrates are remarkably stable—so that, when apolar molecules are dissolved in water to sufficiently high concentration, the freezing point of the mixture is well above 0°C. The hydrates also have large negative heats of formation. Table 5-6 lists the heats of formation of many different apolar hydrates. Even though the molecules differ greatly in composition and size, the heats are all similar—around -15 kcal $mole^{-1}$. This similarity argues that the large and favorable enthalpy change is not directly due to solvent–apolar molecule interactions, but rather to the energy released in organizing the water molecules into cagelike structures of similar size.

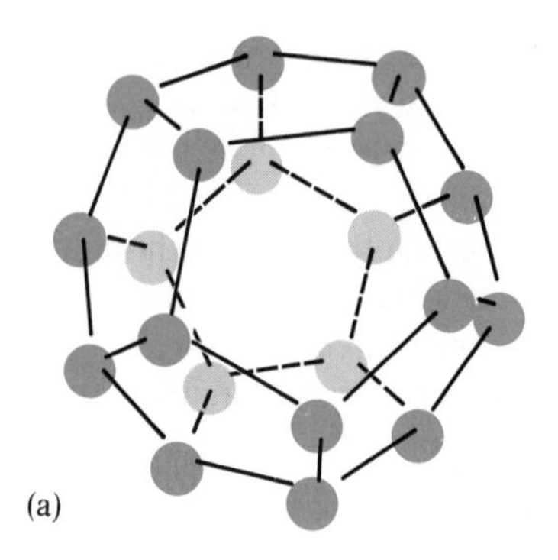

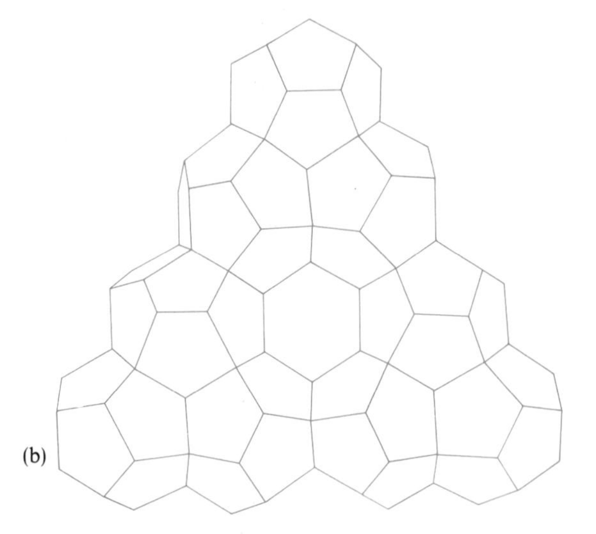

Figure 5-15

Pentagonal dodecahedrons. (**a**) Pentagonal dodecahedron formed by water molecules (balls) enclosing a cavity of ~5Å diameter. [From M. V. Stackelberg and H. R. Müller, *Z. Electrochem.* 58:25 (1954).] (**b**) A cluster of pentagonal dodecahedrons. [From A. F. Wells, *The Third Dimension in Chemistry* (Oxford: Oxford Univ. Press, 1956).]

Table 5-6
Heats of hydration of apolar substances

Molecule	$-\Delta H$ (kcal mole^{-1})
Ar	16.6
Kr	13.9–16.5
CH_4	14.5–17
C_2H_2	15
C_2H_4	15
C_2H_6	15
CH_3SH	16.6

SOURCE: Data from I. M. Klotz, *Brookhaven Symp. Biol.* 13:25 (1960).

Unitary and cratic contributions to free energy of transfer

An approach to the apolar molecule–water interaction was adopted by W. Kauzmann. Rather than focus on the formation of crystalline hydrates, he considered the thermodynamics of transferring an apolar residue from an apolar solvent to water. For example, he asked what the free energy change would be in transferring methane from mole fraction χ in benzene to mole fraction χ in water. This kind of approach basically aims at determining whether the apolar groups would prefer to reside on the protein surface—and thus interact with water—or bring themselves together in the interior of the protein—and thus create an apolar environment for themselves.

Before considering data on free energy of transfer, it is important to consider the distinction between *unitary* and *cratic* contributions to free energy (see Gurney, 1953; Kauzmann, 1959). The reason for this consideration is that we must be sure all data are considered on the same basis. To illustrate the difficulty encountered, consider the reaction

$$A + B \rightleftharpoons AB \tag{5-7}$$

with an equilibrium constant defined as

$$K = (AB)/(A)(B) \tag{5-8}$$

and a standard free energy change of

$$\Delta G^0 = -RT \ln K \tag{5-9}$$

If we choose concentration units of mole liter^{-1}, then ΔG^0 corresponds to the change in free energy when A and B, each at a standard state of 1 mole liter^{-1}, are combined to give AB at 1 mole liter^{-1}. However, the numerical value of K (and consequently that of ΔG^0) will be different if a different standard state (or a different set of concentration units) is employed. This dependence of the free energy change on the choice of the standard state is, of course, true only for reactions in which the number of reactant species is not equal to the number of product species; in these cases, the magnitude of the equilibrium constant depends on the choice of standard state, unlike the situation for a unimolecular action.

The dependence of ΔG^0 on the standard state is a purely statistical effect that arises from the expression for the partial molal entropy; consider the partial molal entropy S_A of A:

$$S_A = S'_A - R \ln \chi_A \tag{5-10}$$

where χ_A is the mole fraction of A (see Box 5-2). (We are considering dilute solutions, and therefore we are ignoring activity-coefficient effects.) The term $-R \ln \chi_A$ is called the *cratic* contribution to the entropy; it is a purely statistical term arising from the mixing of A with solvent molecules. This contribution is independent of the chemical nature of A. The S'_A term is called the *unitary* contribution; it reflects the characteristics of A itself and its interaction with solvent. The standard entropy change for the reaction of Equation 5-7 becomes

$$\Delta S^0 = S'_{AB} - S'_A - S'_B + R \ln \chi_M = \Delta S' + R \ln \chi_M \tag{5-11}$$

Box 5-2 DERIVATION OF EXPRESSION FOR PARTIAL MOLAR ENTROPY

Equation 5-10 is the usual expression for the partial molar entropy. It can easily be derived by starting with the total differential for the Gibbs free energy (derived in introductory physical chemistry textbooks):

$$dG = -S dT + V dP + \sum \mu_i dN_i$$

By cross-differentiation, we obtain

$$S_i = (\partial S/\partial N_i)_{T,P,N_{j \neq i}} = -(\partial \mu_i/\partial T)_{P,N_j}$$

Introductory physical chemistry textbooks derive the following relationship:

$$\mu_i = \mu_i^0(T, P) + RT \ln \chi_i$$

where μ_i^0 is the standard chemical potential at $\chi_i = 1$. Using this relationship, and differentiating the preceding expression, we obtain

$$\begin{aligned} S_i &= -(\partial \mu_i/\partial T)_{P,N_j} = -(\partial \mu_i^0/\partial T)_{P,N_j} - R \ln \chi_i \\ &= S'_i - R \ln \chi_i \end{aligned}$$

where χ_M is the mole fraction of a molecule when its concentration is 1 mole liter^{-1}. Because the water concentration is 55.6 molar in dilute aqueous solution, the cratic contribution to the entropy change is about $-R \ln 56 \cong -8$ cal mol^{-1} deg^{-1}; at 300 K, this corresponds to a cratic contribution to the free energy of about -2.4 kcal mole^{-1}. If a different standard state and different concentration units were chosen, a different contribution would result. The unitary change in entropy, ΔS_u^0, thus is obtained from

$$\Delta S_u^0 = \Delta S^0 - R \ln \chi_M \qquad (5\text{-}12)$$

Consider now the following transfer processes:

$$\text{Hydrocarbon in nonpolar solvent (mole fraction } \chi) \rightarrow \text{Hydrocarbon in water (mole fraction } \chi) \qquad (5\text{-}13)$$

$$\text{Hydrocarbon as pure liquid} \rightarrow \text{Hydrocarbon in water (mole fraction } \chi) \qquad (5\text{-}14)$$

For the process of Equation 5-13, the cratic contributions obviously cancel; however, they do not cancel for the process of Equation 5-14. The unitary change in entropy for Equation 5-14 is obtained from the measured standard entropy change ΔS^0 by $\Delta S_u = \Delta S^0 + R \ln \chi$. Having obtained the unitary change for the second process, we may compare it directly with the entropy change for Equation 5-13.

Data on free energy of transfer

We are now in a position to consider data on free energy of transfer. Table 5-7 gives the thermodynamic parameters for transferring a variety of hydrocarbons from apolar solvents to water. The free energies and entropies given are expressed as unitary

Table 5-7
Thermodynamic parameters for transferring various hydrocarbons from nonpolar solvents to water at 25°C

Transfer reaction	ΔH (kcal mole^{-1})	ΔS_u (cal deg^{-1} mole^{-1})	ΔG_u (kcal mole^{-1})
CH_4 in benzene → CH_4 in H_2O	−2.8	−18	+2.6
CH_4 in ether → CH_4 in H_2O	−2.4	−19	+3.3
CH_4 in CCl_4 → CH_4 in H_2O	−2.5	−18	+2.9
C_2H_6 in benzene → C_2H_6 in H_2O	−2.2	−20	+3.8
C_2H_6 in CCl_4 → C_2H_6 in H_2O	−1.7	−18	+3.7
Liquid C_3H_8 → C_3H_8 in H_2O	−1.8	−23	+5.05
Liquid n-C_4H_{10} → n-C_4H_{10} in H_2O	−1.0	−23	+5.85
Liquid C_6H_6 → C_6H_6 in H_2O	O[§]	−14[§]	+4.07[§]

[§] Measured at 18°C.
SOURCE: Data from W. Kauzmann, *Adv. Protein Chem.* 14:1 (1959).

changes. It is strikingly apparent that, although the hydrocarbons energetically prefer water to an apolar solvent ($\Delta H < 0$), the large negative entropy changes cause the overall free energy change to be positive, thus making the transfer process unfavorable. For example, the transfer of methane from benzene to water results in the liberation of 2.8 kcal $mole^{-1}$ of heat, but the -18 e.u. that accompanies this process gives rise to a ΔG_u^0 of $+2.6$ kcal $mole^{-1}$. Similar results are obtained with other hydrocarbons and other apolar solvents.

How do we interpret these results? A useful way to rationalize them is to assume that in liquid water the apolar hydrocarbon can induce the formation of an "icelike" structure (such as a clathrate structure) that surrounds the hydrocarbon in liquid water much as it does in crystalline hydrates (see Fig. 5-15). This is not unreasonable, especially because the dissolved hydrocarbons can cause an elevation of the freezing point of water. This process thus involves the formation of icelike structure and, therefore, the making of hydrogen bonds. The latter process is exothermic (i.e., energy is released), and this accounts for the negative values of ΔH in the transfer process. However, the ordering of water molecules into a more organized structure naturally decreases the entropy, which accounts for the negative values of ΔS.

Interaction of apolar amino acid side chains with water

Similar data have been obtained for amino acids. We can consider the free energy of transfer of various amino acids from ethanol to water (Table 5-8). In the transfer process, the molecule rearranges from the neutral, apolar form

$$\begin{array}{c} \mathrm{R} \\ | \\ \mathrm{NH_2{-}C{-}COOH} \\ | \\ \mathrm{H} \end{array}$$

in ethanol to the zwitterionic form

$$\begin{array}{c} \mathrm{R} \\ | \\ \mathrm{NH_3^+{-}C{-}COO^-} \\ | \\ \mathrm{H} \end{array}$$

in water. This rearrangement gives a sizable negative contribution to the ΔG of transfer, owing to solvation of the zwitterion by the polar water molecules. The important item to consider is the *difference* ΔG_t between the ΔG of transfer for a particular amino acid and that for glycine. This difference gives a measure of the contribution of just the side chain of that particular amino acid. It is clear from Table 5-8 that ΔG_t increases with increasing side-chain size. Surprisingly enough, the contribution of a —CH_2— group to ΔG_t appears to be similar for different molecules. Table 5-8 shows that the difference in ΔG of transfer between ethane and methane is nearly the same as that between alanine and glycine; moreover, the difference

Table 5-8
Free energy changes for transferring various compounds from ethanol to water at 25°C

Compound	ΔG (kcal mole^{-1})	ΔG_t (kcal mole^{-1})
Glycine	−4.63	0
Alanine	−3.90	+0.73
Valine	−2.94	+1.69
Leucine	−2.21	+2.42
Isoleucine	−1.69	+2.97
Phenylalanine	−1.98	+2.65
Proline	−2.06	+2.60
Contribution of a CH_2 group		
Ethane	+3.02	—
Methane	+2.26	—
Ethane–Methane	—	+0.76
Alanine–Glycine	—	+0.73
Leucine–Valine	—	+0.73

NOTE: For additional data, see Table 2-3.
SOURCE: Data from C. Tanford, *J. Am. Chem. Soc.* 84:4240 (1962).

between leucine and valine is the same as that between alanine and glycine. This suggests that the individual hydrocarbon moieties act more or less independently and additively in these simple small molecules.

Concept of hydrophobic interaction

The data in Tables 5-7 and 5-8 argue that, given a choice, the apolar side-chain moieties of amino acids will prefer to reside in an apolar nonaqueous environment. Therefore, they will tend to gather together to form a cluster of apolar groups, which may be called a hydrophobic interaction or a hydrophobic bond. X-ray crystallographic results on a number of proteins confirm the fact that apolar side chains tend to reside in the "oily" interior of the protein (for an example, see Fig. 2-32). Note that the driving force for the hydrophobic interaction is entropic, and that the enthalpy change actually acts in opposition to this tendency. Therefore, $\Delta H > 0$ when two or more hydrophobic groups, initially solvated, clump together in the interior of the protein. This means that an increase in temperature will tend to drive the equilibrium toward "hydrophobic bonding" rather than toward dissociation,[§] assuming that the

[§] This follows from the van't Hoff relation:

$$d(\ln K)/dT = \Delta H/RT^2$$

where K is the equilibrium constant of a reaction.

temperature range under consideration is small enough that ΔH remains positive. (See Pratt and Chandler, 1977, for a theoretical treatment of the hydrophobic effect.)

Disruption of hydrophobic interactions by urea

The results just given clearly demonstrate the importance of the hydrophobic interaction. Proteins, with their hydrocarbon side chains tucked inside, must derive a considerable amount of their stability from this effect. Moreover, it appears reasonable that the conformation should be sensitive to solvent composition, especially if the changes in solvent affect the partitioning of a hydrocarbon between the hydrocarbon phase and the ambient solvent phase. Consider the case of urea, a popular protein denaturant. Although it was once thought that urea acts by competing for intramolecular peptide hydrogen bonds and thereby disrupts the structure, that notion appears untenable because water itself should be as effective a denaturant. What then is the effect of urea on the partitioning of the hydrocarbon side chains between the aqueous phase and an organic phase?

Table 5-9 lists the free energy of transfer for transferring various amino acids from water to 8 M urea. To assess the side-chain contribution, ΔG_t is also given, where

Table 5-9
Free energy change for transferring various amino acids from water to 8 M urea at 25°C

Amino acid	ΔG (kcal mole^{-1})	ΔG_t (kcal mole^{-1})
Glycine	+0.10	0
Alanine	+0.03	−0.07
Leucine	−0.28	−0.38
Phenylalanine	−0.60	−0.70
Tyrosine	−0.63	−0.73

SOURCE: Data from P. L. Whitney and C. Tanford, *J. Biol. Chem.* 237:1735 (1962).

the value of ΔG for glycine has been subtracted from the other values, just as was done earlier. It is clear that the side chains prefer 8 M urea to water, the degree of preference being correlated with the size of the side chain. Thus, in the presence of 8 M urea, a side chain's preference for the apolar interior of the protein, as opposed to the solvent exterior, diminishes by several hundred calories. Because the apolar residues generally make up 30% to 50% of most proteins, it is easy to see that, even in a small protein of 150 residues, the total change in stability caused by 8 M urea can be enormous.

5-8 IONIC INTERACTIONS

Possibilities for ionic interactions in proteins

The large number of ionic side chains in proteins can obviously form "salt bridges" that can act as an additional noncovalent force to stabilize the native structure. In the neutral pH region, carboxyl groups are the source of negative charges; these occur in the side chains of aspartic and glutamic acid, as well as in C-terminal residues. Positive charges come from the amino groups of side-chain lysines and of N-terminal amino acids, the guanidino groups of arginine side chains, and imidazolium ions of histidine side chains (which ionize in the neutral region).

X-ray crystallographic data have verified the occurrence of ionic pairs in proteins. Some salt bridges found in deoxyhemoglobin are given in Table 5-10. Here we see that both ionic side chains and C-terminal carboxyls are participants. Interestingly enough, these ionic interactions are broken when hemoglobin is in the oxyconformation (Chapter 17).

Table 5-10
Some salt bridges found in deoxyhemoglobin

Positive charge	Negative mate	Remarks
Guanidinium group	Aspartic acid side chain	Links together α_1 and α_2 subunits
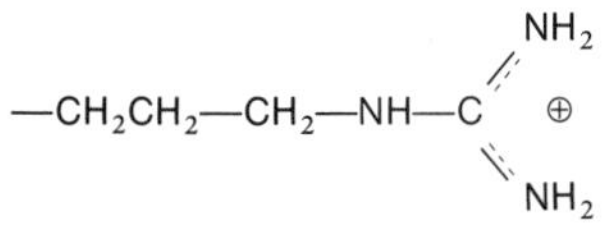 $—CH_2CH_2—CH_2—NH—C(NH_2)_2^{\oplus}$	$—CH_2—COO^-$	
ε-Amino of lysine	C-terminal carboxyl group	Links α_2 with the β_1 subunit, and α_1 with the β_2 subunit
$—CH_2CH_2CH_2CH_2—NH_3^+$	$—COO^-$	
Imidazole ring of histidine	Aspartic acid side chain	Intrasubunit link occurring within each β subunit
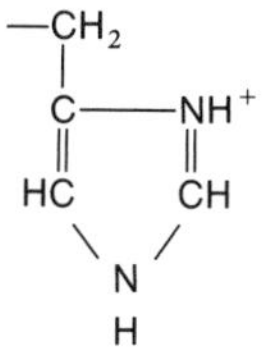 $—CH_2$ attached to imidazolium ring (C, NH^+, HC, CH, NH)	$—CH_2—COO^-$	

SOURCE: Data from M. F. Perutz, *Nature* 228:726 (1970).

Simple treatment of ionic interactions

It is rather difficult to assess the magnitude of the contribution made by ionic interactions to the stability of the native structure. Let us consider the free energy change involved in bringing into contact two oppositely charged ions (see Kauzmann, 1959). We will calculate the difference in free energy ΔG between a final state, in which the charges have approached each other to within the sum of their ionic radii (when they make contact), and an initial state, in which the charges are far enough separated that their interaction energy is negligible. This free energy change is merely the work done on the system in bringing the charges together. For opposite charges this work is negative and is given by

$$\Delta G = -\int_{\infty}^{a} (z_1 z_2 e^2/\varepsilon r^2)\, dr \tag{5-15}$$

$$z_1 z_2 e^2/\varepsilon a \tag{5-16}$$

where r is the variable distance of separation, z_1 and z_2 are the ionic valencies of the two ions, e is the electronic charge, and ε is the dielectric constant. The parameter a is the contact distance and is equal to the sum of the ionic radii.

Equation 5-16 is only an approximate expression, of course. One shortcoming, among others, is that it was derived on the assumption of constant ε, but it is very probable that ε varies considerably with r, because the microscopic environment between the two charges changes as they approach each other. If we assume $z_1 = -z_2 = 1$, and $\varepsilon = 80$ (which is the dielectric constant of water), then ΔG is approximately -1 kcal mole^{-1} for $a = 4$ Å. As the charges approach each other near the surface of a protein, however, some lines of force will pass through the protein itself rather than through the solvent. This will tend to lower the effective dielectric constant and will thus increase the magnitude of the net free energy change. On the other hand, the initial separation of charges in a protein in an unfolded state might not be so great as to give rise to even a negligible electrostatic interaction energy. The change in free energy for forming the salt bridge starting from this unfolded state would therefore be correspondingly reduced.

The enthalpy change associated with ion-pair formation can be calculated from Equation 5-16, assuming that a is temperature insensitive.

$$\begin{aligned} \Delta H &= -T^2[d(\Delta G/T)/dT] = (z_1 z_2 e^2/\varepsilon a) + T(z_1 z_2 e^2/\varepsilon^2 a)(d\varepsilon/dT) \\ &= (\Delta G/\varepsilon)[d(\varepsilon T)/dT] \end{aligned} \tag{5-17}$$

The entropy change is given by

$$\begin{aligned} \Delta S &= -d\Delta G/dT = (z_1 z_2 e^2/\varepsilon^2 a)(d\varepsilon/dT) \\ &= (\Delta G/\varepsilon)(d\varepsilon/dT) \end{aligned} \tag{5-18}$$

Entropic drive for ion-pair formation

Both derivatives (Eqns. 5-17 and 5-18) are negative for aqueous solvent. This means that ΔH for ion-pair formation is actually *positive* in water—that is, the enthalpy change opposes the making of the salt-bridge linkage. Conversely, ΔS is positive, so that the driving force is entropic in origin. Examples in the literature lend support to the qualitative predictions of this simple electrostatic model. For example, the reaction between Mg^{2+} and ATP^{4-} to yield $MgATP^{2-}$ has an entropy change of $+29.4$ cal deg^{-1} $mole^{-1}$ in favor of complex formation, whereas ΔH is $+3.3$ kcal $mole^{-1}$ and thus *opposes* complex formation.

Of course, the picture provided by Equations 5-15 through 5-18 is only a rough approximation. In particular, the solvent has been approximated as a continuous medium with a constant dielectric constant. Variations in dielectric constant with distance between the ions have not been considered; neither has account been taken of specific complexes between the ions and water molecules. Thus, although the picture provided by those equations is instructive, it is only a crude estimate.

It comes as a surprise to find that ionic complex formation in water is entropically driven. One might think that the entropy change would be unfavorable, rather than favorable, because the complex obviously has less translation freedom than the separated ions. Nevertheless, this result can be rationalized in terms of our simple qualitative ideas about water. The ions tend to be heavily solvated, owing to strong ion-dipole interactions with the water molecules. These highly organized and immobilized water molecules are liberated, however, upon formation of the ion pair, thus increasing the "disorder" or configurational freedom in the solvent. This gives rise to a large increase in the overall entropy change associated with complex formation. Moreover, the disruption of the energetically favorable ion–water interactions accounts for the positive enthalpy change associated with complex formation.

Hydrophobic versus ionic interactions

The fact that salt bridges are found in native protein structures certainly implies that they play a role in stabilizing the native structure. As mentioned above, however, it is difficult to assess their relative importance. Salt bridges should strengthen when the dielectric constant of the medium is lowered—for example, when apolar solvents such as dioxane or ethyl alcohol are added to water. When this is done, however, proteins tend to denature, which suggests that the more favorable solvent environment for hydrocarbon side chains provided by the apolar solvent is sufficient to overcome any increase in electrostatic stabilization. This result tends to indicate that hydrophobic interactions outweigh electrostatic ones in the native protein structure.

5-9 DISULFIDE BONDS

Thus far we have considered three kinds of noncovalent forces that are important for determining native protein structure: hydrogen bonding, hydrophobic interactions, and ionic pairing. A *covalent* bond that plays a role in stabilizing protein conformation is the disulfide bridge between cystine residues. These linkages serve to restrict the number of possible three-dimensional configurations the macromolecule can adopt. In this way, they reduce the difference in configurational entropy between the native and denatured states (with all disulfide linkages intact). Without the constraint of the disulfide linkages, the denatured state would have considerably more configurational freedom.

Ways of pairing *N* half-cystines

Let us first consider the number of ways in which an even number of half-cystine residues hypothetically can pair to form disulfide bridges. We will identify the various half-cystines by numbering them 1, 2, 3, ... With two half-cystines, there obviously is only one way to form a linkage. If there are four half-cystines, then there are three ways they can hypothetically pair, schematically represented by

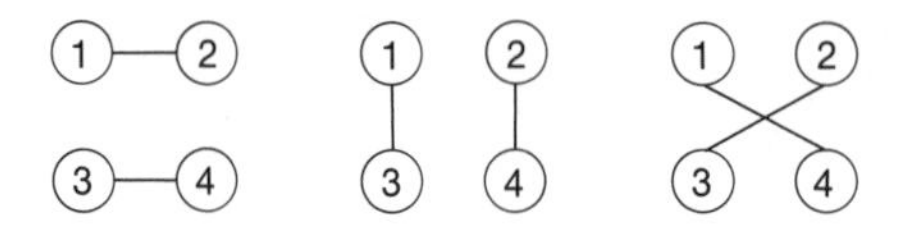

If there are six half-cystines, then there are five different partners with which any one can pair; this leaves four remaining half-cystines, and we have already seen that there are three ways in which they can pair, to form two additional bridges. Thus there are a total of $5 \times 3 = 15$ ways in which the six half-cystines can form three disulfide linkages. In general, if there are N half-cystines (N is even), there are $N - 1$ ways of making the first bridge with a specific residue, $N - 3$ ways of making the second bridge with another specific residue, etc.—so that the total number (W) of hypothetical arrangements is

$$W = (N - 1)(N - 3)(N - 5) \ldots \qquad \text{for even } N \tag{5-19}$$

(Can you modify the equation for odd N?) It is immediately apparent that the number of ways of making a given number of disulfide bridges increases rapidly with N.

Formation of specific disulfide links

Why does a native protein structure form only one particular arrangement of disulfide linkages? There are at least three plausible explanations.

1. Although there are many hypothetical possibilities, only one particular arrangement is geometrically or sterically possible. Thus, the protein adopts that configuration in which that particular arrangement can form.

2. Many of the hypothetical arrangements are indeed possible, but specific events in protein biosynthesis determine the locations of the linkages.
3. The reduced protein folds into its most stable conformation, and the disulfide linkages snap in to reinforce this particular configuration. Other disulfide arrangements are possible, but these are for structures that are not as stable thermodynamically.

Reoxidation of ribonuclease into native protein

The explanations for specific disulfide linkages were investigated by C. B. Anfinsen and colleagues (1961), who chose to work with ribonuclease. This protein ($\sim$14,000 mol wt) has four disulfide linkages and no free SH groups. According to Equation 5-19, therefore, there are 105 ways in which these four disulfide linkages can form among the eight half-cystine residues.

Figure 5-16 is a schematic illustration of the kind of experiment that was conducted. The native molecule first is reduced with β-mercaptoethanol ($HOCH_2CH_2SH$) in the presence of 8 M urea. This produces a reduced, unfolded molecule. The urea then is removed, and the solution is exposed to air (O_2) at mildly alkaline pH (pH $\sim$8) in order to reoxidize the half-cystines. The results obtained are summarized graphically in Figure 5-17, which shows the change in various enzyme properties as a function of time of reoxidation. The circles show the decrease in free SH groups, whereas the squares show the change in optical rotation at 366 nm. When normalized to the same scale, both variables fall on the same experimental curve. This suggests that the disappearance of free SH groups is coincident with the generation of ordered structure, as monitored by optical rotation.

Also shown in Figure 5-17 is the time course of the appearance of enzyme activity. This curve shows a pronounced lag phase, and initially it lingers well behind the curve for the change in physical properties. Eventually, however, fully active molecules are produced, and the lag phase implies merely that they are not generated in a one-by-one (or "all-or-none") fashion. The extent of this lag phase depends upon concentration, thus suggesting that intermolecular as well as intramolecular pairing is occurring. The various disulfide linkages shuffle around until they find the most stable arrangement, which corresponds to the intramolecular bridges of the native conformation.

The active molecules produced during reoxidation are identical to native ribonuclease. Hence, we conclude that the second explanation for specific disulfide bridges is unlikely, because the correct bridges form even in vitro. The validity of the first alternative was tested by allowing reoxidation to occur in the presence of 8 M urea. After complete oxidation, urea is removed, and a product with only 1% activity is obtained. This low level clearly implies the possible formation of one or more disulfide arrangements that are distinct from the native arrangement. (It is perhaps coincidental that the 1% activity is what is predicted if all 105 possible

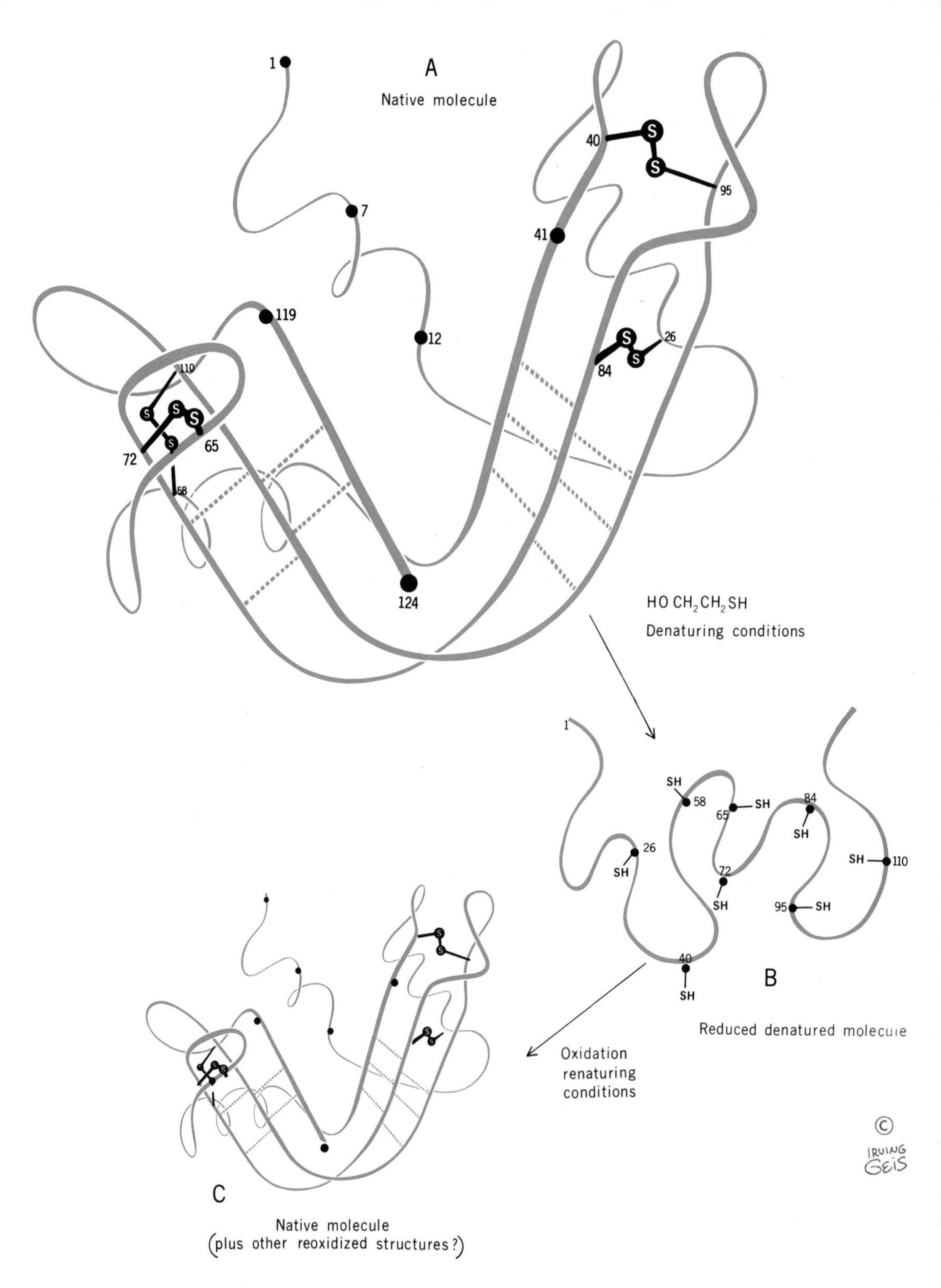

Figure 5-16

Reduction and reoxidation of ribonuclease. [Drawing by Irving Geis.]

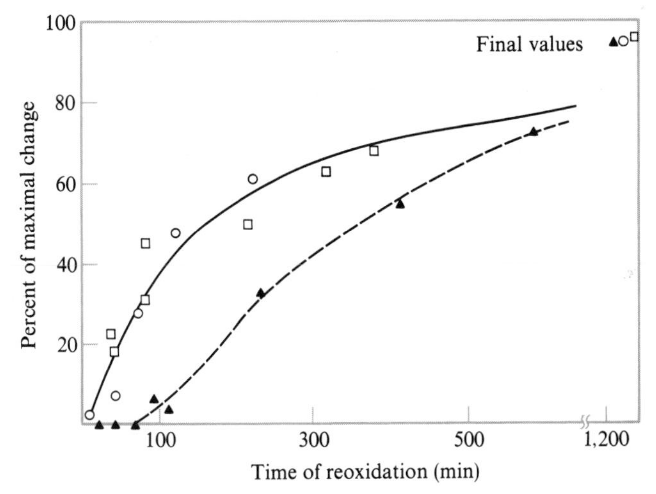

Figure 5-17

Reoxidation of reduced ribonuclease. Solid curve = physical property; dashed curve = enzyme activity; □ = optical rotation; ○ = titration of free SH groups; ▲ = enzyme activity. [After C. B. Anfinsen et al., *Proc. Natl. Acad. Sci. USA* 47:1309 (1961).]

arrangements are formed with equal frequency, with only the single native arrangement possessing activity.) Moreoever, when a thiol reagent is added to the "randomly" oxidized material, reshuffling of the linkages occurs until the native material is generated.

These interesting experiments lend strong support to the third possibility given above. Thus, we conclude that the protein folds to its most stable state, which is further stabilized by the formation of the disulfide bridges. This notion is given further support by the observations that certain proteins, containing no disulfide bridges, adopt a native structure and reversibly unfold.

Reoxidation of proinsulin into native insulin

Several additional proteins have been examined with respect to their ability to reoxidize to the correct structure after disulfide reduction. In these cases, similar results have been obtained—with insulin standing out as a notable exception. Figure 1-11 shows the structure of insulin, which consists of an A chain of 21 residues and a B chain of 30 residues; these chains are joined together by two disulfide bonds. In addition, the A chain possesses an intrachain disulfide link, joining half-cystines 6 and 11. When insulin is reduced, and the chains are mixed together and reoxidized, a good yield of native insulin is not obtained. However, it was eventually discovered that insulin is synthesized as a proprotein, proinsulin (Fig. 1-11). This molecule is enzymatically cleaved after residues 30 and 63, the fragment is removed, and native insulin is generated. When proinsulin is reduced and reoxidized, immunological activity characteristic of native proinsulin returns. Moreover, biologically active insulin can then be generated by treating the reoxidized proinsulin with trypsin. Thus, the disulfide linkages found in proinsulin are formed spontaneously. These linkages then are preserved in insulin and are in fact necessary in order for insulin to attain its native conformation. This raises the question of whether native insulin is in its thermodynamically most stable conformation, at least with respect to the arrangement of disulfide bonds.

5-10 APPLICATION OF PROTEIN STRUCTURE DATA

Number of possible configurations

The preceding discussion gives consideration to the various factors that play a role in determining the three-dimensional structures of native protein molecules. Steric hindrances among near-neighbor atoms in the chain go a long way to reduce the total domain of available configuration space. For example, the steric contour diagram for alanine (Fig. 5-4) shows that the normally allowed area for this residue is less than 10% of the total area in the ϕ, ψ plane. This reduction in configuration space is not nearly great enough, however, to specify even approximately the conformation of a long polypeptide chain. Consider the energy contour diagram for a residue bearing a $—CH_2R'$ type of side chain (Fig. 5-8). Let us assume that each of the local low-energy regions (I, II, and III) represents a distinct conformation, so that a total of three conformations must be considered. Therefore, for a chain consisting of 100 such residues, there are $3^{100} \cong 10^{48}$ distinct configurations that are energetically feasible, at least as far as near-neighbor interactions are concerned. This is an impossibly large number to deal with. Of course, steric overlaps between atoms that are distant in the sequence will reduce this number, but certainly not to the great extent necessary to give a manageable number of conformations. It is therefore remarkable that such factors as hydrophobic interactions, hydrogen bonding, and salt bridges are capable of biasing the chain into essentially one structure.

Comparison of structures of lysozyme and myoglobin

In order to grasp more concretely the role played by these various interactions in determining the three-dimensional structure of proteins, let us consider the crystal structures of lysozyme and myoglobin. Lysozyme is illustrated in Figures 1-12, 2-31, 2-32, and 2-33; a schematic of myoglobin is shown in Figure 1-4a. These proteins share certain general features that are of obvious significance in determining the particular observed structures. For example, the residue conformations tend to fall within the allowed regions (Fig. 5-12a,b for lysozyme); the polar groups generally reside on the surface, and hydrophobic side chains tend to pack together into regions devoid of solvent.

Although there are some general similarities, the two molecules have quite different conformations. This difference emphasizes the important roles of amino

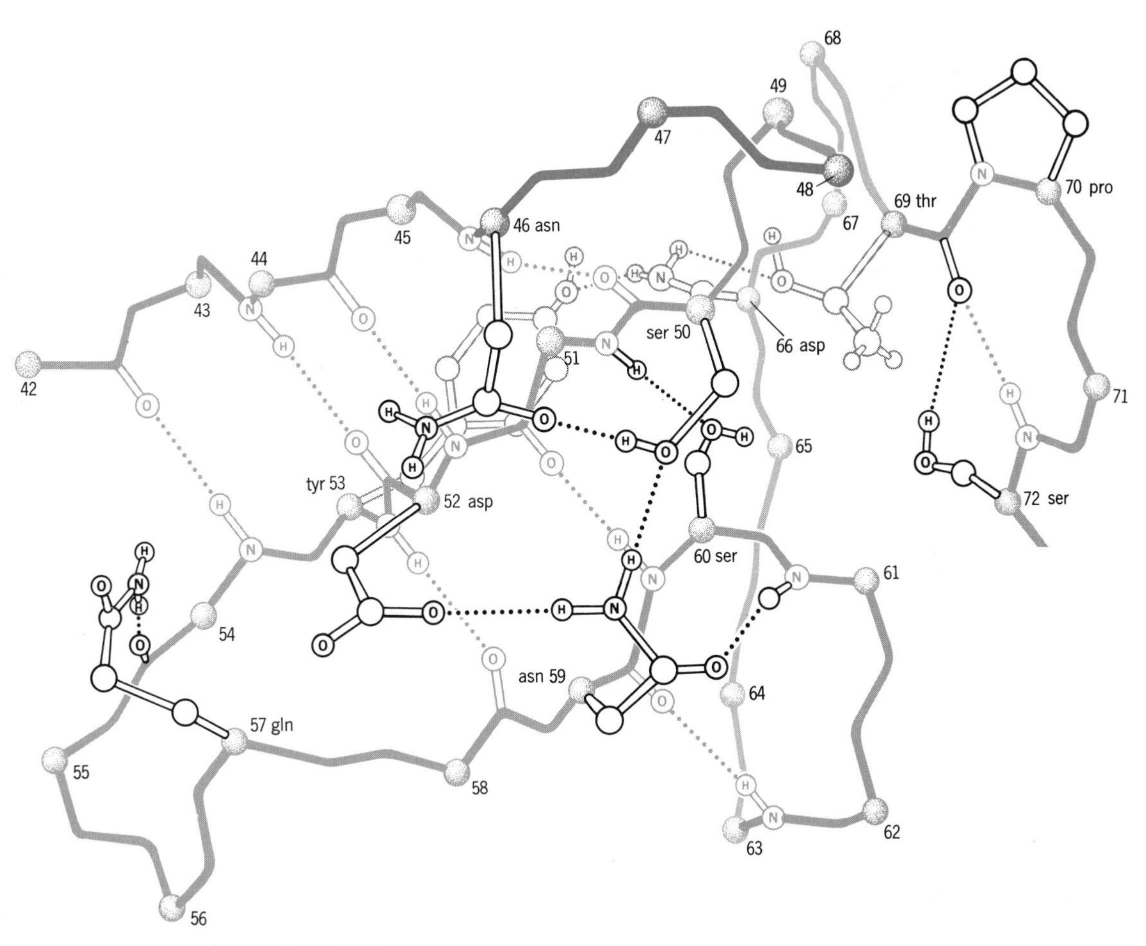

Figure 5-18

β-Sheet region in lysozyme. Hydrogen bonds are indicated by dots. [Drawing by Irving Geis.]

acid sequence and chain length in determining specific structures. First consider lysozyme (129 residues), whose gross morphology is that of a prolate ellipsoid (45 × 30 × 30 Å) with a wedge-shaped piece removed. The wedge-shaped area corresponds to the active site. The molecule contains only a small number of α-helix-like sequences, involving residues 5–15, 24–34, and 88–96. There are also a few other short sequences of residues that adopt ϕ, ψ coordinates approximating those of the α or the 3_{10} (or 3_1) helix.[§] An important structural feature is that of a β sheet, shown in Figure 5-18. This is formed from a straight run of residues 42–48, followed by a

[§] In an earlier notation, the designation 3_{10} referred to 3 residues per turn and 10 atoms in the ring formed by closing a hydrogen bond. In that notation, the α helix is a 3.6_{13} helix. In more modern notation, the designation N_M referrs to the minimal number M of complete turns required to give an integer number N of residues. In this notation, the helix formerly designated 3_{10} is called a 3_1 helix, and the α helix is an 18_5 helix.

doubling back of residues 49–54, so that the two segments run antiparallel to each other and link up via amide-group hydrogen bonds. The chain then turns again, and residues 57–61 form a third segment to the sheet. In addition to amide H-bond stabilization, a scaffolding is provided by hydrogen bonds involving side chains of serine, threonine, asparagine, and glutamine residues. Some of these residues do not occur in the β-pleated sheet sequence itself, but rather in parts of the chain succeeding this region. It is believed that this β structure plays a vital role in stabilizing and determining the enzyme's native conformation. However, it is doubtful that this remarkably cross-hydrogen-bonded region is as important in determining the general structure as are the hydrophobic interactions, because the molecule can be unfolded in aqueous urea.

The three-dimensional structure of myoglobin (153 residues) differs sharply from the preceding structures, especially with regard to the specific backbone conformation. The molecule is roughly a thick oblate ellipsoid (44 × 44 × 25 Å). Its most striking characteristic is the occurrence of 121 of its 153 residues in α-helical sequences that range in length from 7 to 26 residues. These helical sections are folded back on each other to form a "globular" structure; a schematic illustration of these helical segments, lettered from A to H, is given in Figure 1-4a. The globular protein structure forms a cavity for the important heme group, which is completely buried except for its two hydrophilic propionic acid groups. The iron atom of the heme is octahedrally coordinated with six atoms, each donating an electron pair to the iron orbitals; these six atoms occupy the corners of the octahedron. Four coordination positions are occupied by porphyrin ring nitrogens, and a fifth position is occupied by a side chain imidazole nitrogen. The sixth position is occupied by oxygen in oxymyoglobin, and is empty in deoxymyoglobin. Apart from the coordination of the iron with histidine, the porphyrin ring is held in place by hydrophobic interactions involving the large number of hydrophobic residues that line the porphyrin cavity. In contrast to lysozyme, myoglobin essentially lacks the extended β-form type of structure.

The variety of interactions that can take place among amino acid residues provides for a rich diversity of ordered arrangements that can be generated by different amino acid sequences. It is evident, therefore, that the amino acid sequence plays a paramount role in determining the three-dimensional structure, and that the general constraints of steric interactions, and of keeping polar groups on the surface and nonpolar ones on the inside, are not sufficient to prevent proteins from having very different sequences or adopting extremely disparate conformations.

5-11 PREDICTION OF PROTEIN STRUCTURE

A prime long-range objective of researchers in the protein structure field is to predict three-dimensional conformation from the amino acid sequence. From our preceding discussion it is clear that the problem is formidable. Thus, for a protein that consists of only 100 amino acid residues in a single polypeptide chain (ignoring end effects), there are 100 ϕ and 100 ψ backbone torsion angles that must be simultaneously varied in order to find the conformation of lowest free energy in 200-dimensional space. Even if we restrict attention to just those values of ϕ and ψ that are sterically allowed, we saw earlier that, for most residues bearing a side chain (such as L-alanine), there are three distinct regions in the ϕ, ψ plane that can be occupied (see Fig. 5-8) and, for 100 such residues, there are $3^{100} \cong 10^{48}$ conformations that must be considered.

These simple ideas reemphasize the enormous numbers that are encountered in this kind of multidimensional problem. And this is only part of the problem. The considerations of backbone conformation do not get into the serious issue of the relative stabilization of various conformations by the solvent. Also, consideration should be given to side chain conformations and their effects on stabilizing particular backbone conformations.

Procedures for predicting structure

Because of these inherent complexities that are so difficult to surmount, investigators have turned to semiempirical procedures. They also have restricted their objectives to predicting some basic elements of protein secondary structure, such as α-helical and β-sheet regions, rather than the complete secondary and tertiary structure.

Although a variety of procedures are followed, all of them are based on the idea that *in the main* the conformation of a protein is determined by local, short-range interactions. This means that the conformation of a short polypeptide of defined sequence should have a reasonable probability of adopting the same conformation as that same polypeptide sequence within a large protein molecule.

Some experimental evidence supports this idea. M. Sela, C. B. Anfinsen, and colleagues have synthesized the closed "loop" peptide comprised of residues 64–82 of lysozyme (Fig. 5-19; Arnon et al., 1971). This synthetic loop elicits, in rabbits and goats, antibodies capable of reacting with native lysozyme and with the loop peptide prepared from lysozyme. On the other hand, although the loop peptide and native lysozyme can inhibit the immunological interactions with the synthetic material, the performic acid–oxidized peptide (in which the –S–S– linkage is broken) does not act as an inhibitor. These data imply that the loop peptide adopts, at least to a certain extent, the conformation of the same loop in native lysozyme.

In another vein, E. A. Kabat and coworkers have analyzed protein structures with the idea of determining the conformations of the residue Y in tripeptide sequences XYZ (see Kabat and Wu, 1973). Based on available data, they believe that, to a significant extent, the conformation of the specific residue Y is determined by local interactions. This means that the conformation of Y in the sequence XYZ in protein A has a good chance of being the same as that of sequence XYZ in protein B. This forms the basis for a structure predicting scheme.

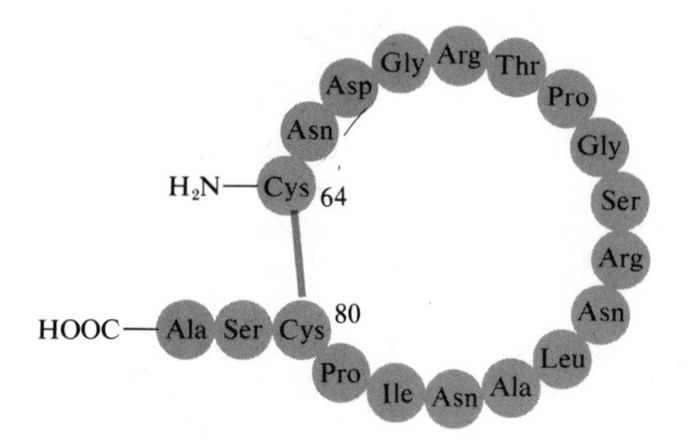

Figure 5-19

Synthetic loop peptide based on residues 64–82 of lysozyme, with alanine replacing the cysteine that occurs at position 76 in the enzyme. [After R. Arnon et al., *Proc. Natl. Acad. Sci. USA* 68:1450 (1971).]

These data indicate that structure prediction might be attempted—not by trying to vary simultaneously all of the ϕ, ψ coordinates—but rather by concentrating on only a few residues at a time along the chain. In this way one hopes to gain at least some estimate of secondary structure.

An example of structure prediction

Many structure-prediction schemes have been put forth over the past few years. For the sake of illustration we will briefly consider the scheme of P. Y. Chou and G. D. Fasman (1974). The choice of this, as opposed to other schemes, is somewhat arbitrary.

The basic idea is to examine a wide number of protein crystal structures with the goal of determining conformational preferences for the various amino acids. The notion is that certain residues are more likely than others to be found, for example, in α-helical sections. Data on synthetic polypeptides do in fact indicate that some residues have a greater tendency to form helical polypeptides (see Chapter 20). The analysis is cast into a quantitative framework by calculating conformational parameters for each of the residues. Table 5-11 gives an example of the data and calculations. This table is based on the analysis of the crystal structures of 15 proteins. The second column gives the fractional occurrence f_α of each residue in an α-helical section ($f_\alpha = n_\alpha/n$, where n is the total number of residues *of a given kind* in the 15 proteins, and n_α is the number of these residues that are in α-helical sections). The third column gives the conformational parameter P_α for each residue. This parameter is simply defined as $P_\alpha = f_\alpha/\langle f_\alpha \rangle$, where $\langle f_\alpha \rangle$ is the average value of f_α (that is, the sum of the f_α values divided by 20, the number of different residues). Note that, when $P_\alpha = 1$ for a residue, then that residue adopts the α-helical conformation at the same frequency

Table 5-11
Frequency of helical, β, and coil residues in 15 proteins with their conformational parameters P_α, P_β, and P_c

Amino acid	f_α	P_α	f_β	P_β	f_c	P_c
Ala	0.522	1.45	0.167	0.97	0.311	0.66
Arg	0.282	0.79	0.154	0.90	0.564	1.20
Asn	0.263	0.73	0.113	0.65	0.624	1.33
Asp	0.351	0.98	0.137	0.80	0.514	1.09
Cys	0.278	0.77	0.222	1.30	0.500	1.07
Gln	0.421	1.17	0.211	1.23	0.368	0.79
Glu	0.549	1.53	0.044	0.26	0.407	0.87
Gly	0.190	0.53	0.138	0.81	0.668	1.42
His	0.446	1.24	0.122	0.71	0.432	0.92
Ile	0.358	1.00	0.274	1.60	0.368	0.78
Leu	0.480	1.34	0.209	1.22	0.311	0.66
Lys	0.383	1.07	0.126	0.74	0.491	1.05
Met	0.429	1.20	0.286	1.67	0.286	0.61
Phe	0.402	1.12	0.219	1.28	0.378	0.81
Pro	0.212	0.59	0.106	0.62	0.682	1.45
Ser	0.282	0.79	0.124	0.72	0.594	1.27
Thr	0.295	0.82	0.205	1.20	0.494	1.05
Trp	0.409	1.14	0.203	1.19	0.386	0.82
Tyr	0.220	0.61	0.220	1.29	0.560	1.19
Val	0.409	1.14	0.282	1.65	0.309	0.66

SOURCE: After P. Y. Chou and G. D. Fasman, *Biochemistry* 13:211 (1974).

as the average residue in the chain. With $P_\alpha > 1$, there is a greater than average tendency to adopt the α-helical conformation, whereas $P_\alpha < 1$ means a less than average frequency of occurrence of the α-helical conformation for that residue. This parameter bears some analogy to the Zimm-Bragg helix-propagation parameter s discussed in Chapter 20; for an infinitely long polypeptide chain, when $s = 1$, the fractional helicity is 50%.

Table 5-11 also lists the parameters f_β, P_β, f_c, and P_c. These parameters are analogous to f_α and P_α, and they apply to β and coil regions.

It is of interest to examine the average values of these parameters (designated with angle brackets) in helical and β-sheet regions of the 15 proteins used to construct Table 5-11. Figure 5-20 shows average conformational parameters for residues in and adjacent to α-helical regions. Thus $\langle P_\alpha \rangle_{nN}$, $\langle P_\alpha \rangle_N$, $\langle P_\alpha \rangle_I$, $\langle P_\alpha \rangle_C$, and $\langle P_\alpha \rangle_{nC}$ are average α-helix conformational parameters for N-terminal nonhelical, N-terminal helical, central (or inner) helical, C-terminal helical, and C-terminal nonhelical residues, respectively. The figure shows that, as one proceeds in either direction from the center of the helix, there is a decrease in the value of $\langle P_\alpha \rangle$, and that just beyond the end of the helix, the value of $\langle P_\alpha \rangle$ drops below unity. This gives rise to the idea

Helical regions

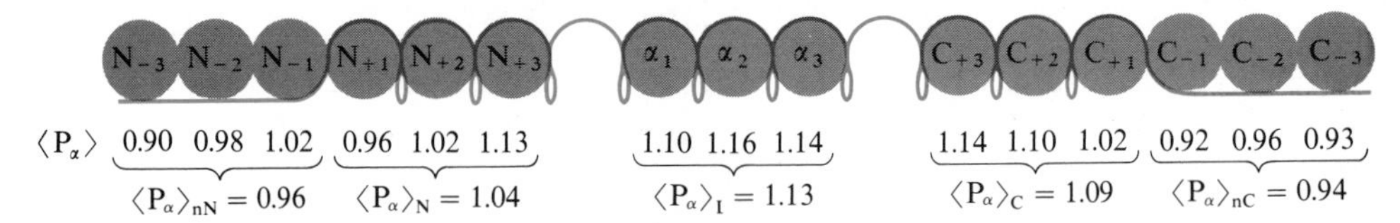

Figure 5-20

Comparison of $\langle P_\alpha \rangle$ *values for helical boundary and central regions.* [After P. Y. Chou and G. D. Fasman, *Biochemistry* 13:211 (1974).]

that the ends of helices are occupied, on the average, by residues with helix-breaking tendencies, and that these residues act to arrest further helix growth. Conversely, the centers of helices are occupied by clusters of residues with a greater than average helix-forming tendency. These clusters might be sites for helix initiation, with helix growth then going outward in both directions.

A similar analysis shows that residues with $\langle P_\beta \rangle$ values below unity tend to occur at the boundaries of β-sheet regions, and that those with relatively high $\langle P_\beta \rangle$ values are in the inner β-sheet region.

These considerations suggest a straightforward scheme for predicting secondary structure. The idea is to examine a given sequence and assign conformational parameters to each residue, and then look for clusters of helix-forming residues flanked at some point by helix breakers, or for clusters of β-sheet-forming residues flanked by sheet breakers. This is easily done by first classifying each residue as a strong helix former, helix breaker, β-sheet former, or β-sheet breaker, in accordance with the values of the conformational parameters. Table 5-12 classifies the residues according to their α-helix and β-sheet making and breaking tendencies. Making use of this table, one can construct a set of simple rules for predicting secondary structure from amino acid sequence.

Comparison of predicted and observed secondary structures

With a prediction scheme based on the ideas outlined above, secondary structures have been predicted for a number of proteins for which crystallographic data are available. Table 5-13 summarizes results for four proteins: concanavalin A, lamprey hemoglobin, thermolysin, and trypsin inhibitor. None of these proteins is contained in the set of 15 that formed the data base for Tables 5-11 and 5-12. The table lists sequence positions where α-helical and β-sheet residues are predicted and observed to occur. In the case of trypsin inhibitor, a small polypeptide of 58 residues, predicted and observed regions of secondary structure are in excellent agreement. Because

Table 5-12
Classification of amino acid residues as formers, breakers, or indifferent for α-helical and β-sheet regions

Helical residues	P_α	Classification	β-Sheet residues	P_β	Classification
$Glu^{(-)}$	1.53	H_α	Met	1.67	H_β
Ala	1.45	H_α	Val	1.65	H_β
Leu	1.34	H_α	Ile	1.60	H_β
$His^{(+)}$	1.24	h_α	Cys	1.30	h_β
Met	1.20	h_α	Tyr	1.29	h_β
Gln	1.17	h_α	Phe	1.28	h_β
Trp	1.14	h_α	Gln	1.23	h_β
Val	1.14	h_α	Leu	1.22	h_β
Phe	1.12	h_α	Thr	1.20	h_β
$Lys^{(+)}$	1.07	I_α	Trp	1.19	h_β
Ile	1.00	I_α	Ala	0.97	I_β
$Asp^{(-)}$	0.98	i_α	$Arg^{(+)}$	0.90	i_β
Thr	0.82	i_α	Gly	0.81	i_β
Ser	0.79	i_α	$Asp^{(-)}$	0.80	i_β
$Arg^{(+)}$	0.79	i_α	$Lys^{(+)}$	0.74	b_β
Cys	0.77	i_α	Ser	0.72	b_β
Asn	0.73	b_α	$His^{(+)}$	0.71	b_β
Tyr	0.61	b_α	Asn	0.65	b_β
Pro	0.59	B_α	Pro	0.62	b_β
Gly	0.53	B_α	$Glu^{(-)}$	0.26	B_β

NOTE: H = strong former; h = former; I = weak former; i = indifferent; b = breaker; B = strong breaker.
SOURCE: P. Y. Chou and G. D. Fasman, *Biochemistry* 13:222 (1974).

this protein is small, local interactions should have a bigger relative effect on determining the structure, and we expect the best results with this protein. Many of the α-helical and β-sheet regions are also accurately predicted in the other three proteins. In fact, 17 of the 18 α-helical sections and 20 of the 22 β-sheet sections are correctly localized, whereas 5 α-helical and 10 β-sheet regions are predicted to occur at regions where none are found. Clearly, the results indicate that predictions of secondary structure can be much more accurate than random chance.

Another common structure in many proteins is the β turn. Figure 2-25 shows that this structure allows a polypeptide chain to reverse its direction. It is possible to calculate a β-turn conformational parameter for each amino acid, analogous to the P_α and P_β parameters discussed above. Use of this parameter, together with P_α and P_β, allows a further extension of the prediction of secondary structure. It is noteworthy that, in a dozen or so proteins recently examined by Chou and Fasman,

Table 5-13

Comparison of experimental and predicted helical and β-sheet regions in four proteins not included in computing the conformational parameters P_α and P_β

	Helical regions		β-Sheet regions	
Protein	X-ray	Predicted	X-ray	Predicted
Concanavalin A	—	38–43	4–9	3–12
	81–85	81–86	25–29	25–29
	—	155–160	48–55	47–55
	—	180–189	59–66	60–67
			73–78	73–80
			92–97	88–96
			106–116	106–113
			125–132	124–134
			140–144	140–144
			173–177	173–177
			190–199	190–200
			209–215	209–215
			—	229–234
Lamprey hemoglobin	12–29	8–24	None	
	30–44	—	—	2–6
	45–52	45–52	—	35–43
	62–66	61–78		
	67–88	80–88		
	92–106	92–100		
	—	104–110		
	111–127	115–128		
	132–148	133–147		
Thermolysin	—	53–58	4–13	4–17
	65–88	67–74	15–32	20–32
	137–152	137–150	37–46	37–50
	159–180	158–180	52–58	—
	235–246	238–246	60–63	61–66
	259–274	261–271	—	75–84
	280–296	281–295	97–106	98–110
	302–313	301–313	112–116	—
			119–123	120–124
			—	127–131
			—	151–157
			—	192–197
			—	221–225
			—	251–260
			—	272–276
Trypsin inhibitor (pancreatic)	3–6	2–7	16–24	16–23
	45–56	45–54	27–36	27–38

SOURCE: After P. Y. Chou and G. D. Fasman, *Biochemistry* 13:222 (1974).

the α-helical, β-sheet, and β-turn conformation make up 80% of the residue conformations, with the remainder being in random conformations.

Chapter 2 mentioned briefly the predictions of secondary structure for adenylate kinase, a small polypeptide chain of molecular weight 21,600. Figure 2-13 compares results obtained by a variety of investigators. The three panels (top to bottom) indicate the results of predictions for bends, β-sheets, and α-helices. The open rectangles indicate the positions in the sequence where these elements of secondary structure occur, and horizontal bars indicate the prediction results for different investigators. The histograms at the bottom of the panels give composites of the results (joint predictions) obtained by the various groups.

From the bottom panel it is clear that the joint predictions are in good accord with the x-ray results. The middle panel shows that adenylate kinase has five β-sheet sections, and that those around residues 12, 91, and 116 are accurately predicted. These three sections form the three central strands of a five-stranded parallel β-sheet in this enzyme. The two outer strands of this sheet are not accurately located on the joint prediction histogram. In the case of bends, the top panel shows that most of these are predicted, although a few are predicted where in fact none occur.

On balance the results with adenylate kinase illustrate that the various semi-empirical prediction schemes, on the average, have a fair degree of success, and that the different schemes can lead to similar results.

It is difficult to guess how far these semiempirical prediction schemes can be carried before they reach inherent limitations. It also must be remembered that correct prediction of secondary structure is a necessary, but not sufficient, step to specifying three-dimensional conformation. Tertiary structure, which depends on the conformations adopted by residues ascribed to "random" conformations (not identifiable as any ordered form such as α-helic, β-sheet, etc.) and on interactions between residues that are far apart in the sequence, is much more difficult to pin down and yet must be put on a rational basis before we can even begin to predict a complete three-dimensional structure from an amino acid sequence.

5-12 MOLECULAR MODEL BUILDING BY COMPUTER

In attempting to predict protein structure, and for many other purposes, it is helpful to have a means for rapidly constructing molecular models of protein structures. Using molecular models to construct manually a wide variety of structures is, of course, completely unfeasible. However, a novel and fascinating approach to this problem has been developed by C. Levinthal and colleagues (1966), who employed a computer to accomplish the model building. The computer rapidly constructs the models from the backbone rotation angles, and the resulting structures can be displayed on an oscilloscope screen. Because the screen gives only a two-dimensional projection, a three-dimensional viewing effect is created by rotating the structures about an axis, so that different parts of the structure become visible to the observer.

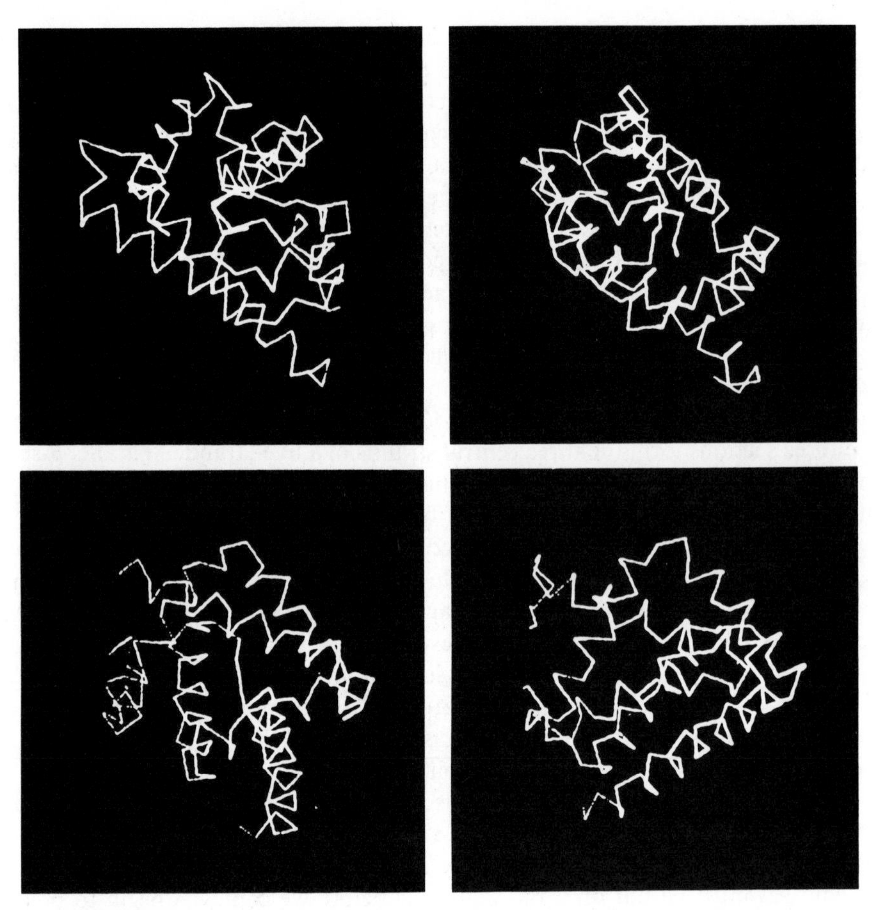

Figure 5-21

Oscilloscope projections of crystal structure of myoglobin. [Courtesy of C. Levinthal.]

Figure 5-21 shows some typical oscilloscope projections of myoglobin's crystal structure. The visual clarity of these structural projections is very impressive. Thus an investigator can rapidly construct and visually examine a wide variety of structures in the course of an attempt to determine preferred ones.

Summary

Various and complex factors determine protein conformation. In the polypeptide chain, bond lengths and bond angles may be considered as relatively fixed, and the amide bond is generally confined to the planar, *trans* conformation. Therefore, the conformation of the polypeptide backbone is specified by the values adopted by the internal rotation angles ϕ and ψ associated with each residue. These rotation angles are limited to certain preferred values determined by steric and other interactions (such as amide–amide dipolar interactions) in the chain. All of these features collectively eliminate a large fraction of possible conformations, but still leave a huge number of possibilities.

Hydrogen bonding, hydrophobic interactions (which arise in consequence of the peculiar structural features of water), ionic interactions, and disulfide bonds are further constraints on polypeptide conformation. The combined effect of all interactions gives rise to formation of a "native" conformation. Amino acid sequence plays a key role in determining the precise native conformation that is adopted. Although a sufficient rational foundation does not yet exist for predicting protein conformation based on amino acid sequence, attempts at such predictions are being made with moderate success.

Further aspects of protein folding are discussed in Chapter 21.

Problems

5-1. Consider a repeating-sequence cyclic polypeptide XYZXYZX consisting of seven amino acid residues. X, Y, and Z represent the three amino acid residues Ala, Pro, and Gly. You wish to make a sequence that will introduce the least strain in closing the circle. The question which you have to answer is "In which order should the amino acid residues be placed?"

Professor A, a brilliant man, says the best solution is Y = Pro, Z = Gly, and X = Ala. A timid graduate student claims that this is not the best solution, but a better one is X = Ala, Y = Gly, and Z = Pro. A third party, basically just a philosopher who knows nothing about amino acids, says that neither of these solutions is truly the best one. There is a better one yet. Who is right and why?

5-2. A protein undergoes a subtle conformational change in which the most significant event is the movement of three closely packed amino acid side chains—leucine, alanine, and phenylalanine—from the ethanollike interior of the protein to the exterior, where they are immersed in solvent. Assuming that this is the only thermodynamically significant event in the conformational change, calculate the equilibrium constant for this change in water and in 8 M urea, at 25°C. Use data given in the chapter.

5-3. Two parts of a protein interact with each other according to a well-characterized force law. In conformation A, these two parts are separated by the distance r_2 and in conformation B this separation is r_1, where $r_2 > r_1$. The force law characterizing this interaction is $F = -\gamma/r^2$, where $\gamma = aT$, and a is a constant independent of temperature. Assuming that this force law describes the only significant interaction occurring in the conformational change, calculate the equilibrium constant, ΔH, and ΔS for the change (in terms of r_1, r_2, γ, and T).

5-4. A student measures the free energy of transfer (ΔG^0_{Tot}) of apolar compound A at mole fraction $\chi = 0.1$ in ethanol to mole fraction $\chi = 0.1$ in water. He also measures the transfer free energy ($\Delta G^{0*}_{\text{Tot}}$) of going from pure liquid A to mole fraction $\chi = 0.1$ in water. He finds that $\Delta G^{0*}_{\text{Tot}} - \Delta G^0_{\text{Tot}} = -RT \ln 56$, and he then claims that this makes perfect sense. Do you agree or not? Why?

5-5. Two polypeptide tendons, tendon 1 and tendon 2, are made of exactly the same length, thickness, and width. One tendon is made largely of long chains of polyalanine, and the other is made largely of long chains of polyglycine. With the tendons in their unstretched, natural state, the constituent polymers in each tendon take up random coil conformations that are dictated, by and large, by the conformational energy diagrams of their individual residues. Interestingly enough, it is found that tendon 1 can be stretched reversibly to a considerably greater length than that possible with tendon 2. On the basis of this observation, and of your knowledge of the preferred conformations of glycyl and alanyl residues, suggest which tendon is made up of polyglycine chains and which is made up of polyalanine chains. State concisely, but carefully, the reasoning behind your choice.

References

GENERAL

Chou, P. Y., and G. D. Fasman. 1978. Empirical predictions of protein conformation. *Ann. Rev. Biochem.* 47:251. [A concise summary of empirical prediction methods.]

Dickerson, R. E., and I. Geis. 1969. *The Structure and Action of Proteins.* New York: Harper and Row. [Includes beautiful illustrations of protein structures. The book is particularly valuable for quick visual insight into protein conformation.]

Flory, P. J. 1969. *Statistical Mechanics of Chain Molecules.* New York: Interscience. [Chapter VII provides a good summary of conformational energy calculations on polypeptides. As in much of the earlier literature, the $\phi = 0°$, $\psi = 0°$ conformation is assigned to the all-*trans* polypeptide chain, which corresponds to $\phi = 180°$, $\psi = 180°$ in the 1970 IUPAC–IUB convention.]

IUPAC–IUB Commission on Biochemical Nomenclature. 1970. Abbreviations and symbols for the description of the conformation of polypeptide chains. *Biochemistry* 9:3471. [Summarizes definitions of polypeptide internal rotation angles and gives related material of interest. Published simultaneously in *J. Biol. Chem.* 245:6489, and *J. Mol. Biol.* 52:1.]

Kauzmann, W. 1959. Some factors in the interpretation of protein denaturation. *Adv. Protein Chem.* 14:1. [A classic paper on the hydrophobic effect; ionic interactions also are discussed.]
Schultz, G., and R. H. Schirmer. 1979. *Principles of Protein Structure.* New York: Springer-Verlag. [This is an extensive discussion with a large bibliography.]

SPECIFIC

Anfinsen, C. B., E. Haber, M. Sela, and F. H. White, Jr. 1961. The kinetics of formation of native ribonuclease during oxidation of the reduced polypeptide chain. *Proc. Natl. Acad. Sci.* 47:1309.
Anfinsen, C. B., and H. A. Scheraga. 1975. Experimental and theoretical aspects of protein folding. *Adv. Protein Chem.* 29:205.
Arnon, R., E. Maron, M. Sela, and C. B. Anfinsen. 1971. Antibodies reactive with native lysozyme elicited by a completely synthetic antigen. *Proc. Natl. Acad. Sci.* 68:1450.
Brant, D. A., and P. J. Flory. 1965. The configuration of random polypeptide chains, II: Theory. *J. Am. Chem. Soc.* 87:2791.
Brant, D. A., W. G. Miller, and P. J. Flory. 1967. Conformational energy estimates for statistically coiling polypeptide chains. *J. Mol. Biol.* 23:47.
Chou, P. Y., and G. D. Fasman. 1974. Conformational parameters for amino acids in helical, β-sheet, and random coil regions calculated from proteins. *Biochemistry* 13:211.
———. 1974. Prediction of protein conformation. *Biochemistry* 13:222.
Doty, P. 1960. Optical rotation and the structure of polypeptides and proteins. In *Fourth International Congress of Biochemistry, Vol. VIII: Proteins* (New York: Pergamon Press), p. 8.
Frank, H. S., and W.-Y. Wen. 1957. Structural aspects of ion-solvent interaction in aqueous solutions: A suggested picture of water structure. *Disc. Faraday Soc.* 24:133.
Gurney, R. W. 1953. *Ionic Processes in Solution.* New York: McGraw-Hill. [Discusses concepts of cratic and unitary contributions to free energy. For example, see pp. 89–92 and subsequent sections.]
Hagler, A. T., and B. Honig. 1978. On the formation of protein tertiary structure on a computer. *Proc. Natl. Acad. Sci.* 75:554.
Kabăt, E. A., and T. T. Wu. 1973. The influence of nearest-neighbor amino acids on the conformation of the middle amino acid in proteins: Comparison of predicted and experimental determination of β-sheets in concanavalin A. *Proc. Natl. Acad. Sci.* 70:1473.
Klotz, I. M., and J. S. Franzen. 1962. Hydrogen bonds between model peptide groups in solution. *J. Am. Chem. Soc.* 84:3461.
Levinthal, C. 1966. Molecular model building by computer. *Sci. American* 214(6):42.
Levitt, M., and C. Chothia. 1976. Structural patterns in globular proteins. *Nature* 261:552.
Némethy, G., and H. A. Scheraga. 1965. Theoretical determination of sterically allowed conformations of a polypeptide chain by a computer. *Biopolymers* 3:155.
Pratt, L. R., and D. Chandler. 1977. Theory of the hydrophobic effect. *J. Chem. Phys.* 67:3683.
Ramachandran, G. N., C. Ramakrishnan, and V. Sasisekharan. 1963. Stereochemistry of polypeptide chain configurations. *J. Mol. Biol.* 7:95.

6

Conformational analysis and forces that determine nucleic acid structure

6-1 GENERAL CHARACTERISTICS OF NUCLEIC ACID STRUCTURE

The physical and chemical properties of nucleic acids differ appreciably from those of proteins and polypeptides. This is a consequence of the entirely different chemical compositions and structures of the two molecules. Whereas the polypeptide backbone is electrically neutral, and forms the attachment point for twenty or so different side chains, the nucleic acid backbone is a highly charged polyelectrolyte, which bears (for the most part) only four different kinds of pendant side groups. Moreover, nucleic acid side chains exhibit a specific complementarity (base pairing), which is absent in amino acids. This complementarity is partly responsible for the helical rodlike structures found in both single-stranded and double-stranded chains. In addition, the charged backbone makes it more difficult for nucleic acids to adopt the compact, globular conformations so often preferred by proteins.

Our objective is to adumbrate the principal important determinants of nucleic acid conformation. We first consider the limitations on possible configurations placed by geometric and steric constraints. It will be immediately apparent that the nucleic acid backbone presents a degree of complexity considerably beyond that encountered with the polypeptides. Next we deal with the specific base-pairing and the nonspecific stacking interactions that play a major role in stabilizing ordered forms of single-stranded and double-stranded chains.

6-2 GEOMETRIES

Figure 6-1 is a schematic diagram of a polynucleotide chain. There are six skeletal bonds in the repeating unit, with the side chain bases attached at the $C^{1'}$ position of the ribose moiety. Figure 3-1 gives the structures of the major bases and many of the

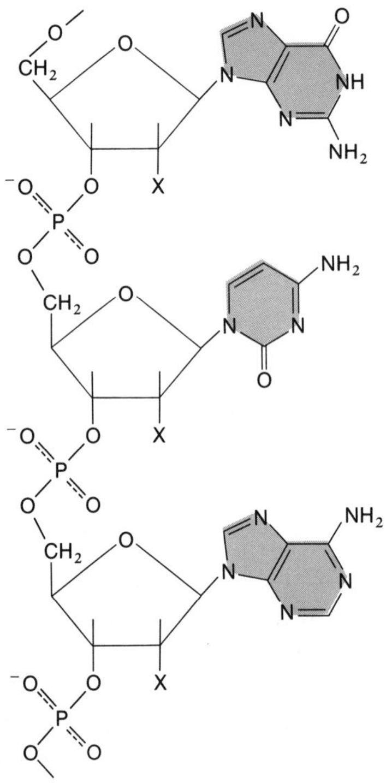

Figure 6-1

Schematic of polynucleotide chain. X = OH in RNA polymers and H in DNA polymers.

minor bases, and indicates the numbering systems for purine and pyrimidine nucleosides. The relatively great structural complexity of the polynucleotide chain is made evident by comparing the polynucleotide backbone with that of the polypeptide illustrated in Figure 5-1.

Table 6-1 lists backbone structural parameters for crystalline polyriboadenylic acid. This table gives bond lengths and various bond angles. These parameters are essentially fixed and should not substantially differ for different RNA polymers.

In order to consider the family of conformations available to the chain, it is expedient to investigate first the conformational characteristics of the individual mononucleoside units (sugar and base) themselves. We can then proceed to a treatment of the steric restrictions in the backbone.

Table 6-1
Polyribonucleotide chain geometries

Bond	Bond length (Å)	Bonds	Bond angle
Backbone geometries			
$C^{4'}$—$C^{3'}$	1.52	$C^{4'}$—$C^{3'}$—$O^{3'}$	110°
$C^{3'}$—$O^{3'}$	1.47	$C^{3'}$—$O^{3'}$—P	119°
$O^{3'}$—P	1.56	$O^{3'}$—P—$O^{5'}$	104°
P—$O^{5'}$	1.57	P—$O^{5'}$—$C^{5'}$	121°
$O^{5'}$—$C^{5'}$	1.46	$O^{5'}$—$C^{5'}$—$C^{4'}$	112°
$C^{5'}$—$C^{4'}$	1.54	$C^{5'}$—$C^{4'}$—$C^{3'}$	112°
Glycosidic geometries			
$O^{1'}$—$C^{1'}$	1.45	$O^{1'}$—$C^{1'}$—N^{9}	112°
$C^{2'}$—$C^{1'}$	1.54	$C^{2'}$—$C^{1'}$—N^{9}	106°
$C^{1'}$—N^{9}	1.47	$C^{1'}$—N^{9}—C^{4}	127°
N^{9}—C^{4}	1.35	$C^{1'}$—N^{9}—C^{8}	127°
N^{9}—C^{8}	1.33		

NOTE: Data are for the polyriboadenylic acid chain.
SOURCE: Data from A. Rich et al., *J. Mol. Biol.* 3:71 (1961).

6-3 GLYCOSIDIC BOND ROTATIONAL ISOMERS AND RIBOSE PUCKERING

The pendant base moiety is attached to the polynucleotide backbone via the glycosidic bond. The relative orientation of the base with respect to the ribose unit is described by right-hand rotations χ about this bond. This angle is assigned a value of 0° for the *cis* conformation of bonds $0^{1'}$–$C^{1'}$ and N^9–C^8. In the case of pyrimidines, the rotation angle is measured with reference to orientation of the $0^{1'}$–$C^{1'}$ and N^1–C^6 bonds.

Two rotational isomers are of particular interest. These are designated *syn* and *anti* (Fig. 6-2). An *anti* conformation is located near $\chi = 0°$, whereas the *syn* domain

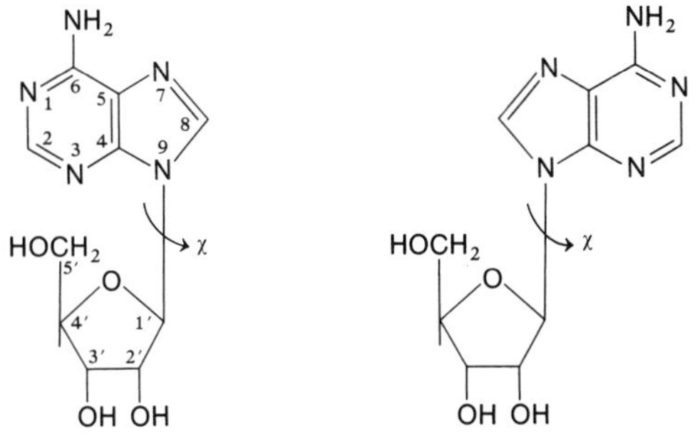

Syn
$\chi \approx 210°$

Anti
$\chi = 0°$

Figure 6-2
Conformers of adenosine.

is around 210°. There is also a "high-*anti*" form at 120°. The degree of preference for one rotameric state over the other is dependent on the puckering of the ribose moiety, which is discussed in the next paragraph. In general, however, calculations suggest—and mononucleoside (and mononucleotide) x-ray data confirm—that pyrimidines tend to adopt the *anti* configuration, whereas purines are able to assume both forms. In double-helical DNA and helical double-stranded polyadenylic acid, the bases adopt the *anti* configuration.

As mentioned, the disposition of the base about the glycosidic bond is coupled to the conformation of the ribose ring. The conformation of the ring is generally found to be such that four of the atoms fall in the same plane, with the fifth displayed above or below this plane. The deviant atom is invariably $C^{2'}$ or $C^{3'}$, and the displacement is usually designated as *endo* or *exo*, depending on whether it occurs on the same or opposite side as $C^{5'}$, respectively. The different forms are usually designated as $C^{2'}$-*endo*, $C^{2'}$-*exo*, $C^{3'}$-*endo*, or $C^{3'}$-*exo*. These puckered forms of the ribose ring are illustrated in Figure 3-15.

6-4 BACKBONE ROTATION ANGLES AND STERIC HINDRANCES

The analysis of the preferred conformations of the nucleic acid backbone is a formidable task. A multitude of steric interactions must be examined as the six rotational angles within each unit are varied, and the possibility of interactions between adjacent units must also be investigated. Thus, at the outset the problem posed is clearly more complicated than the analysis of preferred conformations of polypeptide chains.

To tackle this problem, we follow the procedure of W. K. Olson and P. J. Flory (1972), who demonstrated that a great deal of simplification is possible. Figure 6-3 is schematic illustration of a segment of a polyriboadenylic chain; backbone torsion angles associated with individual bonds are given Greek letter symbols, which are primed or double primed.

An important feature is that the periodic occurrence of the pentose ring effectively ensures that the rotations about the five backbone bonds succeeding the ring are independent of the five that precede the ring. If we therefore take the repeating unit to be the atoms from one $C^{4'}$ to the next (see Fig. 6-3), the conformations of the successive repeating units are independent of each other. We are then left with the mere task of considering the steric consequences of simultaneously varying the six rotation angles (in Fig. 6-3 designated ω', ω'', ψ', ψ'', ϕ', ϕ'') within a single unit!

First-order interactions

By systematically examining the steric interactions encountered as a result of varying each rotation angle separately, we simplify the problem considerably. We assign the *cis* configuration about a given backbone as the 0° rotation angle for that bond.

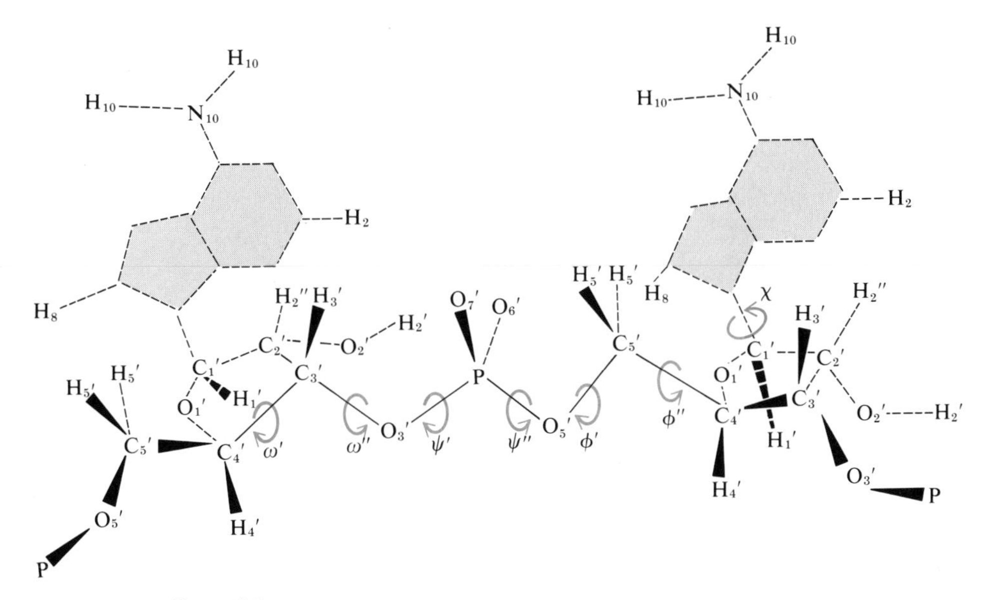

Figure 6-3

Schematic drawing of polyriboadenylic acid chain, showing rotation angles. [After W. K. Olson and P. J. Flory, *Biopolymers* 11:1 (1972).]

First we treat interactions between atoms whose distance of separation depends on a *single* rotation angle. Consider, for example, the steric hindrances that occur as a result of varying only ψ'. These include contacts that $C^{3'}$ can make with $0^{5'}$, $0^{6'}$, and $0^{7'}$; in these cases, the distances of separation between $C^{3'}$ and the oxygens depend only on the value adopted by ψ'. However, contacts between $C^{3'}$ and $C^{5'}$ are not considered in this case, because their distance of separation depends simultaneously upon ψ' and ψ''. The first-order interactions are sufficient to limit considerably the domain accessible to ψ'; this domain consists approximately of regions encompassing $-90°$ to $-30°$, $+30°$ to $+90°$, and $180° \pm 30°$.

Interactions dependent solely on ψ'' are considered in a similar fashion. Figure 6-4a is a graph plotting ψ'' against ψ'. The shaded regions of the graph indicate areas that are sterically forbidden solely by first-order interactions—that is, interactions dependent on a single rotation angle (ψ' or ψ'').

Second-order interactions

We can next ask what parts of the regions allowed by first-order considerations are forbidden by second-order contacts, which are steric conflicts between atoms whose distance of separation depends simultaneously on ψ' and ψ''. An example is the distance between $C^{3'}$ and $C^{5'}$. These second-order contacts eliminate the area enscribed by the solid line in Figure 6-4a. Much of this area falls within the shaded region and is,

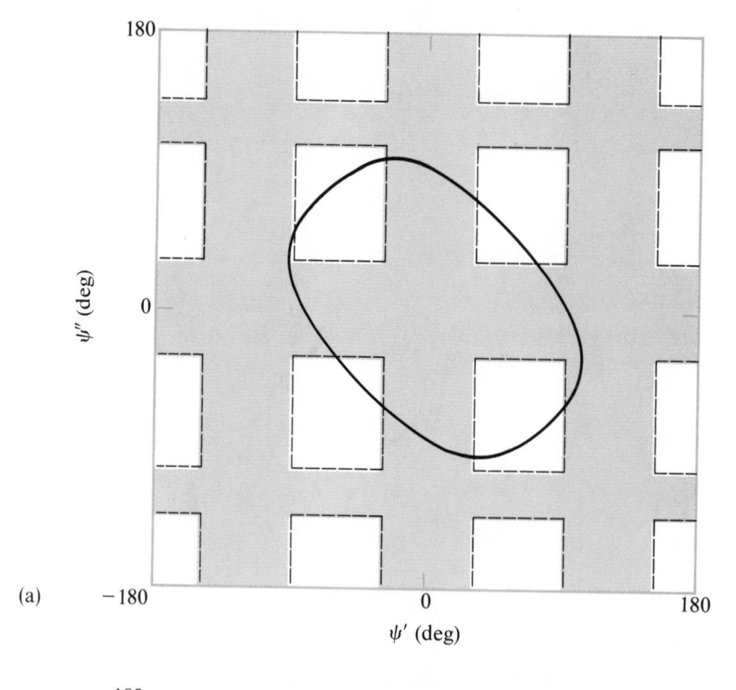

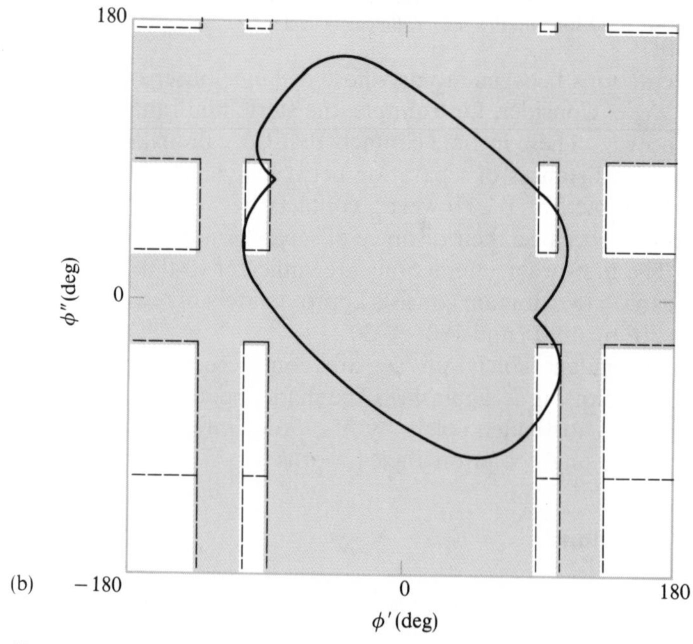

Figure 6-4

Plots of interactions between rotation angles. Regions excluded by first-order interactions are shaded; those excluded by second-order contacts are enclosed by a solid line. **(a)** Plot of ψ'' versus ψ'. **(b)** Plot of ϕ'' versus ϕ'. A small region at $\phi'' = -120°$ may be excluded, and this is indicated by a horizontal dashed line. [After W. K. Olson and P. J. Flory, *Biopolymers* 11:1 (1972).]

therefore, already forbidden by first-order interactions. A small area allowed by first-order interactions is forbidden by second-order effects, however. The remaining unshaded areas (outside of the closed solid line) are free of second-order contacts.

Extension of the analysis

A similar analysis is performed with the rotation angles ϕ' and ϕ'', and the results obtained are summarized in Figure 6-4b. Here again, the unshaded areas represent regions permitted by first-order considerations, and the solid line encloses the region forbidden by collisions between atoms whose separation depends simultaneously on ϕ' and ϕ''. In the case of the angles ω' and ω'', the permitted range of ω'' is somewhat dependent on the value adopted by ω'. The latter angle is restricted to a domain of about 15°, centered at 83° for the $C^{2'}$-*exo* and $C^{3'}$-*endo* forms, and a similar domain centered at 153° for the $C^{2'}$-*endo* and $C^{3'}$-*exo* conformers. The domains accessible to ω'' are centered at about $-145°$ and $-95°$ and have widths in the range of 10° to 20°, the exact value depending on the value of ω'.

By proceeding in the fashion indicated, it is clear that a large fraction of backbone configuration space is excluded on the basis of these first-order and second-order interactions. However, we are still faced with the problem of second-order interactions involving other dihedral angle pairs (such as ψ' and ω'') and of the possible conflicts between atoms whose distance of separation depends on three or more rotation angles. Moreover, we have not yet considered the possibility of contacts involving atoms of the laterally attached bases—both base–base and base–backbone steric conflicts.

It turns out, however, that all of these problems are ameliorated because every one of these additional conflicts occurs within segments of configuration space already excluded by the previously treated first-order and second-order interactions. Figure 6-5 shows, as unshaded regions, those domains in which no *first-order* conflicts occur in the variation of ϕ' and ψ''. The solid line encloses the region of unfavorable second-order steric interactions. It is clear that this line encloses a completely shaded area and, therefore, represents a domain already excluded by first-order contacts.

The same is true of second-order conflicts arising from variations of ψ' and ω''. Moreover, with ω' fixed at either 83° or 153°, there are no conflicts arising from rotations (ω'' and ϕ'') about the two bonds flanking the $C^{4'}$–$C^{3'}$ bond. This assures that rotations within the unit spaning one $C^{4'}$ to the next $C^{4'}$ are independent of the neighboring units. Finally, with the backbone rotation angles fixed in the previously determined sterically allowed domains, there are no steric interactions between successive bases, no base–backbone overlaps, and no hindrances arising from the simultaneous variation of three or more backbone dihedral angles.

These calculations of sterically permitted regions may be refined in a manner analogous to that done in the case of polypeptides—that is, by employing potential energy functions. In the present case, the calculations are rather involved and must include a careful analysis of the electrostatic interactions between various polar

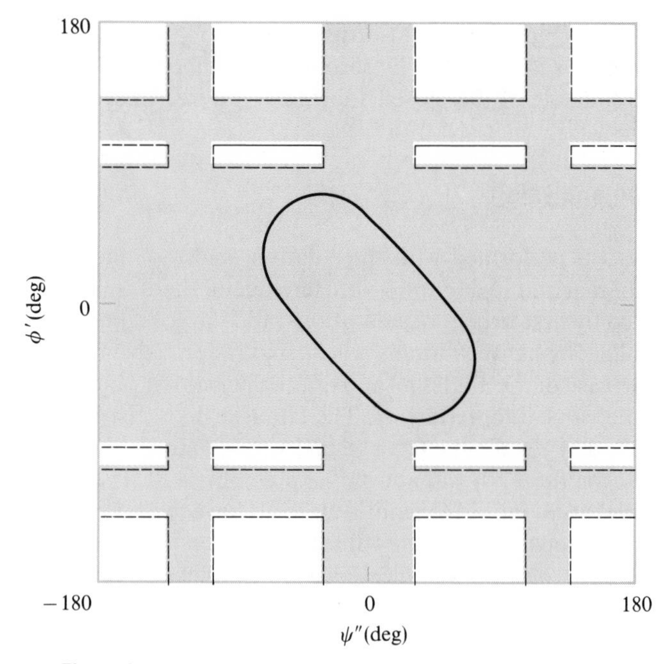

Figure 6-5

Plot of ϕ' versus ψ'', following same conventions as Figure 6-4. [After W. K. Olson and P. J. Flory, *Biopolymers* 11:1 (1972).]

groups. This analysis involves the assignment of partial charges to appropriate atoms of the phosphate group, of the base moiety, and of the ribose unit. The upshot of these calculations is the demonstration that the backbone is biased into those sterically allowed conformations that minimize the coulombic energy, and that these conformations are the more extended ones.

In the case of 2′-deoxyribonucleic acid polymers, the configuration space preferred by the backbone is calculated to be not significantly different than that deduced for the RNA polymers. It turns out that the main difference arises from a change in the energy associated with rotations about the $C^{3'}$–$O^{3'}$ bond.

6-5 FORCES STABILIZING ORDERED FORMS

Although potential energy calculations enable us to estimate the regions of configuration space that the molecule can freely traverse, they fail to account for the markedly organized structures that nucleic acids, like proteins, can adopt under physiological conditions. Thus the degree of freedom permitted by steric and simple electrostatic restrictions, though relatively limited, is still great enough to enable a wide diversity of irregular structures to occur. Moreover, the previous calculations do not provide a basis for understanding the frequently found double-stranded and even triple-stranded structures, where interstrand interactions obviously must be considered. We must turn, therefore, to a consideration of the other kinds of forces that play a paramount role in stabilizing regular nucleic acid structures.

Figure 3-14 illustrates a model of the B-form double-stranded DNA helix, taken from the structure proposed by J. Watson and F. Crick. Two right-handed polynucleotide helices are interwound, with their bases on the inside and the phosphodiester backbone on the outside. The bases are at right angles to the helix axis and stacked upon each other, neighboring bases being separated by 3.4 Å. There are 10 base pairs per turn of the helix, which thus results in a pitch of 34 Å. The strands themselves run in opposite directions (that is, they have opposite polarity), so that at either end of the helix, one strand has a free 5′-phosphate and the other has a free 3′-hydroxyl.

Prior to the elucidation of this structure, it had been shown that the base composition of DNA is such that the purine content equals that of the pyrimidine and, furthermore, that adenine (A) = thymine (T), and guanine (G) = cytosine (C). The Watson–Crick structure provides a rationalization for this early observation. The bases of the opposite strands are positioned in a precise geometry that enables them to form hydrogen bonds. This hydrogen bonding is highly specific and involves the matching of A with T, and G with C. The two strands are therefore complementary. Thus, if in going from the 5′-end to the 3′-end of one chain, we have the sequence AGCTAGGT . . . , the sequence on the opposite strand (reading from the 3′-end to 5′-end) would be TCGATCCA . . . . (It is customary, however, to number sequences from the 5′-end to the 3′-end, so that the latter sequence would be written . . . ACCTAGCT.) This complementary base matching between strands occurs because the enzymatic synthesis of DNA proceeds by the specific manufacture of the complement of a single strand, rather than by the reproduction of the base sequence of that strand. It should also be pointed out that, under a variety of experimental conditions, DNA has been shown to exist in a structure quite consistent with the Watson–Crick helix observed in fibers.

The two obvious features that were not explained in our preceding conformational analysis are the specific base-pairing interaction, and the striking parallel stacking arrangement of the bases. These two interactions are also prominent in other nucleic acid structures, including single-stranded chains such as transfer RNA. It is of considerable interest, therefore, that we examine rather carefully the molecular basis for these two important interactions.

6-6 BASE PAIRING

A diversity of base–base hydrogen-bonding interactions are possible and have been observed.

Figure 3-11 shows the specific base-pairing scheme of Watson and Crick whereby adenine and thymine (or uracil) pair together by two hydrogen bonds, and guanine and cytosine by three bonds.

Obviously we want to know why these particular bases pair together with this kind of base–base specificity and, even given this specificity, why this particular pattern or arrangement of hydrogen bonds should be adopted in the A–T and G–C pair. The answers to these questions are by no means trivial.

It is well known that bases can self-associate to make hydrogen-bonded complexes. There are many examples of such self-associated complexes from crystallographic studies. Figure 6-6 illustrates hydrogen-bonded pairing between adenine bases in crystals. Hydrogen-bonded self-pairing between other bases also has been observed in crystals.

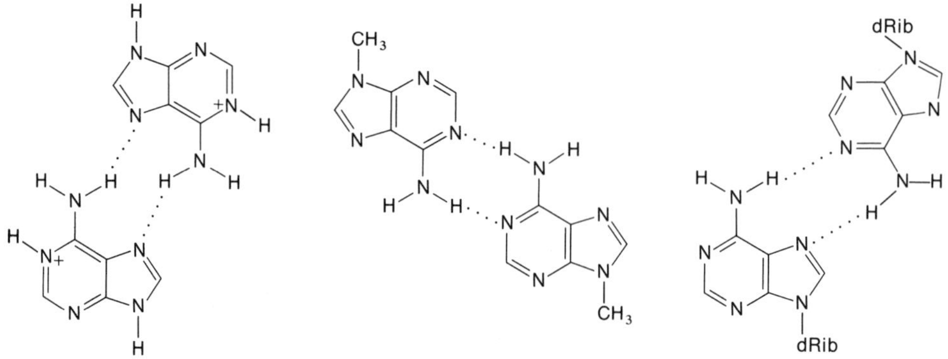

Figure 6-6

Pairing between adenine residues as found in certain crystal structures. [After D. Voet and A. Rich, *Progr. Nucleic Acid Res. Mol. Biol.* 10:183 (1970.]

When monomeric adenine and uracil derivatives are mixed and cocrystallized, A–U pairs are observed to form, but in a crystal *the Watson–Crick type of bonding between normal A and U monomeric derivatives has never been observed.* The kinds of complexes that have been found are illustrated in Figure 6-7. Figure 6-7a shows the Watson–Crick bonding scheme between deoxyadenosine and thymidine. Figure 6-7b shows the bonding scheme observed by K. Hoogsteen (1963) for the complex of 9-methyladenine and 1-methylthymine. This complex involves N^7 of the adenine ring (on the imidazole part of the ring) as an H-bond acceptor. This arrangement is known as the *Hoogsteen* or *imidazole structure.* Figure 6-7c illustrates a reverse-imidazole structure, in which uracil O^2 instead of O^4 is the H-bond acceptor. Molecular orbital calculations by B. Pullman and colleagues (1966) suggest that, for adenine–thymine pairs, the decreasing order of stabilities is the following: imidazole structure, reverse-imidazole structure, and Watson–Crick structure.

The situation differs with respect to the G–C pairs that have been observed in crystals. In this case, essentially only Watson–Crick structures are found. This is probably due to the fact that this structure is stabilized by three hydrogen bonds.

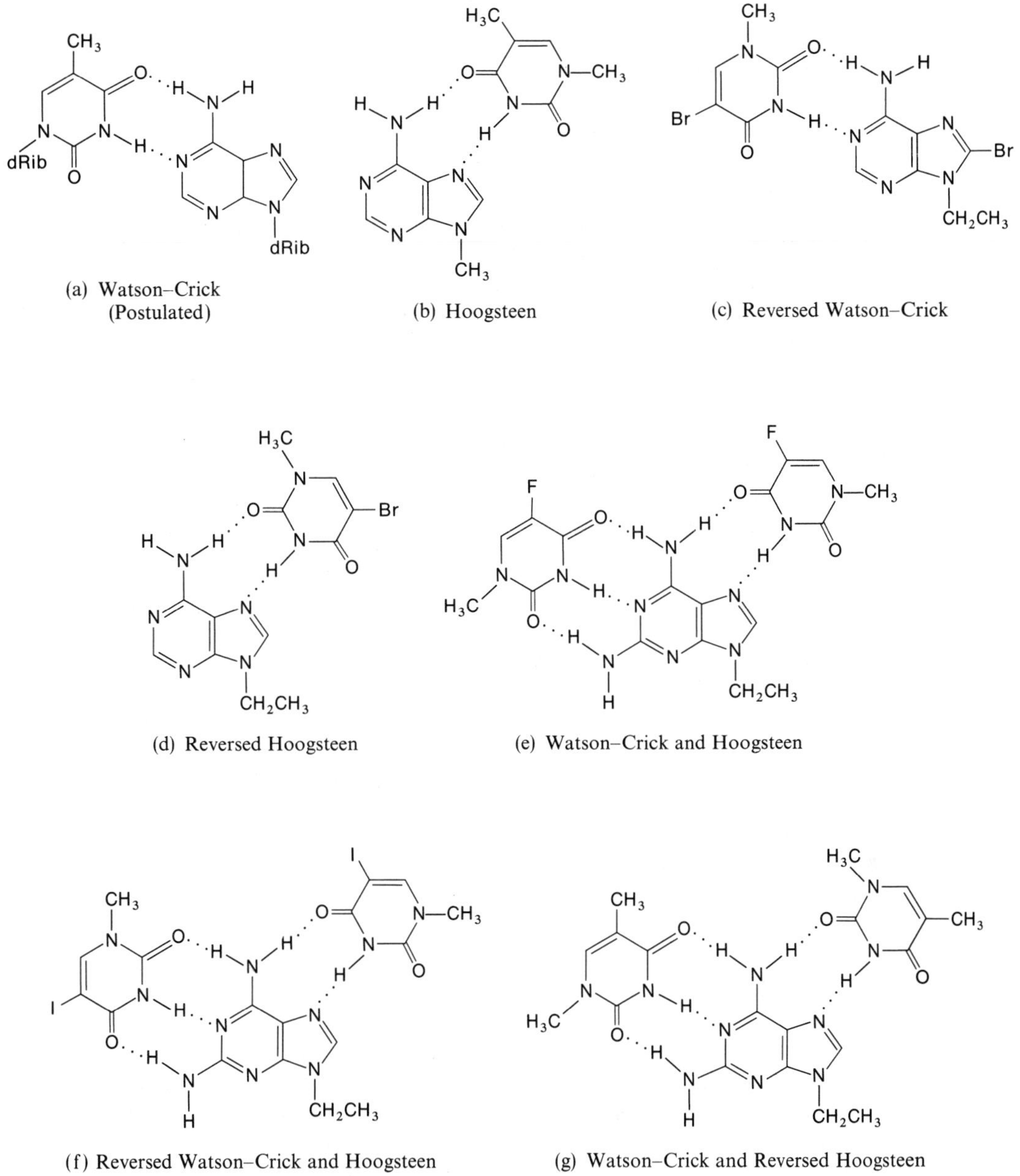

(a) Watson–Crick (Postulated)

(b) Hoogsteen

(c) Reversed Watson–Crick

(d) Reversed Hoogsteen

(e) Watson–Crick and Hoogsteen

(f) Reversed Watson–Crick and Hoogsteen

(g) Watson–Crick and Reversed Hoogsteen

Figure 6-7

Types of hydrogen bonding found between adenine and uracil derivatives.
[After D. Voet and A. Rich, *Progr. Nucleic Acid Res. Mol. Biol.* 10:183 (1970).]

Watson–Crick pairing in the crystal structure of a dinucleotide

Although the Watson–Crick bonding scheme has not been observed with normal A and U monomeric derivatives, the dinucleotide ApU crystallizes as a minihelix in which the Watson–Crick complementary base pairing is found. The self-complementary ApU complex was studied at atomic resolution by Alex Rich and colleagues (1973). The atomic structure is illustrated in the top half of Figure 6-8. It is clear from viewing adjacent ribose units on the ribose-phosphate chain that the complex has a right-handed helical twist. In the bottom panel of the figure, a segment of the RNA 11 helix is given. This structure uses Watson–Crick pairing to generate a helix with 11 base pairs per turn. It is the kind of helix found, for example, with viral RNAs and is clearly similar to the ApU helix. Thus, although A (or U) can pair with itself, and although crystals of mixtures of monomeric A and U derivatives are commonly observed to have A–U complexes stabilized by hydrogen-bonding schemes different than the Watson–Crick arrangement, the Watson–Crick double helix does form with a simple dinucleotide and, as other work suggests, with poly A and poly U. It may be environmental and geometric effects present in oligomers and polymers but not in monomers, that make Watson–Crick pairing the prefered mode of complex formation.

Specific base pairing in solution

We have discussed self-association between bases in the crystalline state, as well as complexes between different bases. These observations illustrate the diversity of interactions that can occur, but they do not give quantitative information on the relative stabilities of the different types of complexes. In particular, it is of interest to determine whether, in a mixture, the bases tend to sort out into specific complexes, or whether a broad distribution of complexes is obtained. This kind of question can be approached by studying base–base interactions with the aid of spectroscopic techniques such as infrared spectroscopy or NMR. Of course, such studies must be done in a solvent that is a relatively weak hydrogen-bond donor or acceptor, in order that the nucleoside association is not obscured.

These kinds of studies have been done by Richard Lord and Alex Rich and their colleagues, among others. Figure 6-9 shows the infrared spectra in the N–H stretching region for guanine, cytosine, and adenine derivatives, as well as for pair-wise mixtures of these bases in deuterochloroform. The solid lines in the left panels give the observed spectra for G and C separately, at 0.0008 M base concentration, and together in an equimolar mixture (giving 0.0016 M total base concentration). If no interaction occurred, then the observed spectrum should be equal to the sum of the two individual spectra, as indicated by the dashed line in the bottom left panel. The observed spectrum is quite different, however, thus providing evidence for an interaction between G and C. On the other hand, the curves in the right-hand panel clearly indicate that,

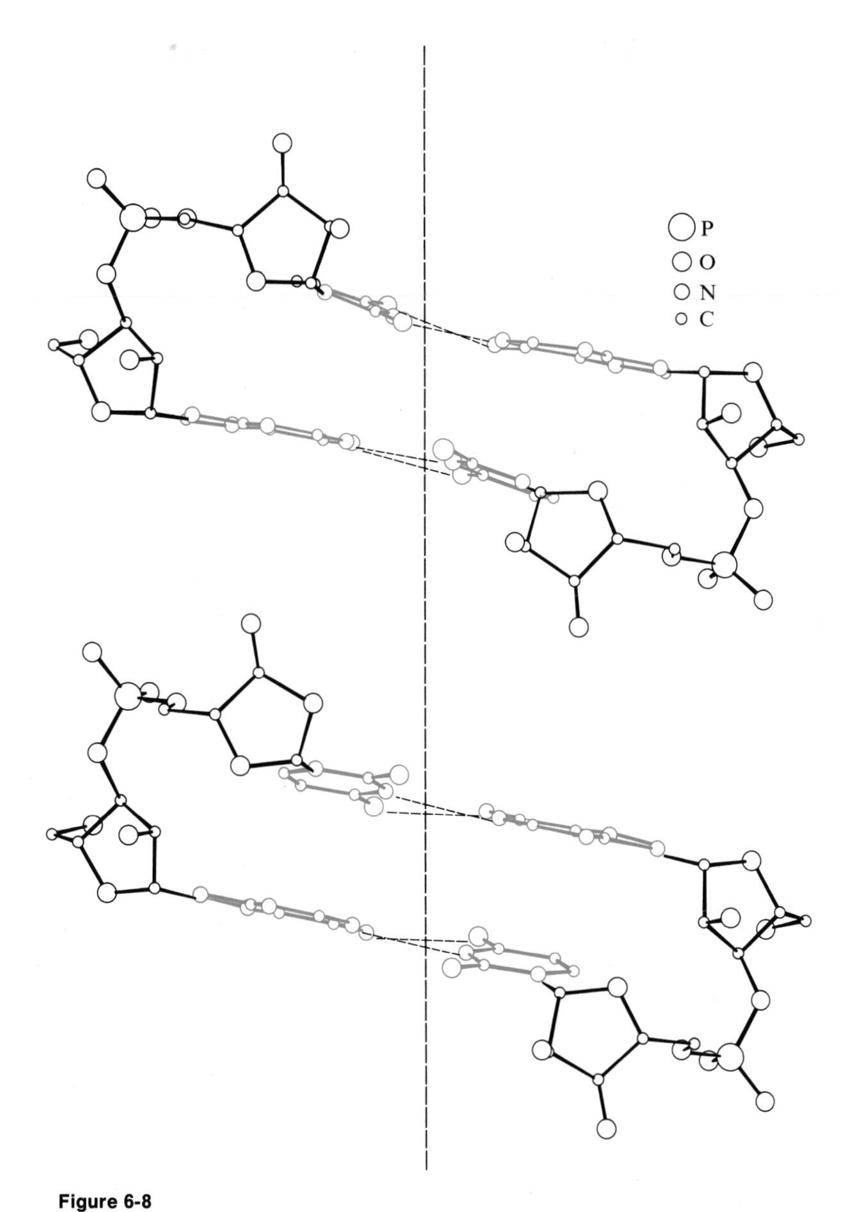

Figure 6-8

Structures of ApU (top) *and RNA* 11 (bottom) *helices*, viewed roughly perpendicular to the helix axis (*vertical dashed line*). [After J. M. Rosenberg et al., *Nature* 243:150 (1973).]

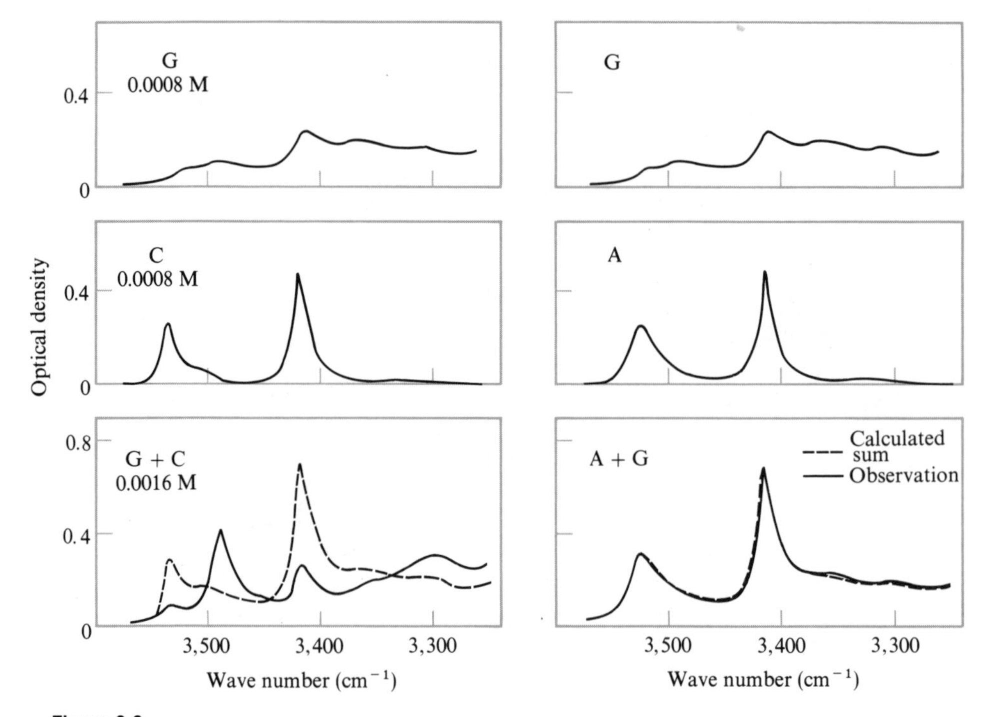

Figure 6-9

Infrared spectra for guanine, cytosine, and adenine derivatives separately and in mixtures. The spectra cover the N–H stretching region. [After R. C. Lord and G. J. Thomas, Jr., *Devel. Appl. Spectroscopy* 6:179 (1968).]

when the same kind of experiment is repeated with G and A, no evidence for interaction is seen.

Table 6-2 summarizes data from investigations analogous to that shown in Figure 6-9, in which pair-wise interactions of the four major bases, as well as the minor base inosine (I), were tested. The results are reported as qualitative observations of infrared spectra in mixtures containing, in most cases, 0.004 M of each base. These qualitative results indicate a remarkable specificity in the base-pairing reaction, with many presumably possible complexes showing no evidence of occurring. The pattern of specific pairing of A with T (or U) and G with C is observed, however. In addition, a strong I–C association is found, as well as a somewhat weaker A–I complex.

In the case of the A–U pairing, other data indicate that both Watson–Crick and imidazole (and/or reverse-imidazole) structures are formed. However, regardless of the details of the structures formed, it is clear that the base–base interaction specificity observed for monomers in solution is exactly that postulated by Watson and Crick for the DNA double helix.

Table 6-3 gives some thermodynamic data on the association of the bases at 25°C in deuterochloroform. It is seen, for example, that the association constants

Table 6-2
Qualitative specificity of base pairing measured by infrared spectroscopy

	Second base				
First base	C	I	G	T	U
A	0	M	0	S	S
U	0	0	0	0	
T	0	0	0		
G	S	0			
I	S				

NOTE: Estimated relative strengths of association are indicated by S = strong, M = medium, 0 = no reaction.
SOURCE: Data from Y. Kyogoku et al., *Biochim. Biophys. Acta* 179:10 (1969).

Table 6-3
Thermodynamic parameters for base-pair formation in deuterochloroform at 25°C

Base pair	K (M^{-1})	$-\Delta H$ (kcal mole^{-1})	$-\Delta S$ (cal deg^{-1} mole^{-1})
U—U	6.1	4.3	11.0
A—A	3.1	4.0	11.4
A—U	100	6.2	11.8
C—C	28	6.3	15
G—G	10^3–10^4	(8.5–10)[§]	(15)[#]
G—C	10^4–10^5	(10–11.5)[§]	(15)[#]

§ Value obtained by assuming $\Delta S = -15$ cal deg^{-1} mole^{-1}.
# Value assumed.
SOURCE: Data from Y. Kyogoku et al., *Biochim. Biophys. Acta* 179:10 (1969).

for the self-association of a base and for bonding to its Watson–Crick mate differ by anywhere from 15-fold to 1,000-fold or more. This preference for the Watson–Crick partner appears to be an energetic one, as reflected by the significant differences in ΔH for the various base pairs.

In order to gain further insight into the factors that determine the strength and specificity of a given association, we can investigate a series of base derivatives that are altered in a variety of minor ways from the parent species. A series of such derivatives for the A–U association is shown in Figure 6-10. Association constants are written over arrows for the interaction of 9-ethyladenine with 1-cyclohexyluracil and their various derivatives. Self-association constants are indicated next to each compound. These data provide important clues to the details of the pairing reaction. For example, the inability of 3-methyluracil to associate with adenine is apparently because a cyclic dimer cannot be made. The enhanced association of 5-bromouracil with adenine is probably due to the greatly increased NH acidity caused by the bromine atom's electron-withdrawing power; the increased NH acidity must outweigh the countereffects of the simultaneously decreased oxygen basicity. (The increased NH acidity implies that the hydrogen is less strongly attached to the nitrogen and hence more able to interact with another electronegative group. The decreased oxygen basicity means that it has a lower affinity for a hydrogen ion.) Can you rationalize some of the other results shown?

At this point we should emphasize that it is not entirely clear why the specificity is so pronounced at the mononucleoside level. Thus, it appears that, although many base pairs show a geometric complementarity, this bias is not sufficient to ensure that a strong complex will form. It must be argued, therefore, that the specific A–U and G–C pairs possess an "electronic" complementarity in addition to a geometric one, and that a quantum mechanical explanation doubtlessly is required. The reason that

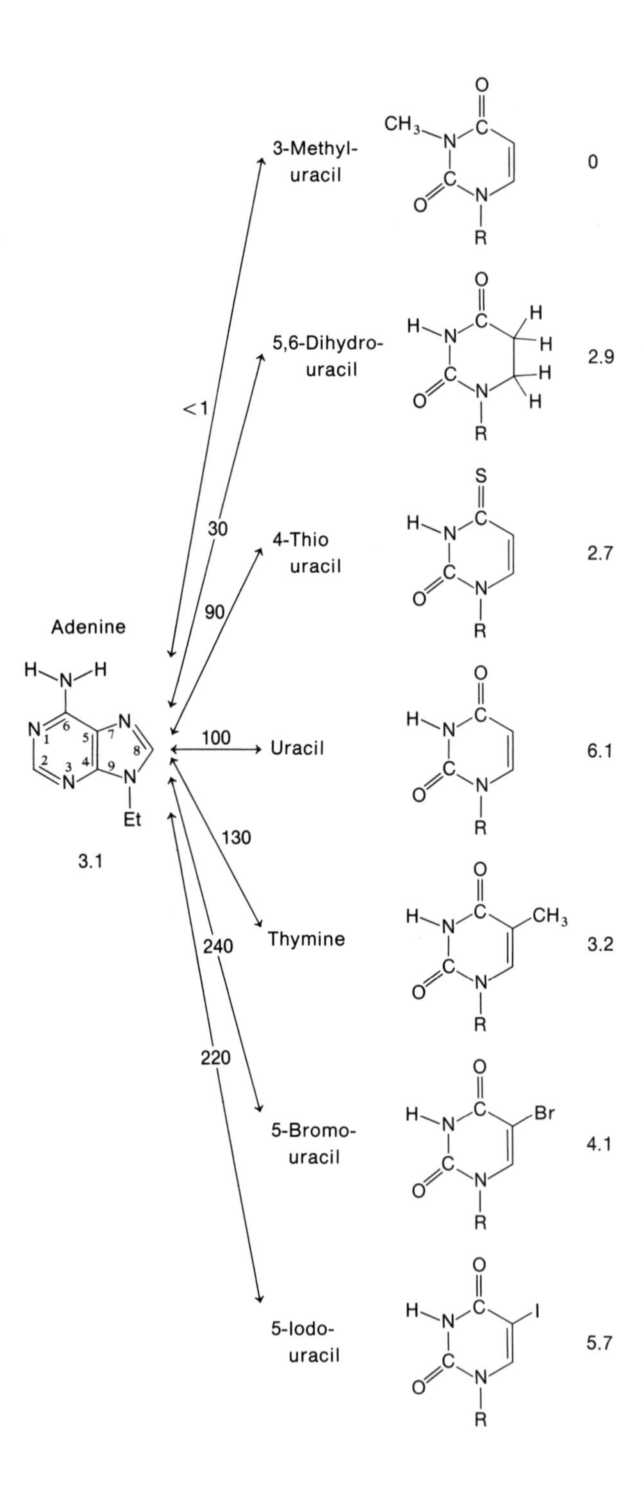
Adenine
3.1
3-Methyl-
uracil
0
5,6-Dihydro-
uracil
2.9
4-Thio
uracil
2.7
Uracil
6.1
Thymine
3.2
5-Bromo-
uracil
4.1
5-Iodo-
uracil
5.7
<1
30
90
100
130
240
220

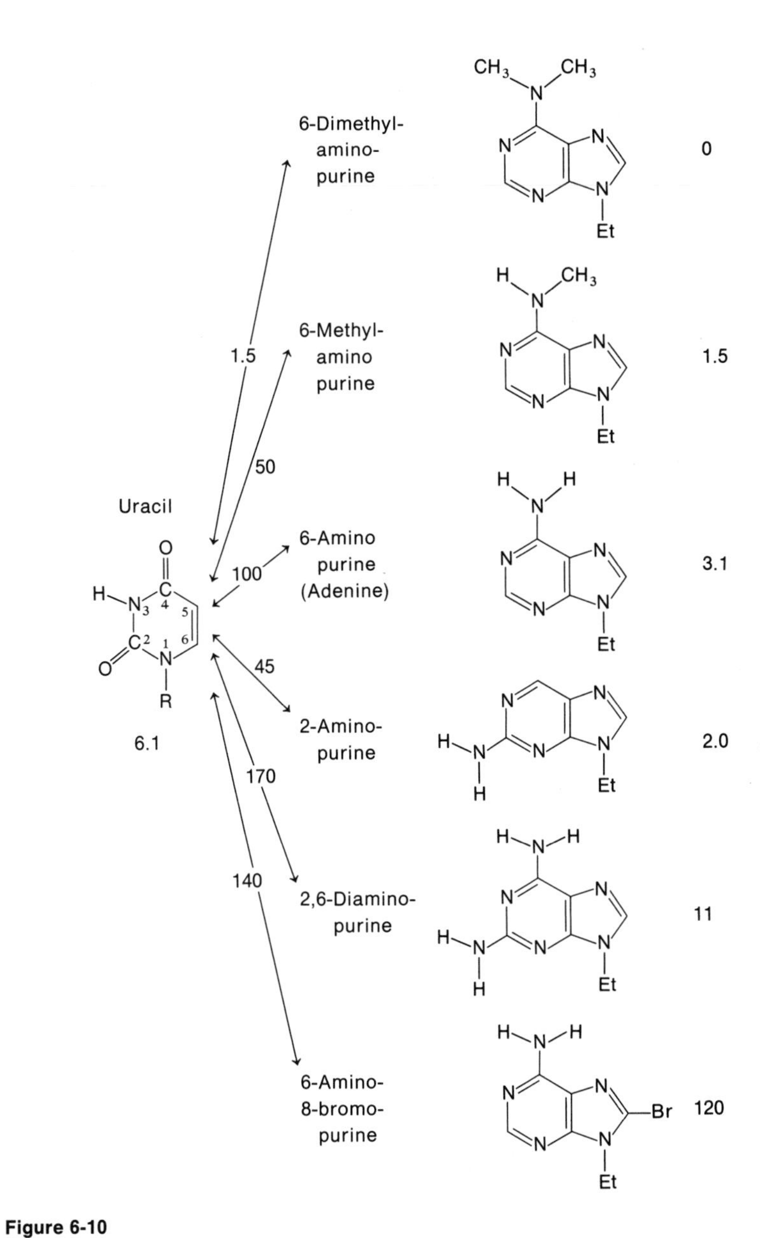

Figure 6-10

Interactions of 9-ethyladenine and 1-cyclohexyluracil derivatives. Association constants (M^{-1}) in $CDCl_3$ at 25°C are given over arrows; self-association constants are shown next to each compound. [After R. C. Lord and G. J. Thomas, Jr., *Devel. Appl. Spectroscopy* 6:179 (1968).]

such an explanation has not yet come forth is due to the approximations that enter into any quantum treatment; these approximations make it impossible to account for the relatively small difference in energy between a "specific" and a "nonspecific" complex.

The results of these experiments suggest that A–U and G–C specificity of base pairing is an inherent property of the monomeric bases themselves, a property we ascribe to an electronic complementarity. However, a pair such as A–U can form by more than one hydrogen-bonding scheme (imidazole structure, reverse-imidazole structure, or Watson–Crick structure), and the formation of a double helix with the geometries of the Watson–Crick structure must be due to geometric and environmental factors. (For some novel base-pairing interactions, see Box 6-1.)

6-7 BASE STACKING

Examination of the crystal structures of helical nucleic acids reveals that the planes of successive base moieties are virtually parallel and overlap considerably, the dihedral angle between successive planes being less than 20°. (These features are apparent in the double helical structure of DNA shown in Figure 3-12.) Is this stacked arrangement of bases a consequence of the geometric restrictions imposed on the backbone rotation angles by the specific geometry required by the Watson–Crick base pairing, or is this arrangement energetically favorable in and of itself and thus, in double helices, serves to stabilize further the Watson–Crick geometry?

Stacking of bases in single-stranded poly A in solution

The conditions for base stacking have been extensively investigated by examining the solution properties of single-stranded polynucleotides, especially polyadenylic acid. Particularly useful in this regard are the optical properties. Figure 6-11 compares the absorption spectrum of AMP with that of a solution of poly A containing an equivalent amount of nucleotide residues. The integrated absorption of poly A is clearly less than that of its constituent monomers, an effect known as hypochromism. (See Chapter 7 for further discussion of optical phenomena.) In addition, the wavelength of maximum absorption for poly A at neutral pH is 3 nm lower than that of AMP. This sharp difference in absorption gradually diminishes, however, as the temperature is raised.

Figure 6-12 shows some data obtained by M. Leng and G. Felsenfeld (1966). The graph plots the absorption at about 260 nm against temperature for oligomers of A and for poly A. It is clear that the optical density passes between two extremes—a hypochromic form at very low temperatures ($\sim -20°$), and a form at high temperatures ($\sim 90°$) that shows the absorption characteristics of AMP. Over the same temperature range, the absorption of AMP barely changes.

Box 6-1 NOVEL BASE-PAIRING REACTIONS

The barbiturates provide an interesting side topic here; they are all similar to uracil. These drugs are popular sedatives that also have been shown to cause a reduction in oxygen consumption in mammals, and inhibition of respiration in mitochondria. The structures of barbital, phenobarbital, thiopental, and uracil are shown here.

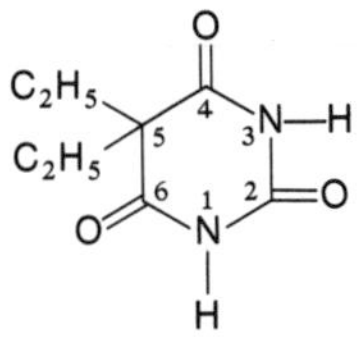

Barbital
(1,000 M^{-1})

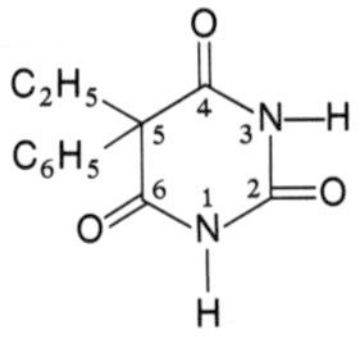

Phenobarbital
(1,200 M^{-1})

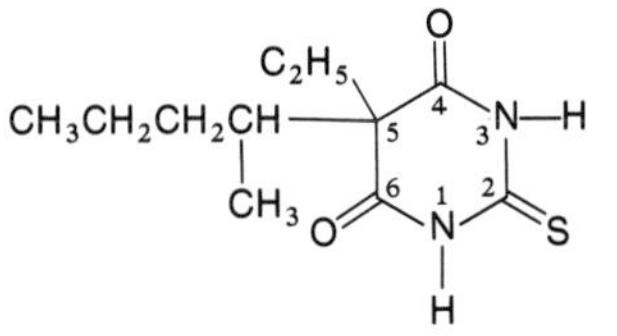

Thiopental
(600 M^{-1})

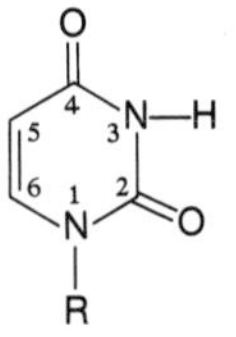

Uracil
(100 M^{-1})

The chief difference among the various barbiturates lies in the side chains attached at position 5, and the occurrence of oxygen or sulfur at position 2. Because of their similarity to uracil, the barbiturates might be expected to form specific complexes with adenine, and this is indeed the case. Numbers in parentheses beneath each structure indicate the association constants for reaction with A in deuterochloroform (Kyogoku et al., 1968). No significant complexation occurs with C, G, or U, and the self-association constants all fall within the range of 1 to 10 M^{-1}.

The interaction between A and the barbiturates is considerably stronger than that between A and U. Two factors are responsible for this. First, the barbiturate NH is far more acidic: pK = 9.4 for uridine, compared to pK = 7.8 for barbital, 7.3 for phenobarbital, and 7.4 for thiopental. Second, there are twice as many NH and C=O groupings in the barbiturates as in uracil, so that a purely statistical effect enhances the association of barbiturates with A by a factor of two. Note that the association of adenine with thiopental falls below that of phenobarbital and barbital, presumably because the electronegativity of sulfur is less than that of oxygen.

The drugs are believed to be active in nonpolar environments, possibly in membranes. Thus the results obtained in chloroform may not be an altogether inaccurate representation of a physiological situation in which the drugs interact with nucleotide bases. Of course, whether or not these data have any bearing on the physiological role of barbiturates is an open question, but the results do indicate how physical studies into the mechanism of a chemical process can lead one into a study of the mechanism of a physiological phenomenon.

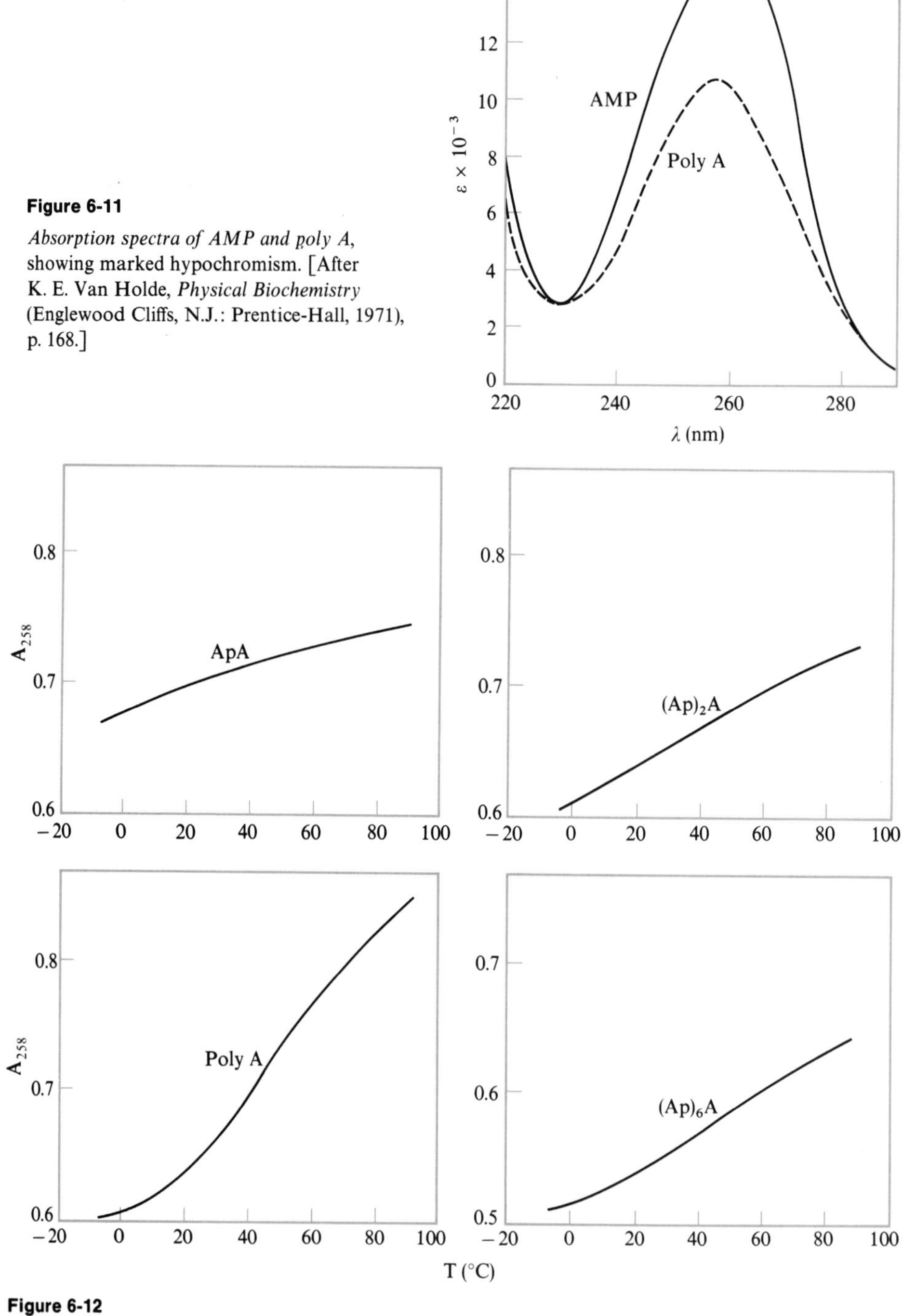

Figure 6-11

Absorption spectra of AMP and poly A, showing marked hypochromism. [After K. E. Van Holde, *Physical Biochemistry* (Englewood Cliffs, N.J.: Prentice-Hall, 1971), p. 168.]

Figure 6-12

Absorbance at 258 nm plotted against temperature for poly A and various oligomers. [After M. Leng and G. Felsenfeld, *J. Mol. Biol.* 15:455 (1966).]

When other spectroscopic properties (such as circular dichroism) are monitored, analogous results are found—that is, poly A undergoes temperature-dependent changes not observed in AMP. The interpretation of all of these data is that a fraction of the bases in the chain are stacked upon each other, the degree of stacking gradually decreasing with increasing temperature. (See Chapter 7 for a discussion of the physical basis for hypochromism in stacked structures.)

The data in Figure 6-12 are amenable to a particularly straightforward interpretation. We can assume a simple two-state equilibrium in which the chain of N nucleotide residues behaves as an array of N-1 dinucleotide units that stack independently of each other. If K is the equilibrium constant for the reaction and f_μ is the fraction of units that are unstacked, then

$$K = f_u/(1 - f_u) \tag{6-1a}$$

and

$$f_u = (A_T - A_{min})/(A_{max} - A_{min}) \tag{6-1b}$$

where A_{max} is the absorbance at the high-temperature limit, A_{min} at the low-temperature limit, and A_T at the temperature T. The K values so obtained may be plotted according to the usual van't Hoff analysis of ln K versus $1/T$ in order to determine the thermodynamic properties (Fig. 6-13). This graph shows the excellent linear dependence of ln K on reciprocal temperature. From these data are obtained values of $\Delta H^0 = 13$ kcal per mole (of stacks) and $\Delta S^0 = 40$ e.u. per mole for the unstacking reaction. (However, it should be pointed out that significantly different thermodynamic parameters have been obtained recently from a calorimetric analysis; see Box 6-2.) Roughly similar values are found for chains of much shorter length, thus further substantiating the hypothesis that each stack forms semiindependently.

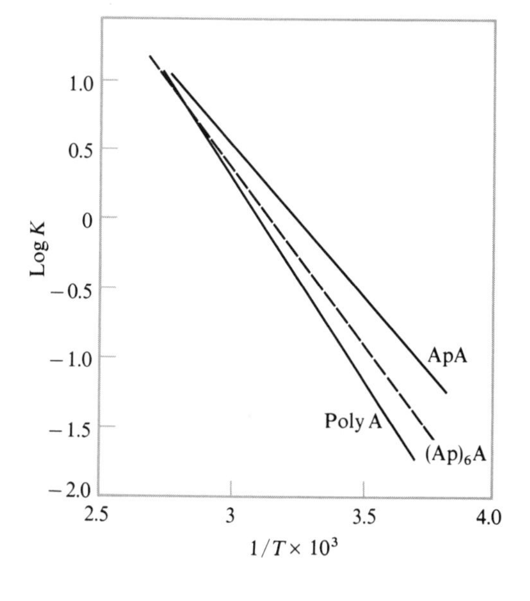

Figure 6-13

Plot of log K *against* $1/T$ *for adenylate oligomers and polymers.* [After M. Leng and G. Felsenfeld, *J. Mol. Biol.* 15:455 (1966).]

Box 6-2 DISCREPANT EVIDENCE ON THERMODYNAMICS OF UNSTACKING

K. J. Breslauer and J. M. Sturtevant (1977) have carried out differential scanning calorimetry on adenylate heptamer. They obtain $\Delta H = 3.4$ kcal for the breaking of one mole of stacks. They also note that van't Hoff enthalpies derived from melting curves obtained by optical techniques show a spread of about 5 to 13 kcal (mole of stacks)$^{-1}$. The discrepancies may be due to difficulties in obtaining accurate base lines in the optical measurements, but clearly more investigation is required. (See Box 21-1 for a discussion of calorimetry.)

Stacking of mononucleosides in solution

The propensity of bases to stack is not confined to those that are nearest neighbors in a polynucleotide chain. Considerable evidence has been marshalled to show that mononucleosides tend to aggregate in aqueous solution. Both infrared and NMR spectra indicate that this association is not by hydrogen bonds. Furthermore, the resonances of specific hydrogens in the complexes are all shifted upfield, as expected for ring-current effects on the positions of these resonances. (See Chapter 9 for a discussion of NMR.) Thus, the rather strong tendency of the units to stack in poly-adenylic acid is an intrinsic property of the bases themselves.

One of the most useful ways to explore the thermodynamic properties of the monomer-stacking reaction is by means of classical vapor pressure measurements on dilute solutions of mononucleosides. Assuming ideal solution behavior at the low concentrations employed, the lowering of the solvent vapor pressure caused by introduction of the base is measured. The fractional lowering is equal to the mole fraction of all species in the solution according to Raoult's law:

$$(P_0 - P_v)/P_0 = \chi \tag{6-2}$$

where P_0 and P_v are solvent vapor pressures in the absence and presence of solute, respectively, and χ is the mole fraction of all solute species in solution. Thus, if n_0 = number of moles of solvent, n_1 = number of moles of solute monomer, n_2 = number of moles of solute dimer (stacked pair), n_3 = number of moles of solute trimer (three molecules stacked together), and so on, then

$$\chi = \left(\sum_{i \geqslant 1} n_i\right) \Big/ \sum_{i \geqslant 0} n_i \tag{6-3a}$$

$$\cong \left(\sum_{i \geqslant 1} n_i \right) \Big/ n_0 \tag{6-3b}$$

because $n_0 \gg \sum_{i \geqslant 1} n_i$ for dilute solutions. The experimentally determined mole fraction χ then can be used to calculate the total molality of all solute species or, equivalently, the colligative molality

$$m = \sum_{i \geqslant 1} m_i \tag{6-4a}$$

where m_i is the molality of the ith species.

The parameter m is useful when it is compared with the known total molality m' of all monomer units in solution, irrespective of their states of association:

$$m' = \sum_{i \geqslant 1} i m_i \tag{6-4b}$$

The ratio m/m' is known as the practical osmotic coefficient ϕ_p. Clearly, $\phi_p \leqslant 1$ and, the smaller it becomes, the greater the degree of association of monomer species into aggregates. (See Chapter 25 for a discussion of osmotic pressure.)

It is a simple matter to relate the practical osmotic coefficient to a model for the association of monomer base units B. We follow the treatment of T. N. Solie and J. A. Schellman (1968). For the indefinite association of monomer B_1, we have

$$\begin{array}{ll} 2\,B_1 \rightleftarrows B_2, & K_2 \\ B_1 + B_2 \rightleftarrows B_3, & K_3 \\ B_1 + B_3 \rightleftarrows B_4, & K_4 \\ \vdots \quad \vdots \quad \vdots & \vdots \end{array} \tag{6-5}$$

where K_i is the association constant for the formation of B_i. For the case where only dimerization occurs, it is easy to show the relationship

$$(1 - \phi_p)/(2\phi_p - 1)^2 = K_2 m' \tag{6-6}$$

Similarly, for the case where indefinite association occurs but every K_i is the same (all $K_i = K$), we have

$$(1 - \phi_p)/\phi_p^2 = Km' \tag{6-7}$$

(See Problem 6-1.) Equations 6-6 and 6-7 provide straightforward ways to use the practical osmotic coefficients in treating models for the association. Figure 6-14a plots $(1 - \phi_p)/(2\phi_p - 1)^2$ against m' for deoxyadenosine. It is clear that marked curvature sets in by $m' = 0.03$, thus indicating the existence of higher-order aggregates. Of course, the limiting slope as $m' \to 0$ gives the association constant for dimerization.

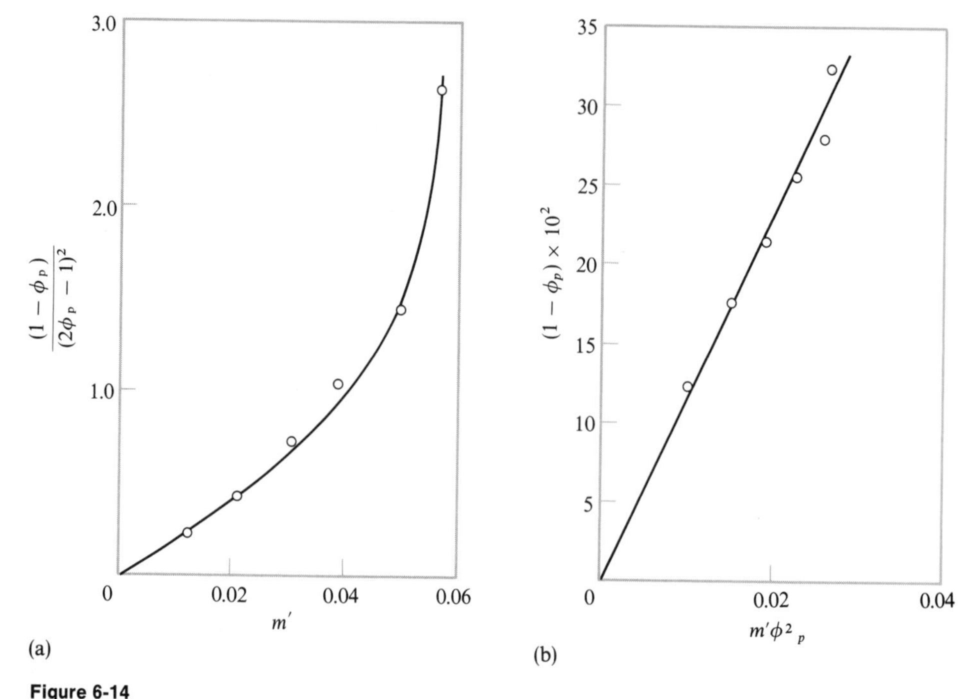

Figure 6-14

Osmotic plots for deoxyadenosine. **(a)** Plot of $(1-\phi_p)/(2\phi_p-1)^2$ versus m'. **(b)** Plot of $(1-\phi_p)$ versus $m'\phi_p{}^2$. [After T. N. Solie and J. A. Schellman, *J. Mol. Biol.* 33:61 (1968).]

When the data are replotted as $(1-\phi_p)$ versus $m'\phi_p^2$ (Eqn. 6-7), the resulting plot is quite linear (Fig. 6-14b). This suggests that the formation of each higher aggregate occurs with a fairly similar association constant.

Stacking preference of purines over pyrimidines

Table 6-4 tabulates some association constants obtained by the methods just described. In general, the association constants are small and tend to indicate that the strength of interaction is proportional to the size of the rings of the stacking partners. The stability constants show the progression purine–purine > purine–pyrimidine > pyrimidine–pyrimidine.

The free energies given in Table 6-4 cannot be compared directly with those for bases in a polynucleotide chain because the former ones contain the extra RT ln 55.6 contribution discussed earlier (Chapter 5). Thus, a more appropriate comparison could be made by adding $-RT \ln 55.6 = -2.4$ kcal mole^{-1} to each of the values

Table 6-4

Association constants and free energies for presumed stacking of bases in aqueous solution at 25°C

Base	K (molal^{-1})	ΔG (kcal mole^{-1})
dA	12	-1.50
Purine riboside	3.5	-0.74
I	3.0	-0.64
dT	0.91	$+0.06$
dC	0.91	$+0.06$
U	0.70	$+0.21$
dA–dT	3 to 6	-0.70 to -1.10
dG–dC	4 to 8	-0.80 to -1.20
dA–dC	3 to 6	-0.70 to -1.10
dA–dU	3	-0.70
dT–dC	0.91	$+0.06$

SOURCE: Data from T. N. Solie and J. A. Schellman, *J. Mol. Biol.* 33:61 (1968).

in the table. In the case of deoxyadenosine, therefore, a value of -3.9 kcal mole^{-1} at 25°C is obtained; the value of ΔG^0 for stacking of adenine bases in poly A is -1 kcal mole^{-1} at 25°C (see Fig. 6-13) under somewhat different conditions. Can you suggest reasons why the two values are not the same?

The role of solvent in stabilizing stacks

What provides the driving force for the stacking reaction? Clearly, one possibility is the induced dipole interactions between the pi electron clouds of the stacked bases. Another important clue to a key source of stabilization is that the stacked conformation is most favored in aqueous solvents. This is suggested, for example, from studies of hypochromicity changes in DNA in various solvents. However, a perhaps even more direct demonstration of this fact comes from studies of photochemically induced thymidine dimer formation in the dinucleotide thymidyl–thymidine, T_PT. It is well known that irradiation of DNA at 280 nm can cause covalent linking of adjacent thymidine bases in the same strand. The accumulated evidence suggests that the bases must be in the stacked conformation in order for dimerization to occur. When T_PT is irradiated with a fixed amount of radiation, it is found that the amount of dimer formed is very sensitive to the solvent employed. The greatest amount of dimer T_D is formed in water, with various lesser amounts in other solvents. The amount of dimer formed in these other solvents is proportional to the estimated ability of T_PT to stack in these media. Table 6-5 lists some solvents in increasing order of their

Table 6-5
Formation of thymine dimer in various solvents

Solvent§	Percentage T_D formed
t-Butanol	5
Ethanol	5.5
Methanol	6
n-Butanol	7.5
Ethylene glycol	19
Formamide	21
Glycerol	33
Water	35

Listed in increasing order of ability to promote stacking of T_pT.
SOURCE: Data from W. Wacker, *Angew Chem. Internat.* 4:150 (1965).

ability to induce stacking. It is seen that this order perfectly corresponds to the percentage of T_D formed in these various solvents.

These studies indicate the important role of the aqueous medium in biasing nucleic acids toward ordered conformations. This role is similar to that of the same medium in determining protein structure, where it is evident that hydrophobic interactions play a crucial part in determining ordered conformations. Thus, the seemingly innocuous water molecule is of great biological importance.

Water clearly plays a role in the stabilization of stacks. But we cannot unequivocally say that hydrophobic interactions along the lines of those for apolar molecule interactions in water (Chapter 5) are the primary source of stabilization. These interactions are endothermic and, consequently, are entropically driven. Base stacking is exothermic. To gain further insight into forces stabilizing stacks, D. M. Crothers and D. I. Ratner (1968) characterized the thermodynamics of stacking in solution between deoxyguanosine and actinomycin. The structure of actinomycin is shown in Figure 23-24; it has a phenoxazone ring, which can stack with the guanine ring, as crystallographic data from Henry Sobell have shown. Crothers and Ratner found that, in aqueous phosphate buffer at 25°C, complex formation has $\Delta H^0 = -10.3$ kcal mole^{-1} and $\Delta S^0 = -19.8$ cal mole^{-1} deg^{-1} but, when methanol is added to the aqueous buffer, both ΔH^0 and ΔS^0 become more negative. For example, with 25% methanol, $\Delta H^0 = -14.05$ kcal mol^{-1} and $\Delta S^0 = -33.7$ cal mol^{-1} deg.$^{-1}$ Therefore, in the pure aqueous solution, there is hidden a positive entropy and a positive enthalpy of complex formation, which are lost as methanol is added. Quite possibly this hidden contribution to the enthalpy and entropy arises from hydrophobic effects involving reordering of water molecules (see Chapter 5). Another view of stacking interactions is given in Box 6-3.

Box 6-3 SURFACE TENSION EFFECT

O. Sinanoğlu and S. Abdulnur (1964) developed an interesting approach to the energetics of base stacking. They put forth the rather ingenious notion that the surface tension, among other factors, could be used to compute the work needed to convert solvated, disordered bases into a helix that excludes solvent from between the bases. Thus, the problem reduces to computing the work needed to create cavities in solution for individual solvated bases and for the stacked bases in a helix. The differential work involved in creating a cavity of area dA is merely $dw = \gamma dA$, where γ is the surface tension. Obviously, it takes more work to create separate cavities for each individual base than to create a single larger hole that will contain all the bases with solvent excluded from between them. Hence, the tendency (as far as solvent–solvent interactions are concerned) will be to "squeeze" the bases into a single cavity.

But why should water show such a strong relative tendency to promote stacking? A possible answer lies in the fact that water has a relatively high surface tension: about 72 dynes cm^{-2} as compared, for example, to ethanol with about 22 dynes cm^{-2} at room temperature. Thus the difference in energy between creating the individual cavities and creating the single larger cavity is greater for water than for the other solvents. Other solvents known to denature DNA (methanol, glycol, formamide, glycerol, ethanol, *n*-propanol, and *n*-butanol) are all significantly inferior to water in their ability to promote helix formation by the surface tension effect. However, these solvents also have lower dielectric constants and therefore increase the electrostatic destabilization of a nucleic acid helix.

6-8 TERTIARY STRUCTURE IN NUCLEIC ACIDS

Our analysis thus far has given us a good picture of the geometric constraints and forces that play a role in determining the secondary structure of nucleic acids. One might conclude from this analysis that the only significant structures of nucleic acids are random coils and the linear single-stranded and double-stranded helical forms. This is a false impression, however, fostered in part by too heavy an emphasis given to model systems, with the assumption that tertiary interactions are primarily restricted to proteins.

Transfer RNAs are well studied examples of nucleic acids having considerable tertiary structure. These single-stranded polymers of usually 73 to 93 nucleotides can all be folded into the basic cloverleaf pairing arrangement of secondary structure (Figure 3-6). The transfer RNAs (tRNAs) are a family of similar chains, each of which is specific for a particular amino acid. This specificity comes about because the 3′-terminal ribose of each tRNA can be esterified with one particular amino acid by means of a specific enzyme. The amino acid that can be esterified to a given tRNA corresponds to a triplet sequence of bases on the tRNA—the anticodon—which is the complement to the codon. The insertion point of a given amino acid in a growing polypeptide chain is determined by the interaction between the anticodon of the tRNA and the codon of the messenger RNA.

There are many data that corroborate a tRNA secondary structure of cloverleaf hydrogen bonding with intervening loops. Thus, the strong tendency to form paired and stacked helical segments gives a strong bias to the structure adopted. However, there is a great body of solution data that clearly indicate that a considerable degree of tertiary folding also exists. The nature of this folding has been greatly clarified by the three-dimensional crystallographic structural determination of yeast phenylalanine-specific tRNA by A. Rich, S. H. Kim, A. Klug, and colleagues.

Figure 3-17 is a schematic diagram of the folding of the cloverleaf into a tertiary structure. The molecule is L-shaped with the helical stems of the TψC loop and the 3′-acceptor end forming one branch of the L, and the helical stems of the anticodon loop and dihydrouridine loop forming the other branch. The TψC and dihydrouridine loops themselves form the corner of the L. The two ends of the L are the anticodon loop and the 3′-acceptor stem, which are separated by about 75 Å. From these drawings, it is clear that the anticodon loop is on the "outside" of the molecule, where it can readily participate in base pairing. Although the existence of this conformation in solution has not been "proven," most data are consistent with it.

The tertiary structure of tRNA is stabilized in part by an interesting and previously unknown set of hydrogen-bonding interactions involving triple base pairs (Fig. 3-18). These unusual interactions enable different segments of the chain to cross-bond, and thus to stabilize a more compact configuration. It is also clear that multivalent cations (such as Mg^{2+}, polyamines) play a role in stabilizing the structure, although the specific details of these interactions are not completely elucidated.

The results with transfer RNA show that highly organized three-dimensional conformations can be adopted by nucleic acids. It is almost certain that much larger RNAs, such as ribosomal RNAs, also have complex three-dimensional structures. Complex forms of DNA also exist. At this time, our understanding of the tertiary folding of nucleic acids other than transfer RNA is primitive, and the problem of predicting a unique three-dimensional structure from the nucleotide sequence is just as formidable (if not more so) as the analogous problem for proteins. (See Chapter 24 for further discussion of nucleic acid tertiary structure.) Nevertheless, it is clear that geometric and steric constraints, hydrogen bonds, hydrophobic interactions, and electrostatic forces are generally the repetoire of interactions with which a polynucleotide, like a protein, builds up its tertiary structure.

Summary

The backbone conformation of a polynucleotide chain is determined by six internal rotation angles associated with each repeating unit. The large number of variables complicates the sterochemical analysis of chain conformations but, through systematic procedures, considerable simplification is possible, and it can be shown that steric effects limit significantly the available conformations.

Two key stabilizing forces in nucleic acids are derived from base-pairing and base-stacking interactions. Many different kinds of base pairs are possible, and some of these have been observed in crystallographic studies of monomers. In the case of monomeric adenine and uracil derivatives, the formation of base pairs of the design specified by Watson and Crick (for base pairing in double-helical DNA) has never been observed in crystallographic studies, although alternate pairing schemes have frequently been noted. However, the Watson–Crick helix has been observed for the self-complementary ApU dinucleotide studied in the crystal state at atomic resolution. Base pairing between monomers in solution has also been studied; preferential pairing of A with U and of G with C has been found, although more than one bonding scheme probably operates in the formation of the A–U specific complexes. The combined evidence suggests that Watson–Crick helical polymers form because of "electronic" complementarity in A—U and G—C pairs and because of geometric constraints in double-helical polymers.

Base-stacking interaction occurs in aqueous solution and has been observed in single-stranded polynucleotides, such as poly A, and between free monomeric nucleosides. In general, purines stack more readily than do pyrimidines. The interaction is associated with a negative ΔH, unlike hydrophobic interactions, although other data suggest that hydrophobic effects play a significant role in stabilizing base stacks.

Nucleic acids also form specific tertiary structures, as exemplified by transfer RNA. In transfer RNA, unusual base pairing schemes and other interactions make possible a compact, specific structure.

Problems

6-1. Using the definition of the practical osmotic coefficient ϕ_p and Equations 6-4 and 6-5, prove the validity of Equations 6-6 and 6-7.

6-2. An investigator studies the vapor pressure of adenosine (A) over three kinds of adenosine solutions. Solution I consists of 8 M urea in water as solvent; solution II consists of CCl_4 as solvent; solution III consists of pure water as solvent. He measures the adenosine vapor pressure in order to ascertain the activity of adenosine (or the free adenosine concentration) in these various solutions. He then compares this with the total amount of adenosine $(A)_0$ that he added to the solution. He does this over a wide range of variations in $(A)_0$ and finds that, in solution I, the relationship between $(A)_0$ and actual free adenosine concentration (A) in solution is

$$(A)_0 = (A)$$

That is, in solution I the free monomer concentration equals the total concentration. For solution II he finds

$$(A)_0 = (A) + C(A)^2$$

where C is a constant, and he cannot improve the fit of the equation to the data by adding any terms for higher powers of (A) than $(A)^2$. For solution III he finds

$$(A)_0 = C_1(A) + C_2(A)^2 + C_3(A)^3 + C_4(A)^4 + \cdots$$

where $C_1 = 1$, C_2 is a constant, and

$$C_k = (k/2^{k-1})C_2^{k-1} \qquad \text{for } k = 3, 4, \ldots$$

From these data alone, deduce the types of specific interactions that probably occur in these three solutions. Back up your suggestions by showing that the equations given here agree exactly with expectations based on your suggestions.

6-3. An investigator studies base stacking of mononucleosides by measuring the parameter ϕ_p at three values of the total mononucleoside concentration. He finds that the value he calculates from these data for K (the stacking equilibrium constant) is almost exactly the same whether he uses a model in which stacking proceeds to the dimer stage only, or whether he assumes an indefinite association model, in which higher-order aggregates are formed (with each equilibrium having the same K). He believes that something is fishy but cannot seem to find the flaw. If there is a flaw, point out why the data are inconsistent. Otherwise, explain the conditions under which you would get almost the same value of K with data at three different concentrations. (NOTE: $K = 0$, a trivial reason, is not allowed as an explanation.)

6-4. Double-stranded nucleic acids undergo a helix–coil transition somewhat analogous to the thermal melting of poly A. However, DNA melting curves are extremely sharp. The midpoints of these cooperative transitions are characterized by a temperature T_m. Based on information contained in this chapter, what factors would you consider to be important influences on the steepness and T_m of a DNA melting transition? Why? Can you think of a physical explanation for the cooperativity of the transition?

6-5. Consider a reagent such as urea. It is known to be a good protein denaturant. What effect, if any, do you expect 1 M to 8 M aqueous urea solutions to have on the stability of ordered single-stranded and ordered double-stranded nucleic acid structures? Why?

References

GENERAL

Olson, W. K., and P. J. Flory. 1972. Steric configurations of polynucleotide chains, I: Steric interactions in polynucleotides: A virtual bond model. *Biopolymers* 11:1. [This and accompanying articles in the same volume provide a critical introduction to polynucleotide chain conformation and stereochemistry.]

Ts'O, P. O. P., ed. 1974. *Basic Principles in Nucleic Acid Chemistry*. 2 vols. New York: Academic Press. [A good summary of fundamental aspects of nucleic acid chemistry and physics, written by experts in the field. Chapter 6 of vol. 1 summarizes intrinsic properties and interactions of monomer units.]

Voet, D., and A. Rich. 1970. The crystal structure of purines, pyrimidines and their intermolecular complexes. *Progr. Nucleic Acid Res. Mol. Biol.* 10:183. [Includes an excellent summary of hydrogen-bonded base complexes studied in the crystal state, and of related issues.]

SPECIFIC

Breslauer, K. J., and J. M. Sturtevant. 1977. Calorimetric investigation of base-stacking in Ribo A_7. *Biophys. Chem.* 7:205.

Crothers, D. M., and D. I. Ratner. 1968. Thermodynamic studies of a model system for hydrophobic bonding. *Biochemistry* 7:1823.

Hoogsteen, K. 1959. The structure of crystals containing a hydrogen-bonded complex of 1-methylthymine and 9-methyladenine. *Acta Cryst.* 12:822.

———. 1963. The crystal and molecular structure of a hydrogen-bonded complex between 1-methylthymine and 9-methyladenine. *Acta Cryst.* 16:907.

Kim, S. H., G. J. Quigley, F. L. Suddath, A. McPherson, D. Sneden, J. J. Kim, J. Weinzierl, and A. Rich. 1973. Three-dimensional structure of yeast phenylalanine transfer RNA: Folding of the polynucleotide chain. *Science* 179:285.

Kim, S. H., F. L. Suddath, G. J. Quigley, A. McPherson, J. L. Sussman, A. H. J. Wang, N. C. Seeman, and A. Rich. 1974. Three-dimensional tertiary structure of yeast phenylalanine transfer RNA. *Science* 185:435.

Kyogoku, Y., R. C. Lord, and A. Rich. 1968. Specific hydrogen bonding of barbiturates to adenine derivatives. *Nature* 218:69.

Leng, M., and G. Felsenfeld. 1966. A study of polyadenylic acid at neutral pH. *J. Mol. Biol.* 15:455.

Pullman, B., P. Claverie, and J. Caillet. 1966. Van der Waals–London interactions and the configuration of hydrogen-bonded purine and pyrimidine pairs. *Proc. Natl. Acad. Sci. USA* 55:904.

Rich, A., and S. H. Kim. 1978. The three-dimensional structure of transfer RNA. *Sci. American* 238(1):52.

Robertus, J. D., J. E. Ladner, J. T. Finch, D. Rhodes, R. S. Brown, B. F. C. Clark, and A. Klug. 1974. Structure of yeast phenylalanine tRNA at 3 Å resolution. *Nature* 250:546.

Rosenberg, J. M., N. C. Seeman, J. J. P. Kim, F. L. Suddath, H. B. Nicholas, and A. Rich. 1973. Double helix at atomic resolution. *Nature* 243:150.

Sinanoğlu, O., and S. Abdulnur. 1964. Hydrophobic stacking of bases and the solvent denaturation of DNA. *Photochem. Photobiol.* 3:333.

Solie, T. N., and J. A. Schellman. 1968. The interaction of nucleosides in aqueous solution. *J. Mol. Biol.* 33:61.

Appendix A
Review of elementary matrix algebra

A matrix is an array (or table) whose elements are numbers or symbols; for example,

$$\begin{pmatrix} 8 & 7 \\ 23 & 28 \end{pmatrix} \qquad \begin{pmatrix} a_{11} & a_{12} \\ a_{21} & a_{22} \end{pmatrix}$$

These are 2×2 matrices (having two rows and two columns); the symbol a_{12} denotes the element in the first row and the second column. In general, a_{ij} represents the element in the ith row and the jth column. For the matrix on the left, $a_{22} = 28$, $a_{21} = 23$, and so on. In this text, a matrix is denoted by a bold-faced letter with a "tilde" underscore. For example, we could represent the matrix at the right above as $\underset{\sim}{\mathbf{A}}$, where

$$\underset{\sim}{\mathbf{A}} = \begin{pmatrix} a_{11} & a_{12} \\ a_{21} & a_{22} \end{pmatrix}$$

A *row* matrix (also called a row vector) contains only one row:

$$\underset{\sim}{\mathbf{A}}_{\mathrm{r}} = (a_{11}, a_{12})$$

This matrix $\underset{\sim}{\mathbf{A}}_{\mathrm{r}}$ is a 1×2 row matrix (with one row and two columns). A *column* matrix (or column vector) contains only one column:

$$\underset{\sim}{\mathbf{A}}_{\mathrm{c}} = \begin{pmatrix} a_{11} \\ a_{21} \end{pmatrix}$$

The matrix $\underset{\sim}{\mathbf{A}}_c$ is a 2×1 column matrix (with two rows and one column). In general, a matrix may be of any size $n \times m$ (with n rows and m columns).

MATRIX MULTIPLICATION

Two matrices may be multiplied together to give a third matrix:

$$\underset{\sim}{\mathbf{a}}\underset{\sim}{\mathbf{b}} = \underset{\sim}{\mathbf{c}}$$

The element c_{ij} in matrix $\underset{\sim}{\mathbf{c}}$ is defined as

$$c_{ij} = \sum_k a_{ik} b_{kj}$$

That is, the elements in the ith *row* of a are multiplied by counterparts in the jth *column* of b to produce c_{ij}. For example, if

$$\underset{\sim}{\mathbf{a}} = (a_{11}, a_{12})$$

and

$$\underset{\sim}{\mathbf{b}} = \begin{pmatrix} b_{11} & b_{12} \\ b_{21} & b_{22} \end{pmatrix}$$

then

$$\begin{aligned} \underset{\sim}{\mathbf{a}}\underset{\sim}{\mathbf{b}} &= (a_{11}, a_{12}) \begin{pmatrix} b_{11} & b_{12} \\ b_{21} & b_{22} \end{pmatrix} \\ &= (a_{11}b_{11} + a_{12}b_{21}, a_{11}b_{12} + a_{12}b_{22}) \end{aligned}$$

Therefore,

$$\underset{\sim}{\mathbf{c}} = (c_{11}, c_{12})$$

where

$$c_{11} = a_{11}b_{11} + a_{12}b_{21}$$

$$c_{12} = a_{11}b_{12} + a_{12}b_{22}$$

It is clear that multiplication is possible only when the number of columns of $\underset{\sim}{\mathbf{a}}$ equals the number of rows of $\underset{\sim}{\mathbf{b}}$. In general, an $n \times m$ matrix can multiply only an $m \times p$ matrix, where n and p are arbitrary; the resulting matrix is $n \times p$. Thus, in the example just given, $n = 1$ and $m = 2$ for $\underset{\sim}{\mathbf{a}}$, and $m = 2$ and $p = 2$ for $\mathbf{b}$; multiplication of $\underset{\sim}{\mathbf{a}}$ times $\underset{\sim}{\mathbf{b}}$ yields $\underset{\sim}{\mathbf{c}}$, which is $n \times p$ or 1×2 (a row matrix).

As another example, let

$$\mathbf{a} = \begin{pmatrix} a_{11} & a_{12} \\ a_{21} & a_{22} \end{pmatrix}$$

$$\mathbf{b} = \begin{pmatrix} b_{11} & b_{12} \\ b_{21} & b_{22} \end{pmatrix}$$

$$\mathbf{a} = \mathbf{ab}$$

$$= \begin{pmatrix} a_{11}b_{11} + a_{12}b_{21} & a_{11}b_{12} + a_{12}b_{22} \\ a_{21}b_{11} + a_{22}b_{21} & a_{21}b_{12} + a_{22}b_{22} \end{pmatrix}$$

Thus, when a 2×2 matrix multiplies a 2×2 matrix, the product also is a 2×2 matrix.

When a 1×2 matrix multiplies a 2×1 matrix, the result is a 1×1 matrix that is just a number (a scalar) and is no longer considered a matrix. For example:

$$\mathbf{a} = (a_{11}, a_{12})$$

$$\mathbf{b} = \begin{pmatrix} b_{11} \\ b_{21} \end{pmatrix}$$

$$\mathbf{ab} = a_{11}b_{11} + a_{12}b_{21}$$

If the matrix is square (number of rows equal to number of columns), then the matrix may be raised to any power whatever. For example, for a square matrix $\mathbf{M}$,

$$\mathbf{M}^3 = \mathbf{MMM}$$

$$\mathbf{M}^N = \prod_{i=1}^{N} \mathbf{M}$$

(which is just $\mathbf{M}$ multiplied by itself N times). This procedure is possible only with a square matrix, because only in that case are the rules of matrix multiplication fulfilled. Note also that, if $\mathbf{M}$ is $n \times n$, then $\mathbf{M}^N$ also must be $n \times n$.

MATRIX INVERSION

For a square matrix $\mathbf{a}$, you can find an inverse matrix $\mathbf{a}^{-1}$ such that

$$\mathbf{a}^{-1}\mathbf{a} = \mathbf{I}$$

where $I_{ij} = 0$ for $i \neq j$, and $I_{ij} = 1$ for $i = j$. The matrix $\mathbf{I}$ is called the unit matrix.

For example, if $\underset{\sim}{\mathbf{a}}$ is a 2×2 matrix, then

$$\underset{\sim}{\mathbf{I}} = \begin{pmatrix} 1 & 0 \\ 0 & 1 \end{pmatrix}$$

Note that $\mathbf{I}$ has the property that any matrix multiplied by it is unchanged by the multiplication. Thus $\underset{\sim}{\mathbf{I}}$ is analogous to the number 1 in scalar algebra, so that $\underset{\sim}{\mathbf{a}}\underset{\sim}{\mathbf{I}} = \underset{\sim}{\mathbf{a}}$. The calculation of $\underset{\sim}{\mathbf{a}}^{-1}$ is relatively straightforward. For example, let

$$\underset{\sim}{\mathbf{a}} = \begin{pmatrix} a_{11} & a_{12} \\ a_{21} & a_{22} \end{pmatrix}$$

Then the rule for calculating $\underset{\sim}{\mathbf{a}}^{-1}$ is

$$\underset{\sim}{\mathbf{a}}^{-1} = \begin{pmatrix} a_{22}/\alpha & -a_{12}/\alpha \\ -a_{21}/\alpha & a_{11}/\alpha \end{pmatrix}$$

where

$$\alpha = a_{11}a_{22} - a_{21}a_{12}$$

which is the value of the determinant of $\underset{\sim}{\mathbf{a}}$. (Clearly, the determinant must be nonzero in order for $\underset{\sim}{\mathbf{a}}^{-1}$ to exist.)

MATRIX DIAGONALIZATION

For a 2×2 matrix $\mathbf{a}$, there is a matrix $\mathbf{T}$ such that

$$\underset{\sim}{\mathbf{T}}^{-1}\underset{\sim}{\mathbf{a}}\underset{\sim}{\mathbf{T}} = \begin{pmatrix} \lambda_1 & 0 \\ 0 & \lambda_2 \end{pmatrix}$$

where λ_1 and λ_2 are the eigenvalues of $\underset{\sim}{\mathbf{a}}$, and $\underset{\sim}{\mathbf{T}}$ is called a transformation matrix. The eigenvalues λ_1 and λ_2 are easy to find. Let

$$\underset{\sim}{\mathbf{a}} = \begin{pmatrix} a_{11} & a_{12} \\ a_{21} & a_{22} \end{pmatrix}$$

Then the eigenvalues of $\underset{\sim}{\mathbf{a}}$ are obtained by solving the determinant

$$\begin{aligned} 0 &= \begin{vmatrix} a_{11} - \lambda & a_{12} \\ a_{21} & a_{22} - \lambda \end{vmatrix} \\ &= (a_{11} - \lambda)(a_{22} - \lambda) - a_{12}a_{21} \\ &= \lambda^2 - (a_{11} + a_{22})\lambda + a_{11}a_{22} - a_{12}a_{21} \end{aligned}$$

Using the quadratic formula to solve for λ, we obtain

$$\lambda = \{(a_{11} + a_{22}) \pm [(a_{11} + a_{22})^2 - 4(a_{11}a_{22} - a_{12}a_{21})]^{1/2}\}/2$$

There are two roots, λ_1 and λ_2; let λ_1 be associated with the + sign, and λ_2 with the − sign. Thus we take $\lambda_1 > \lambda_2$. The rule for calculating $\underset{\sim}{\mathbf{T}}$ is simple. If

$$\underset{\sim}{\mathbf{a}} = \begin{pmatrix} a_{11} & a_{12} \\ a_{21} & a_{22} \end{pmatrix}$$

then

$$\underset{\sim}{\mathbf{T}} = \begin{pmatrix} (\lambda_1 - a_{22})/a_{21} & (\lambda_2 - a_{22})/a_{21} \\ 1 & 1 \end{pmatrix}$$

$$\underset{\sim}{\mathbf{T}}^{-1} = \begin{pmatrix} a_{21}/(\lambda_1 - \lambda_2) & (a_{22} - \lambda_2)/(\lambda_1 - \lambda_2) \\ -a_{21}/(\lambda_1 - \lambda_2) & (\lambda_1 - a_{22})/(\lambda_1 - \lambda_2) \end{pmatrix}$$

where λ_1 and λ_2 are the eigenvalues of $\underset{\sim}{\mathbf{a}}$. Two other useful facts for a 2×2 matrix $\underset{\sim}{\mathbf{a}}$ are

$$\lambda_1 + \lambda_2 = a_{11} + a_{22}$$

$$\lambda_1\lambda_2 = a_{11}a_{22} - a_{12}a_{21}$$

These relationships, together with the expressions just given for $\underset{\sim}{\mathbf{T}}$ and $\underset{\sim}{\mathbf{T}}^{-1}$, suffice to derive the expression for the transformation matrix in Equation 20-53 and its inverse in Equation 20-54. It also is a simple matter to generalize these results to $n \times n$ matrices, where $n > 2$.

For a clear and useful introduction to matrix algebra, see Chapter 1 of F. B. Hildebrand, *Methods of Applied Mathematics* (Englewood Cliffs, N.J.: Prentice-Hall, 1965).

Appendix B
Answers to problems

Chapter 1

1-1. The diameter of the cylinder is $2\pi r$. In the projection, one helical turn becomes the hypotenuse of a right triangle with sides of $2\pi r$ and P.

1-2. The regulatory scheme keeps the concentration of protein high enough to favor the stable octamer without allowing overproduction.

Chapter 2

2-1. The five bands seen in a B + C mixture imply that the proteins are tetramers. The bands are B_4, B_3C, B_2C_2, BC_3, and C_4. To explain the occurrence of only three bands in A + C or A + B mixtures, suppose that the A_2 dimer is very stable relative to the heterodimers AC and AB. Isolated A would still be a tetramer $(A_2)_2$. In A + B mixtures the three bands would correspond to $(A_2)_2$, A_2B_2, and B_4. Equivalent bands would be seen in A + C mixtures.

2-2. Adding a monomer to a dimer forms two protein–protein contacts, whereas only one contact is formed when two monomers become a dimer or when a trimer and a monomer combine to give a tetramer. Similarly, adding the last monomer to complete a disk forms four protein–protein contacts, while adding monomers to less complete disks or to completed disks forms fewer contacts. Assuming each contact has about the same favorable formation energy, trimers (pentamers, heptamers, etc.) and disks will be favored over other oligomers with additional or fewer monomers.

2-3. The primary structure is probably a nearly perfect duplicated sequence (like ferredoxin, perhaps). Within the tertiary structure it is likely that the N- and C-termini are nearly in contact.

2-4. An α helix of 85 residues would be only 128 Å long. Thus the protein cannot be a pure α helix. An 85-residue chain in the β-sheet conformation can extend 295 Å. If this folded into a two-stranded antiparallel sheet, the resulting structure would be about 150 Å long. It is also possible to generate structures this length by various combinations of helices and sheets. However, in any of these structures, most or all of the amino acid side chains would be exposed to the solvent. If the protein contains a normal complement of hydrophobic residues, it is very unlikely that any of these extended structures would be stable in aqueous solution. On the ribosome such structures would be stabilized by involving the side chains in extensive contacts with other proteins (or with ribosomal RNA). However, when such a protein is placed in aqueous solution, it is almost sure to assume a more collapsed structure formed at the expense of some sheets or helices.

2-5. Note the pattern of alternating polar and nonpolar residues in the sequence. If the monomer folded into a two-stranded antiparallel β sheet, this would have one entire face polar and the other nonpolar. Such a monomer would readily stick to membranes through its nonpolar face. It would also readily dimerize through hydrophobic interactions to make a sandwich structure with a nonpolar interior.

Chapter 3

3-1. The probability of finding a particular sequence in n contiguous bases is $1/4^n$. Within a 10^4-base DNA, this sequence will occur $10^4/4^n$ times. We must pick n such that this chance of accidental occurrence is 0.01 or less. This means that $0.01 \simeq 10^4/4^n$ or $n \simeq 10$.

3-2. In the B form the DNA is 13.1×10^{-3} cm long and has 3.79×10^4 turns. In the A form the DNA is 9.70×10^{-3} cm long and has 3.44×10^4 turns.

3-3. For a discussion of this problem, see U. H. Settler et al., *J. Mol. Biol.* 131:21 (1979).

3-4. The first half of the molecule forms into a single-stranded circle. The second half continues and twists around this circle to form the double strand. The authors are unware if such a structure has ever been seen experimentally.

3-5. The DNA can still form Watson–Crick base-paired hairpin loops although these will now probably have DNA B geometry rather than RNA 11. While some of the same tertiary base interactions may still form, others will not. For example, hydrogen bonds involving the ribose 2′-OH are impossible for DNA. Interactions that explicitly exploit modified nucleotides will also not be possible for DNA.

Chapter 4

4-1. See Figure 9 in C. Sardet, A. Tardieu, and V. Luzzati, *J. Mol. Biol.* 105:383 (1976).

4-2. This is not an easy problem. One can try to block random collisions by cross-linking the membrane at temperatures below the lipid phase transition(s). If activity assays are available, one can attempt to reconstitute purified cross-linked protein pairs and measure activity. The most effective potential strategy is to try to dilute the concentration of proteins and see if cross-linking still occurs. In practice this is difficult to do. For further discussion, see K. Peters and F. M. Richards, *Ann. Rev. Biochem*, 46:523 (1977).

4-3. There can be a symmetry match between the protein and nucleic acid C_2 axes.

4-4. Irregularities destabilize highly ordered helical and oligomeric structures. By default, less ordered gels are left as the remaining option.

4-5. Complexes between nucleic acids and basic (positively charged) proteins will be destabilized by salt. These complexes bind fewer counterions than do the separated components. Complexes between nucleic acids and acidic (negatively charged) proteins will be stabilized by counterions. They will bind more cations than will the separated components. Such types of complexes are found in nucleoproteins isolated from extreme halophiles. They presumably represent an adaptation to the high intracellular salt concentration in these organisms.

Chapter 5

5-1. Recall that the residue preceding proline has an altered conformational energy diagram, except for glycine, which is not greatly affected if it is succeeded by a proline. (For the case of an alanine residue followed by a proline, see Fig. 5–11.) In the case of an Ala·Pro sequence, region I is unavailable to the alanine residue, and this region has some of the less extended conformations that might be adopted in making a bend. Therefore, in order to accommodate bends, as is necessary in forming a cyclic peptide, it is better to have glycine, rather than alanine, before proline. The glycine residue, because it lacks a side chain, can still adopt all of its less extended conformations even if it is followed by a proline. Therefore, the desired sequence is X = Ala, Y = Gly, and Z = Pro.

5-2. For this problem, use Tables 5-8 and 5-9. Call conformation A the one in which the three residues are buried in the interior of the protein, and conformation B the one in which the three residues are exposed to the solvent. From Table 5-8, $\Delta G_t = 0.73$, 2.42, and 2.65 kcal/mole for Ala, Leu, and Phe, respectively. The sum of these free energy changes is $\Delta G_{AB} = 5.80$ kcal/mole and this is, to a first approximation, the free energy change associated with the transition from A to B in water. Therefore, $K = (A)/(B) = \exp(-\Delta G_{AB}/RT) = 1.8 \times 10^4$. According to Table 5-9, the free energies of transfer are reduced by 0.07, 0.38, and 0.70 kcal/mole, for Ala, Leu, and Phe, respectively, when the transfers are made to aqueous urea. Therefore, ΔG_{AB} is reduced by 1.15 kcal/mole and $K = 2.6 \times 10^3$, for the conformation change in urea.

5-3. This is a straightforward application of the approach used in Section 5-8. Let ΔG_{AB} be the free energy change for going from conformation A to conformation B. Therefore,

$$\Delta G = -\int_{r_2}^{r_1} (F)\,dr = \gamma(r_2^{-1} - r_1^{-1}), \text{ where } \gamma = aT$$

$$K = \exp(-\Delta G/RT) = \exp[-\gamma(r_2^{-1} - r_1^{-1})/RT]$$

$$\Delta H = -T^2 d(\Delta G/T)/dT = -T^2 d[a(r_2^{-1} - r_1^{-1})]/dT = 0$$

$$\Delta S = (\Delta H - \Delta G)/T = \gamma(r_1^{-1} - r_2^{-1})/T$$

5-4. Refer to Equations 5-13 and 5-14. Because $\chi = 0.1$ in both ethanol and water, ΔG^0_{Tot} has no cratic contribution to the free energy of transfer. But $\Delta G^{0*}_{\text{Tot}}$ has a cratic contribution because $\chi = 1$ in ethanol and $\chi = 0.1$ in water. Therefore, $\Delta G^{0*}_{\text{Tot}} = \Delta G^0_{\text{Tot}} + RT \ln 0.1$ and $\Delta G^{0*}_{\text{Tot}} - \Delta G^0_{\text{Tot}} = RT \ln 0.1$. Therefore, the student is incorrect.

5-5. From Section 5-4 and from Figures 5-7 and 5-8, we know that glycine residues in a chain encourage more compact, less extended conformations. In fact, for the same number of residues, the root-mean-square end-to-end distance of a polyglycine random coil is about a factor of two smaller than that of a polyalanine random coil (Section 5-4; see also Chapter 18). Therefore, if the unstretched tendons are exactly the same length, the one containing polyglycine must have about twice the number of residues. When the tendons are stretched, the residues will tend to adopt the extended, zig-zig (*trans*) conformation where $\phi = \psi = 180°$. The one containing polyglycine will extend further because the chains in the unstretched state are more compact to begin with and there are more residues in the chains. Therefore, tendon 1 contains polyglycine, and tendon 2 contains polyalanine.

Chapter 6

6-1. Using the definitions and notation in Section 6-7, the derivation of Equation 6-6 (dimerization case) is straightforward. Start with the definition of K_2 and express K_2 in terms of m' and ϕ_p. Because $m_2 = m' - m$, $m_1 = 2m - m'$, and $m = \phi_p m'$,

$$\begin{aligned} K_2 &= m_2/m_1^2 = (m' - m)/(2m - m')^2 \\ &= m'(1 - \phi_p)/m'^2(2\phi_p - 1) \end{aligned}$$

or

$$K_2 m' = (1 - \phi_p)/(2\phi_p - 1)$$

This is Equation 6-6.

To derive Equation 6-7, start with the definition of ϕ_p and note that $K = m_2/m_1{}^2 = m_3/(m_2)(m_1) = m_4/(m_3)(m_1) = \cdots$

$$\begin{aligned} \phi_p &= (m_1 + m_2 + \cdots + m_n + \cdots)/(m_1 + 2m_2 + \cdots + nm_n + \cdots) \\ &= (1 + Km_1 + Km_2 + \cdots)/(1 + 2Km_1 + 3Km_2 + \cdots) \\ &= (1 + Km)/(1 + Km' + Km) \\ &= (1 + Km'\phi_p)/(1 + Km' + Km'\phi_p) \end{aligned}$$

This simplifies to

$$Km' = (1 - \phi_p)/\phi_p{}^2$$

This is Equation 6-7.

6-2. For solution I: No base–base interactions.

For solution II: Dimerization occurs, probably by an interaction, such as hydrogen bonding, which prevents species bigger than dimer from forming. If $2A \rightleftarrows A_2$ and $K = (A_2)/(A)^2$, then

$$(A)_0 = (A) + 2(A_2) = (A) + 2K(A)^2 = (A) + C(A)^2$$

where $C = 2K$.

For solution III: Indefinite association occurs, where the association constant for each step is the same. Therefore, if $K = (A_2)/(A)^2 = (A_3)/(A_2)(A) = \cdots$, then

$$(A)_0 = (A) + 2(A_2) + 3(A_3) + \cdots = (A) + 2K(A)^2 + 3K^2(A)^3 + \cdots$$
$$= C_1(A) + C_2(A)^2 + C_3(A)^3 + \cdots$$

where $C_1 = 1$, $C_2 = 2K$, $C_3 = 3K^2$, and $C_4 = 4K^3$. Thus

$$C_2 = 2K \qquad \text{and} \qquad C_k = (k/2^{k-1})C_2^{k-1}$$

The most likely mechanism for indefinite associations is base stacking.

6-3. Referring to Equations 6-6 and 6-7, it should be possible to distinguish between the two modes of association, so that it is likely that something is wrong with the experiment. Note that the left-hand sides of Equations 6-6 and 6-7 will be equal when $\phi_p \approx 1$. At this point, $1 - \phi_p$ is very small so that the entire expression on the left-hand side of each equation is near zero. This would mean that $m' \ll K_2^{-1}$(Eqn. 6-6) or that $m' \ll K^{-1}$ (Eqn. 6-7); that is, the mononucleoside concentration is too low to observe any association. However, this consideration should be obvious to the experimental investigator; if not, then it could explain the results.

6-4. DNA structural transitions are described in more detail in Chapters 22 and 23. We expect the melting curves to depend on base composition because G–C base pairs are more stable than A–T hydrogen-bonded complexes, and we also expect the base-stacking interactions, which stabilize the helix, to depend on base composition and even sequence. Because duplex formation involves the union of identically charged strands, we also expect a substantial salt dependence to the melting transition.

Cooperativity will arise if the first base pair (nucleation) is difficult to form compared to subsequent base pairs. Apart from the unfavorable entropy of bringing two strands into register so that they can make the first pair (nucleation), the first pair also does not have a stacking interaction. However, if subsequent base pairs (propagation) form contiguously, each of these will have the benefit of a base-stacking interaction for every additional base pair. Thus, propagation is more favorable than nucleation, and cooperativity results.

6-5. We know from the discussion of Chapter 5 that urea increases the solubility, in aqueous solution, of nonpolar side chains of amino acids. In this way, it can disrupt hydrophobic interactions. To the extent that base stacks are stabilized by hydrophobic interactions, we expect urea to destabilize single-stranded stacked structures and nucleic acid double helices. There is some experimental evidence to this effect.

Index to Part I